10

Springer Series on Fluorescence

Methods and Applications

Series Editor: O.S. Wolfbeis

For further volumes:
http://www.springer.com/series/4243

Springer Series on Fluorescence

Series Editor: O.S. Wolfbeis

Recently Published and Forthcoming Volumes

Advanced Fluorescence Reporters in Chemistry and Biology III
Applications in Sensing and Imaging
Volume Editor: A.P. Demchenko
Vol. 10, 2011

Advanced Fluorescence Reporters in Chemistry and Biology II
Molecular Constructions, Polymers and Nanoparticles
Volume Editor: A.P. Demchenko
Vol. 9, 2010

Advanced Fluorescence Reporters in Chemistry and Biology I
Fundamentals and Molecular Design
Volume Editor: A.P. Demchenko
Vol. 8, 2010

Lanthanide Luminescence
Photophysical, Analytical and Biological Aspects
Volume Editors: P. Hänninen and H. Härmä
Vol. 7

Standardization and Quality Assurance in Fluorescence Measurements II
Bioanalytical and Biomedical Applications
Volume Editor: Resch-Genger, U.
Vol. 6, 2008

Standardization and Quality Assurance in Fluorescence Measurements I
Techniques
Volume Editor: U. Resch-Genger
Vol. 5, 2008

Fluorescence of Supermolecules, Polymeres, and Nanosystems
Volume Editor: M.N. Berberan-Santos
Vol. 4, 2007

Fluorescence Spectroscopy in Biology
Volume Editor: M. Hof
Vol. 3, 2004

Fluorescence Spectroscopy, Imaging and Probes
Volume Editor: R. Kraayenhof
Vol. 2, 2002

New Trends in Fluorescence Spectroscopy
Volume Editor: B. Valeur
Vol. 1, 2001

Advanced Fluorescence Reporters in Chemistry and Biology III

Applications in Sensing and Imaging

Volume Editor: Alexander P. Demchenko

With contributions by

A.M. Brouwer · A.P. Demchenko · A. Harmelin · E.A. Jares-Erijman · M. Joselevich · J. Klohs · V.B. Kovalska · V. Kalchenko · J. Lee · F.C. Leskow · L. Lin · Y. Li · M. Neeman · J. Piehler · T. Plass · T.N. Raja · M. Rudin · C. Schultz · A. Samanta · C. Spagnuolo · S.M. Yarmoluk · K.D. Volkova · S.O. Yesylevskyy

Springer

Volume Editor
Prof. Dr. Alexander P. Demchenko
Palladin Institute of Biochemistry
National Academy of Sciences of Ukraine
Kyiv 01601
Ukraine
alexdem@ukr.net

ISSN 1617-1306 e-ISSN 1865-1313
ISBN 978-3-642-18034-7 e-ISBN 978-3-642-18035-4
DOI 10.1007/978-3-642-18035-4
Springer Heidelberg Dordrecht London New York

Library of Congress Control Number: 2010934374

Cover design: WMXDesign GmbH, Heidelberg, Germany

Printed on acid-free paper

Springer is part of Springer Science+Business Media (www.springer.com)

Series Editor

Prof. Dr. Otto S. Wolfbeis

Institute of Analytical Chemistry
Chemo- and Biosensors
University of Regensburg
93040 Regensburg
Germany
otto.wolfbeis@chemie.uni-regensburg.de

Aims and Scope

Fluorescence spectroscopy, fluorescence imaging and fluorescent probes are indispensible tools in numerous fields of modern medicine and science, including molecular biology, biophysics, biochemistry, clinical diagnosis and analytical and environmental chemistry. Applications stretch from spectroscopy and sensor technology to microscopy and imaging, to single molecule detection, to the development of novel fluorescent probes, and to proteomics and genomics. The *Springer Series on Fluorescence* aims at publishing state-of-the-art articles that can serve as invaluable tools for both practitioners and researchers being active in this highly interdisciplinary field. The carefully edited collection of papers in each volume will give continuous inspiration for new research and will point to exciting new trends.

Preface

The key element of any fluorescence sensing or imaging technology is the fluorescence reporter. It is a molecular or nanoscale device that transforms the information on molecular interactions and dynamics into measurable signal of fluorescence emission. Due to these reporters, fluorescence technologies feature the unique combination of very attractive properties, such as extreme sensitivity up to individual molecules, ultrahigh resolution in time, potential for multiplexing applications, and remote accessibility with the possibility of obtaining the high-resolution images. These properties are in the background of innumerable applications. Within this content, the aim of three volumes under a common title: "Fluorescence Reporters in Chemistry and Biology" is to provide the comprehensive overview of fluorescence reporters, concentrating on those of them that represent organic fluorophores or exhibit similar spectroscopic behavior. Being historically the first, organic dyes maintain their leading positions both in basic studies and technology developments. In logical sequence, Part I was devoted to fundamentals and design of organic fluorophores and to optimization of their valuable properties. Part II addressed nanoscale constructions that are designed based on organic fluorophores and the systems that behave like these reporters but display strongly improved functioning – the clusters of only few noble metal ions and the conjugated polymers. The present Part III is the final part highlighting the applications of fluorescence in sensing and imaging, from the nanoscopic properties of materials to the biological whole-body imaging.

Heterogeneity in structural and dynamic properties is peculiar to any condensed state system. Special attention is required to those systems, in which such heterogeneity reveals on a very short length scale comparable with the size of molecular reporters. The introductory chapter addresses this important issue and critically analyzes different approaches based on physical theory, empirical correlations, and computer simulations. A strong attempt is made toward reducing ambiguity in interpretation of experimental data.

Improving old and designing new materials require new developments in fluorescence techniques. Static and dynamic properties of synthetic polymers and of their change during the synthesis, processing, and aging are the subject of the special chapter. The readers interested in unique properties of ionic liquids as the media of prospective "green chemistry" will obtain the important information on them derived from molecular probing. The systems with nanoscale anisotropy in molecular interactions and dynamics include liquid crystals, molecular-scale thin films, micelles, and bilayers formed of surfactants and lipid molecules. Strong electric field gradients are the characteristic of them. Physical modeling and structural data with atomic-level resolution help in understanding the spectroscopic response in proteins and nucleic acids.

The most extensive and efficient applications of fluorescence techniques are in the studies of biological macromolecules in different aspects – for sensing of other molecules as the targets and for obtaining the distributions of sensed molecules within the living cells and tissues. Operating with organic fluorophores within living cells is a challenging task and it can be addressed by the design of biosynthetically produced protein and peptide tags. Three chapters of this book address the design of organic fluorophores and their covalent modification for fitting these tags inside the cells.

Fluorescence reporters possess extending possibilities for applications on tissue and whole-body levels. They became realized with the design of bright fluorescent dyes absorbing and emitting light in the near-IR range of spectrum. In vivo imaging of vascular targets and recognition of hematopoetic and cancer cells belong to successful applications of organic dyes as fluorescence reporters. Concluding chapters of this book demonstrate recent progress in these practically important areas.

In line with other books of these series, this volume demonstrates the advancement in a rapidly developing interdisciplinary field of research and development. Therefore, it addresses an interdisciplinary audience starting from photophysicists and organic chemists to specialists in material science, chemical technology, and also to researchers working with living objects on molecular and cellular levels. This knowledge will provide additional stimulus to continuous progress in industrial technologies and biotechnologies.

Kyiv, Ukraine
January 2011

Alexander P. Demchenko

Contents

Part I
General Aspects

Interfacial Behavior of Fluorescent Dyes

Power and Weakness of Nanoscopic Description

Alexander P. Demchenko and Semen O. Yesylevskyy

Abstract Our macroscopic world and the world of atoms and small molecules are separated by length scales differing by seven or more orders of magnitude. Describing the latter world with fluorescence probes in terms of structure and dynamics has both merits and difficulties due to peculiarities and limitations of fluorescence method. Demonstrating unique resolution in time and very high sensitivity to interaction energies, this method generally lacks structural resolution on the level of atomic details. Therefore, presentation of fluorescent probe by its molecular structure or its derivatives (size, charge distribution, dipole moment, etc.) and of its tested molecular environment in terms of continuous medium (such as micropolarity, microfluidity, or proticity) became the common method of analysis. This description that combines molecular-level parameters and reduced to molecular-level macroscopic parameters can be termed "nanoscopic". The strong demand towards rational description of systems with molecular and nanoscale heterogeneity (surfaces of liquids and solids, liquid–liquid and liquid–solid interfaces, nanoparticles and porous nanocomposites) requires critical analysis of methodology when applied to these systems. This will be the subject of the present chapter.

Keywords Fluorescence reporters · Nanocavities and nanocomposites · Nanoscale polarity · Nanoscale viscosity · Solvatochromy · Surfaces and interfaces

---

A.P. Demchenko (✉)
A.V. Palladin Institute of Biochemistry, National Academy of Sciences of Ukraine, Leontovicha st. 9, Kyiv 01601, Ukraine
e-mail: alexdem@ukr.net

S.O. Yesylevskyy
Institute of Physics, National Academy of Sciences of Ukraine, Prospect Nauki, 46, Kyiv 03039, Ukraine
e-mail: yesint3@yahoo.com

A.P. Demchenko (ed.), *Advanced Fluorescence Reporters in Chemistry and Biology III: Applications in Sensing and Imaging*, Springer Ser Fluoresc (2011) 10: 3–62,
DOI 10.1007/978-3-642-18035-4_1,

Contents

1 Introduction

Determination of composition, structural arrangement, and dynamics of molecules and their groups of atoms at liquid–liquid and liquid–solid interfaces are extremely important for understanding various phenomena related to adsorption and catalysis and for technologies of chemical synthesis and separation/purification of reaction products. The properties of nanoscale porous materials and of nanoparticles, possessing very high surface-to-volume ratios, can be to a dramatic extent determined by interactions at their solvent interface. Structure and stability of synthetic polymers and biopolymers (proteins, polysaccharides, and DNA) are governed by interactions with the solvent and with the adsorbed low-molecular ligands. Understanding the behavior of these materials requires molecular-size tools integrated into these composite systems and serving as the reporters. Fluorescence probing methodology can provide such tools that are simple, highly sensitive, and nondestructive.

Fluorescence reporting focuses on *nanoscopic* properties of matter. It uses different organic dyes, luminescent metal complexes, labeled macromolecules, and different kinds of nanoparticles to evaluate local properties of their environment and of their intermolecular interactions. Sensing local field effects, the fluorescent reporter probes simultaneously the local polarity of the host medium, specific chemical interactions, and geometrical or morphological constraints. Meantime, the description of the probed system on the level of atomic details here is not available (exceptions are the formations of strong complexes and of covalent bonds by the probes, which is generally outside the probing methodology). This limitation is simply due to the size of reporters that is larger than atomic. Another limitation comes with the restricted geometry of probe location and orientation in structurally inhomogeneous systems that induces new difficulties in the understanding of their properties. Using fluorescence method, we possess a very limited number of parameters (they are intensity, anisotropy, lifetime, and the changes in excitation and/or emission spectra) that could provide informative reporter signal.

Because of these peculiarities, analysis of fluorescence data depends strongly on physical modeling leading to simplification of molecular systems or on empirical correlations relating spectroscopic parameters and intermolecular interactions [1]. Both approaches lead to quasi-continuous characterization of reporter surrounding. They allow exploration of such terms as micropolarity, microfluidity, or proticity as the *nanoscopic* analogs of parameters characterizing *macroscopic* condensed media. Such correlation of parameters that refer to macroscopic scale and molecular scale is not easy and requires different assumptions and approximations that are rarely discussed in original works.

When the averaged properties of the solvent as the "bulk medium" are of primary importance, then the quasi-continuous models (such "*continuum solvation models*") that ignore the solvent molecular structure are effective in description. Meantime, local field effects that deviate from that described in these models and a restricted geometry of probe location and orientation can be important parameters for the full understanding of spectroscopic behavior in such complex systems. Orientation of amphiphilic molecules at the polar–nonpolar interface and formation of specific interactions (such as charge-transfer complexes or H-bonds) may result in additional geometrical or morphological constraints. Then the models based on exact molecular structures (such as "*explicit solvation models*") should be applied.

The sensitivity of fluorescence emission from dye molecules to weak intermolecular interactions and their dynamics has been recognized as an important means to probe local field effects not only in homogeneous systems but also in the systems with molecular and nanoscale heterogeneity. This chapter is focused on mesoscopic description of the systems with this type of heterogeneity based on the data obtained in fluorescence probing. We analyze different methodologies in this description.

2 Essentials of Mesoscopic (Nanoscopic) Analysis Based on Computational Approach

Mesoscopic objects are the objects dealt in physics of condensed systems that are larger than atoms but small enough to observe fluctuations of statistically averaged variables. Fluorescent dyes always probe their *local* molecular environment, whatever is the dimension of studied object, in which they are incorporated. The system composed of the dye and of its local environment is always mesoscopic. Because of nanometer size of such systems they are often called *nanoscopic*. The description of nanoscopic systems and, particularly, those properties that give rise to spectroscopic changes should require some combination of classical and quantum-mechanical variables. Electronic excitations and redistributions of electronic density that accompany them are described by the laws of quantum mechanics. Meantime, there is a possibility to use intermolecular potentials derived from classical mechanics to describe intermolecular interactions in the ground and excited states. This allows considering the change of energies of electronic transitions as information on these interactions.

Why we are not satisfied by just empirical correlations between spectroscopic and macroscopic-like properties and need going deeper in the analysis of molecular and electronic structures? Our first aim is to reduce ambiguity in interpretation that commonly exists even in neat solvents. For instance, the strong shifts in fluorescence spectra could be due to the change of polarity or of H-bonding potential in the dye environment. But it can be also due to some photophysical reaction in the dye coupled with the dynamics in this environment. Even more difficult is the analysis of structure and dynamics in heterogeneous systems. Imagine that we study the properties of two interacting media of macroscopic dimensions. We can describe them in *macroscopic* terms, such as polarity and viscosity. However, this does not allow for the understanding of the properties of interface, such as the sorption of amphiphilic molecules (e.g., detergents), aggregation of nanoparticles, and the interfacial catalysis. On the other hand, *atomistic* description of these systems is very hard to achieve experimentally due to limitations of the methods that are commonly used for structural analysis, such as X-ray diffraction and NMR. But even if we do so by overcoming the problems of sample crystallization or isotope enrichment, we get a huge amount of extra structural information that is hard to analyze. We will then search the possibilities for *nanoscopic* description.

An "in silico" experiment with molecular dynamic (MD) simulations [2, 3] has a much broader applicability. This approach is based on application of classical mechanics and allows computing the forces between all atoms in the system and equilibrating the structure in chosen thermodynamic ensemble. This provides the atomistic detail in structure together with realistic dynamics of individual atoms. Meantime, it is often hard to operate with such large massive of information. Numerous atomic details can mask the general picture of physical and chemical processes that depend on statistical behavior of molecular ensembles rather than on detailed atom–atom interactions. Therefore, the technique of MD simulations is moving toward the *coarse-grained models*. In these models, the groups of adjacent atoms are combined into the "beads", which interact with each other by means of empirical potentials. Since the number of beads is much smaller than the number of individual atoms, significant speed up of computations could be achieved. The coarse-grained models could describe slow collective motions of complex molecules (such as large proteins) or macromolecular assemblies like the membranes of liposomes [4]. The coarse graining provides the *nanoscopic* description of the studied systems instead of purely atomistic treatment.

The difference in the characteristic times for different components of the studied system is crucial for adequate interpretation of MD results. Very slow rotational degrees of freedom of large solutes, such as proteins, DNA, and membranes, are never sampled adequately in MD simulations. However, the dynamics of solvent molecules is usually so fast that the solvent could be considered in local thermodynamic equilibrium for each given position and orientation of the solute. This allows averaging the solvent properties effectively and obtaining integral characteristics, such as local effective dielectric constant, local charge density, local hydrogen bonding propensity, local ionic concentrations, local electric field, etc. Such local properties could be computed around active sites of the protein and even in the vicinity of individual amino acids exposed to the solvent. As a result, unnecessary details of fast solvent motion are averaged out, while the solute is still described with atomic resolution. This represents another facet of the nanoscopic description of the system.

One of the most significant drawbacks of classical MD simulations that deal with atoms as classical bodies but not the electrons is the inability to handle the excited states, redistributions of the electronic density inside the molecules, and chemical reactions. Thus, the development of hybrid simulation techniques, which combine MD with the quantum mechanics [5, 6], has boosted in recent years. Such combined techniques allow computing electronic properties of small critical subsystems (such as organic dye molecules together with their binding sites) in realistic dynamic environment, which is described in the terms of classical mechanics. The interactions of fluorophores with their environment become dependent on their electronic state in this approach.

All MD or hybrid simulations provide the trajectories of individual atoms and, in the case of the hybrid simulations, the electronic densities of the quantum subsystems. These quantities should be integrated over the statistical ensemble to obtain useful characteristics of the system, such as viscosity, polarity, diffusion coefficient, interaction or solvation free energies, etc.

In contrast to the methods that provide atomic structural resolution, fluorescence spectroscopy allows achieving response to intermolecular interactions already in integrated manner. A *nanoscopic* level of details can be addressed here by describing the studied object as an integral but nonhomogeneous system with some of its essential properties presented on molecular level. The other properties appear as integrated over elementary interactions and their dynamics, which requires introduction of quasi-continuous description in such terms as "polarity" or "viscosity". Essentials in spatial resolution here are not lost if they are determined by the structural location and orientation of reporting fluorophore [7]. The nanoscopic view could also be achieved in MD simulations by performing limited statistical averaging, which preserves the intrinsic heterogeneity of the system. For example, the averaging in the plane of the lipid bilayer provides nanoscopic quantities, such as the profiles of polarity, electric field, dipole momentum, etc., across the bilayer [7].

Atomistic simulations of the compounds in the excited electronic states are rather challenging. Excitation is an electronic process that can only be described adequately by the methods of quantum chemistry, which are extremely intensive computationally. Therefore, the reasonable compromise is needed between the accuracy of the quantum description and the computational speed of the classical one. The rational choice of different computational schemes can be made in several distinct cases:

- *An excited state is long living compared to the characteristic times of rotational and translational diffusion. There is no significant time-dependent electronic charge transfer inside the excited molecule and between this molecule and its environment.* In this case, the classical MD could be used because the charge distribution in the excited molecule could be approximated adequately by the point charges on individual atoms. This distribution is computed on the quantum level when the empirical parameters of the force filed are determined (the parameterization stage). Only one such computation is needed, which makes it realistic even for relatively large molecules. As a result, the molecules in the ground and excited states differ only by their point charges. However, there are a number of complications in this approach. For example, the energies of bond stretching and angle bending in the excited state may be very different from the ground state, while empirical force field parameters for these energy terms are usually determined in the ground state. This can lead to systematic errors in the local dynamics and mobility of the studied compounds, although no systematic studies of these issues seem to be performed.
- *An excited state may be short living, but the perturbations of charges of surrounding molecules, caused by the excitation, are localized in the vicinity of the chromophore.* In this case, the hybrid quantum mechanics/molecular mechanics (QM/MM) methods could be used to reduce the computational burden of the quantum dynamics while keeping the accurate quantum description of the excited states. In the hybrid techniques, a small part of the system in the vicinity of the excited molecule is described on the quantum level, while the rest of the system is purely classical (Fig. 1). The quantum part usually includes the chromophore itself and the groups, which interact with it directly. Time-dependent intra- and intermolecular charge transfer could occur in the quantum subsystem as well as any chemical reactions. On each simulation step, the forces from the classical and quantum subsystem are updated in a consistent manner. This requires one quantum calculation per time step, which is feasible only for relatively small quantum subsystems.

  The consistent coupling between the quantum and classical parts is one of the most challenging problems in QM/MM calculations, especially if these regions

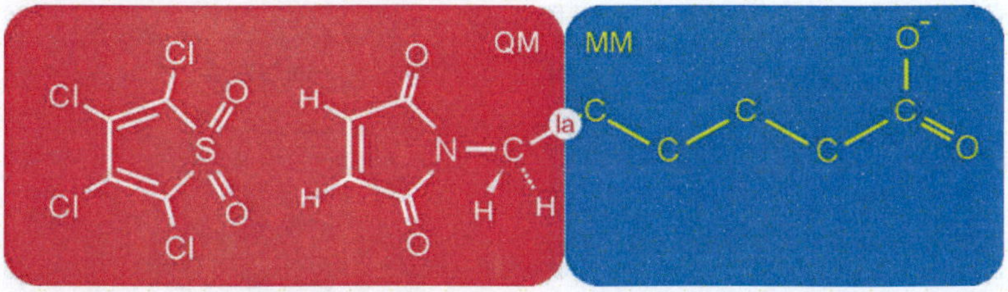

Fig. 1 The scheme of QM/MM simulation. Simulation of the Diels–Alder cyclo-addition reaction is shown as an example. The reacting groups are treated at quantum level, while the rest of the system is purely classical. The *white circle* corresponds to so-called linking atom, which is an auxiliary particle used to connect quantum and classical subsystems seamlessly. Picture from http://www.dddc.ac.cn/embo04/practicals/qmmm/qmmmvacuum.html

are covalently linked. Despite many technical and theoretical difficulties, the QM/MM functionality is now present in the widely used MD packages such as Gromacs [8] and Amber [9]. The questions, which can be addressed with the QM/MM technique, range from the optical properties of small molecules in different solvents [10] to the charge-transfer phenomena and the mechanisms of enzymatic reactions in proteins [11, 12].

- *Electronic excitation causes significant change of the electronic structure and the charge distribution, which is not localized.* For example, the long-distance charge transfer or fast diffusion of the excited molecules. In this case, the methods of the so-called quantum dynamics allow studying dynamic behavior of the system and its evolution in time on a quantum level. The most widely used variant of the quantum dynamics is the Car–Parrinello method [13], which incorporates electronic degrees of freedom directly into the equations of motions of atomic nuclei. The quantum dynamics is extremely intensive computationally because the quantum calculations of the whole system should be performed on each step. This method is usually limited to isolated molecules or small clusters with at most few dozens of atoms. Recent studies of such relatively complex heterocyclic molecules as coumarin dye show that purely quantum dynamics could be used to study their excited state in the gas phase and in the solution [14].

The treatment of *molecular environment* is one of the most important issues in the simulations of the fluorescent probes and other compounds in the excited state. The influence of the solvent on the spectral properties is often the major point of interest in such studies. As it was already stated before, an explicit solvation model includes all solvent molecules surrounding the solute (see Fig. 2a). The dynamics of the solvent should be modeled for rather long period of time, which should be sufficient for obtaining reliable statistical averages. This is usually prohibitively time consuming for quantum dynamics simulations. Alternative approach is to account for the solvent implicitly. In this case, the molecule of interest is placed into the cavity formed in the surrounding continuous dielectric medium (Fig. 2c).

In the MD simulations, the variety of methods based on the Generalized Born concept is used [16]. In these methods, each atom is represented by the sphere,

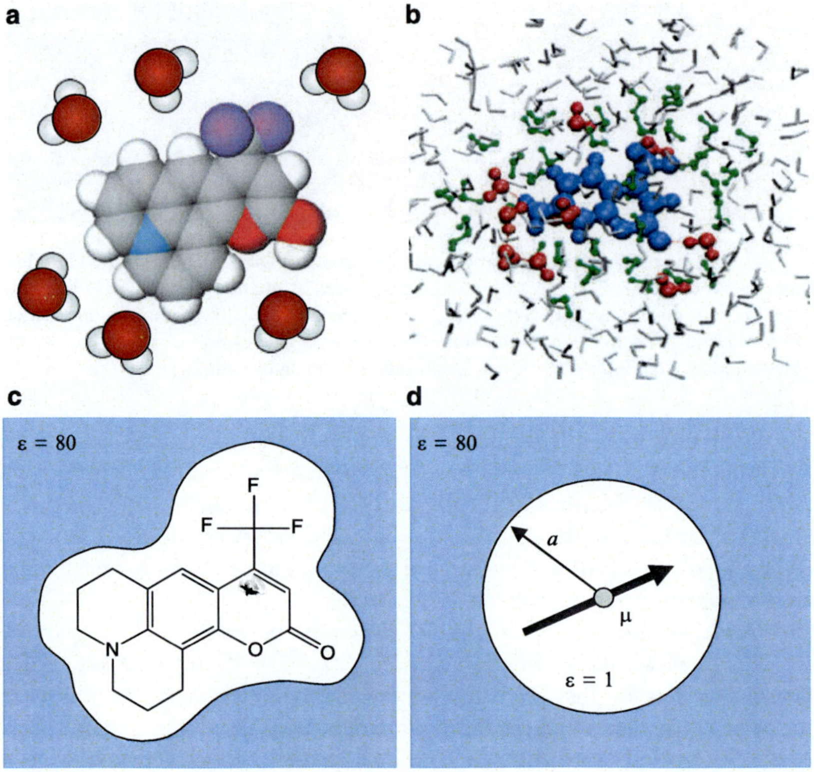

Fig. 2 Schematic view of various solvation models. (**a**) Explicit solvation. (**b**) Hierarchical solvation in quantum dynamics [15], different solvation shells are shown in *different color*. (**c**) Implicit solvation. (**d**) Simple Debye–Onsager model. The coumarin 153 probe in water is used as an example

which is cut out of the continuous dielectric medium and the electrostatic energy of such exclusion is computed. The radii of the spheres, called the Born radii, are either assigned based on the chemical identity of the atoms or updated during the simulation using various complicated algorithms. In quantum simulations, the COSMO implicit solvation model is widely used [17]. In this model, the cavity with the shape of the studied molecule is cut out in the continuous dielectric medium (Fig. 3). It screens the electronic charge density according to the dielectric constant of the medium and provides quite accurate approximation of the solvation effects. Particularly, it was shown that COSMO solvation model allows computing fluorescence spectra of 3-hydroxychromone fluorescent dyes in different solvents quite accurately [18].

Recently, improved implicit solvent models were used to study the dynamics of electronic excitations. Particularly, the effects of volume polarization and the penetration of the electronic density through the walls of dielectric cavity [19] were taken into account. It was shown that these effects play an important role in the excitations of small test molecules [20].

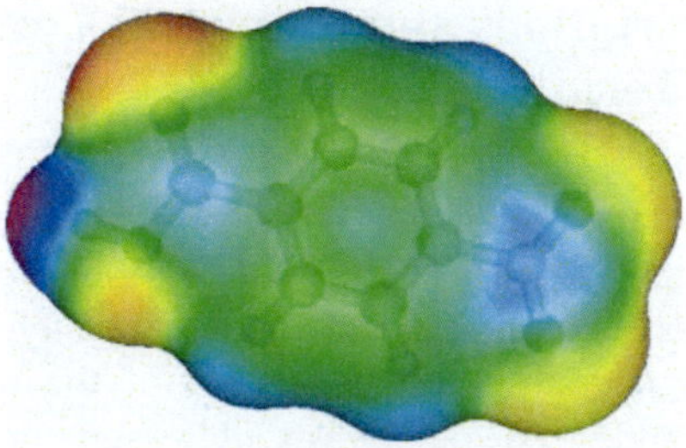

Fig. 3 COSMO surface of 4-nitrobenzoic acid colored according to the screened surface charges induced by the enclosed molecule. *Red* corresponds to the positive charge, *blue* to the negative. Picture from http://commons.wikimedia.org/wiki/File:Nitrobenzoic_acid.png

Another direction of improvement of implicit solvent models is the accounting for time-dependent solvation effects. It was shown that introduction of the time-dependent continuous polarization into the quantum dynamics allows reproducing the experimental time-dependent Stokes shift in coumarin 153 and the charge transfer in *N*,*N*-dimethylaniline extremely well [21].

There are attempts to develop the methodology, which keeps an atomic representation of the solvent, but improves the computational efficiency significantly in comparison to purely explicit solvent model, which is especially important for quantum dynamics simulations. The hierarchical approach is utilized in such methods (Fig. 2b). Solvent molecules from the first solvation shell, which contact directly with the solute, are modeled explicitly at the quantum level with full precision. Few more distant solvent shells are still treated at the quantum level but the approximation of frozen charge density is used, which speeds up the simulations. Even more distant solvent molecules are considered flexible, but their geometries are taken from the precomputed snapshots. Finally, the outer solvent shell is represented by the rigidly fixed solvent molecules [15].

It is possible to conclude that the computational approaches to describe the interactions of fluorescent dyes with their environments become more and more popular in recent years. Rapid advances in computer hardware and computational software allow in silico simulations of either very large systems or intricate time-dependent quantum phenomena, such as electronic charge transfers. The general trend of modern computational methods is the combination of different techniques, which describes the system on different hierarchical levels in order to obtain the general mesoscopic picture with an emphasis on quantum behavior of critical sites of interest. Such decomposition keeps the balance between the accuracy of quantum chemistry and the speed of classical empirical MD simulations and thus is very promising for large heterogeneous systems. Despite the exceptional spatial and time resolution of computational techniques, they depend strongly on experimental reference information, which is used to develop the force fields and to prepare the realistic starting structure for simulations. Thus, the "golden standard" of modern mesoscopic studies includes both experimental and computational approaches to the same system, which complement each other.

3 Analytical Descriptions and Empirical Correlations Exploring the Term "Polarity"

The term "polarity" is often used in a very general sense describing solvating capability of the medium. Whereas chemical definition of polarity is based on distribution of compounds between aqueous and highly hydrophobic phases (e.g., using the index log P, which is the logarithm of the octanol–water partition coefficient), physicists and physical chemists addressing the properties of materials use two different concepts based on *physical modeling* or on *empirical correlations* that will be discussed below.

3.1 Physical Modeling of Polarity Effects

Physical definition of polarity is quite different from that used by chemists and physical chemists. Here, polarity is considered as the composition of two basic effects in accordance to two general ways to stabilize a given molecule in a particular environment. One is the *electronic polarizability* of the medium that provides ultrafast response that can be described as the function of the square of refraction index, n^2, and is frequently presented as $f(n^2) = (n^2 - 1)/(2\,n^2 + 1)$. The other is the *nuclear polarizability* that describes the presence of molecular dipoles interacting with the probing dye and their much slower motion in the electric field created by this dye. This effect is expressed as a function of dielectric constant, ε. When the effects of electronic polarizability are small (e.g., in fluorescence spectra in highly polar liquids), the Onsager polarity function $f(\varepsilon) = 2(\varepsilon - 1)/(2\,\varepsilon + 1)$ can be used as a simplified estimate of *polarity*. In addition, there are the effects of generation of induced dipoles that need a more complicated description in both ε and n^2 terms. The interactions giving rise to electronic and nuclear polarizability in the fluorophore environment are considered "universal" in contrast to "specific" interactions that could be the charge-transfer (CT) complexes and H-bonds [22].

All mesoscopic models used for describing the solvation effects operate with the concept of Onsager *reactive field*. Here, the multitudes of real weak (Van der Waals) intermolecular interactions are represented in an integrated manner as the macroscopic electric field acting on a dye molecule at the site of its location. It may be considered as the field created and sensed by the solute dipole in its dielectric environment. Its electric dipole moment polarizes the solvent so that the solute itself experiences an electric field, the reaction field. The reaction field strength is proportional to the solute dipole moment in the ground and excited states. The energy shift on transition from vacuum ($h\nu_0$) to dielectric environment ($h\nu$) is proportional to the product of reactive field vector R and $\Delta\mu = \mu_e - \mu_g$, the change of dipole moment on electronic excitation:

$$h\Delta\nu = h\nu_0 - h\nu = \Delta\mu R/h.$$

The simplest models consider the dye as a point dipole located in the center of spherical cavity of radius corresponding to the dye dimension and dielectric constant equal to 1 (Onsager sphere radius a) (see Fig. 1d). The frequently used Lippert–Mataga equation [23] is based on approximation, in which all polarization effects except the generation of reactive field are neglected and the dipoles of the ground and the excited states (μ_g and μ_e) are oriented in the same direction. As the spectroscopic parameter, the model uses the Stokes shift, which is the difference between the positions of dye absorption and emission maxima on the wavenumber scale (in cm^{-1}). It describes how the general solvent effects expressed as a function of n^2 and ε can produce the relative shifts between absorption and fluorescence emission spectra:

$$\bar{\nu}_A - \bar{\nu}_F{}^- = \frac{2}{hc}\left(\frac{\varepsilon - 1}{2\varepsilon + 1} - \frac{n^2 - 1}{2n^2 + 1}\right)\frac{(\mu_e - \mu_g)^2}{a^3} + \text{const.} \tag{1}$$

Here, c is the speed of light and h is Planck's constant. The function $\Delta f(\varepsilon,n)$

$$\Delta f = \frac{\varepsilon - 1}{2\varepsilon + 1} - \frac{n^2 - 1}{2n^2 + 1},$$

that is called *orientational polarizability* is the difference between two terms inside large parentheses of (1). The refractive index contribution accounts for the ability of the electrons belonging to polarizable groups in the environment to polarize in order to stabilize the dipole moment of the fluorophore in the excited state, and this factor decreases the Stokes shift. Such polarization is instantaneous and occurs during the absorption process. It shifts the absorption spectrum to lower energies. In contrast, the dielectric constant term accounts for the relaxation process that involves the rotational or translational motions of groups of atoms or whole molecules. It develops in time and results in shifts of fluorescence spectrum in the direction of decrease of the energy difference between the ground and excited states. The constant term in (1) accounts for small additional spectral shifts due to excitation and emission to higher vibrational levels. The Δf values calibrated in different aprotic solvents are commonly used as the measures of polarity [24, 25].

There were a number of attempts to improve the Lippert–Mataga equation (see [26]). Thus, Bakhshiev considered the induced polarization terms and the nonparallel orientation of ground-state μ_g and excited-state μ_e dipole moments [27]. Somewhat different Δf values are used for calibration when the generation of induced dipoles resulting in more complex functions of ε and n are accounted.

Specific interactions, such as CT complexes and H-bonds, cannot be integrated easily into this treatment. But there is a frequently used possibility to consider the complexes formed by these interactions as discrete molecular entities that do not reorganize during electronic excitation. Such approach is not possible with the structures reorganized in the excited state (e.g., excimers and exciplexes), which require different description. In all these treatments, the involvement of specific interactions can be detected as the deviations from linearity in Stokes shift versus Δf plots (the Lippert plots).

The experimental data show that polarity is an efficient variable describing the interaction forces of the probe with the solvent and its components. Their description must involve several molecular properties (electronic and nuclear polarizabilities, abilities to form specific bonding), and therefore it is difficult to provide a strict definition of polarity and provide its measurement in a straightforward manner.

Accounting for polarity in quantum and QM/MM simulations is also nontrivial. The polarity is an integral property averaged over many solvent molecules during sufficiently large period of time, which should be larger than the characteristic time of rotational diffusion of the solvent molecules. Such averaging is problematic in quantum simulations due to huge amount of computational time required. However there are several approximations, which allow effective accounting for polarity effects. The method of effective fragment potentials (EFP) subdivides the system into the solute, which is treated at full precision according to selected basis set, and the solvent, which is treated at the simplified semiclassical level (Fig. 4). Such semiclassical description is, nevertheless, more precise than purely classical treatment in traditional QM/MM but is also more expensive computationally. Recently, the influence of the polar water environment on the excited state of Coumarin 151 dye was successfully modeled using this technique [14]. Another approach to the polarity effects in quantum calculations is based on the continuous implicit solvent model with volume polarization effects [19, 20].

Theoretical description of the polarity effects on the nanoscopic length scale also progressed in recent years. Particularly, the theory of the electronic polarization of

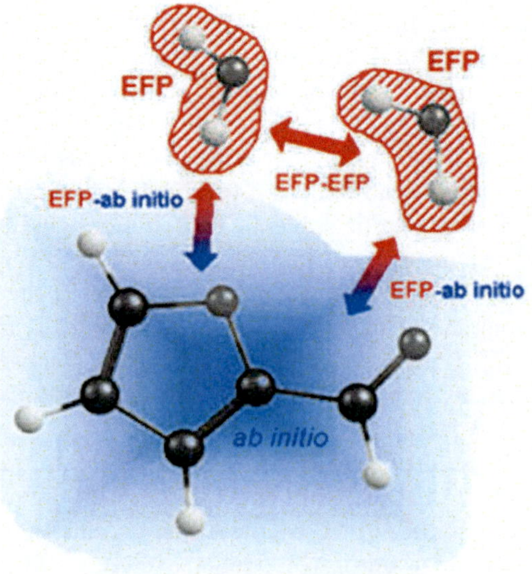

Fig. 4 The scheme of the EFP method. The solute (*blue*) is treated at the quantum ab initio level, while the solvent (*red*) is treated at the simplified semiclassical level

the molecules based on the concept of nonlocal dielectric function was built [28]. This theory also includes the contribution from spontaneous quantum fluctuations in polarization, which are rarely taken into account in traditional approaches. Although such theoretical works are of little practical importance now, they may lead to the development of innovative computational techniques in the future.

Extended versions of physical theory allow closer approximations of real systems accounting real shape of the dye molecule and its charge distribution including not only dipolar but also octupolar and higher order moments and also anisotropy in solvent polarization. Meantime, the basic theory is sufficient for many applications. The most important advantages in this analysis are the following:

- The shifts in both absorption and fluorescence spectra are described in a consistent way as the differences in energy of the ground and excited states.
- The time-dependent shifts of fluorescence spectra can be described within the same formalism in terms of relaxation of surrounding dipoles leading to changes in energies of these interactions. Thus, dynamics in the medium can be characterized.
- The physical parameters of reporting dyes are included into the analysis so that they can nicely reproduce the most important spectral features, including frequencies, intensities, and band shapes, based on a few molecular parameters that are kept fixed in all solvents. Therefore, this analysis can serve as a guideline for developing new probing dyes. The responses of dyes exhibiting specific interactions can be recognized as deviations from "universal" behavior.
- The models are extendable for describing the mechanisms of excited-state reactions, such as intramolecular charge transfer (ICT) and excited-state intramolecular proton transfer (ESIPT), in which the motions along "the solvation coordinate" play an important role. These effects can provide additional tools for characterizing the dye environment.

3.2 Effects of Local Electric Fields

Electric fields, either external or that produced by nearby atomic charges, influence the energies of electronic transitions resulting in spectral shifts [29] providing the tools for characterizing electrostatic interactions on a molecular scale. Local electrostatic fields generated by closely located charges are sensed in the same way as the effects of polarity being an additional component to the reaction field [30]. In these cases and, particularly, when the densely packed molecules form charged interfaces, such as in Langmuir–Blodgett films [31] and biomembranes [7], reorientation of the dye with respect to the field can distinguish the electric field effects. A group of voltage-sensitive styryl or naphthyl styryl dyes has been successfully applied for the visualization of the changes of biomembrane potentials [32].

The electrochromic dye senses the integrated electric field at the site of its location whenever this field is applied externally in a macroscopic device or internally, on a molecular level, produced by nearby charges. The "mesoscopic" approach considering the dye π-electronic system as a point dipole, electric field as a vector $\vec{F}$ that averages all the fields influencing this system and its surrounding as the medium with effective dielectric constant ε_{ef} can be used for the description of electrochromism in the simplest dipole approximation. The direction and magnitude of the shift, $\Delta\nu_{obs}$, is proportional (in the first approximation) to the electric field vector $\vec{F}$and the change of dipole moment associated with the spectroscopic transition $\Delta\vec{\mu}$:

$$h\Delta\nu_{obs} = -(1/\varepsilon_{ef})|\Delta\vec{\mu}|\left|\vec{F}\right|\cos(\theta),$$

where θ is the angle between $\Delta\vec{\mu}$ and $\vec{F}$ vectors. It follows that in order to show maximal sensitivity to electrostatic potential, the probe dye should be located in low-polar environment (low ε_{ef}) and oriented parallel ($\cos\theta = 1$) or antiparallel ($\cos\theta = -1$) to the electric field. It is also important that the correlation between the electric field strength and the spectroscopic effect within the applied approximation is linear, which allows, in principle, easy calibration of this effect in absolute values.

It is essential to note that electrochromism (the response to external electric field) and solvatochromism (the response to dielectric interactions with molecular environment) are based on the same physical mechanism [30]. Practically, this means that the effects of dipoles chaotically surrounding the dye in a polar liquid (effects of polarity) are basically not distinguishable from that of dipoles arranged in organized ensemble and generating the electrostatic potential. In experiments with the charged interfaces, the distinguishing feature can only be based on the fact of high structural anisotropy of this interface and the possibility of locating the probe also anisotropically, orienting it along, reverse or perpendicular to the interface, as it was shown for lipid bilayers [33]. Therefore, the effects that do not depend on dye orientation should be attributed to the effects of polarity and those that show such dependence should be to electrostatic potentials.

Because of stringent demands on the dye location and orientation, the correlation of spectroscopic effects with the applied field strength is possible only if its location and orientation relative to the field are fairly known.

It is important to note that the concept of *dielectric constant*, which is used in nearly all continuous models of environment, becomes invalid for atomistic simulations, such as MD or QM/MM techniques. In these simulations, the Coulomb forces are treated explicitly without the need for involving continuous medium with certain dielectric constant. In the classical subsystem, all pair-wise interactions of the point charges are computed either explicitly or by means of Particle-Mesh Evald summation [9]. In the quantum subsystem, the electron density and the nuclei are treated explicitly. As a result the dielectric effects appear naturally from the motions of individual atoms and the redistributions of their electronic density.

The *electronic polarization* is described directly as a redistribution of the electronic density in the quantum subsystem. This term of polarization is naturally missed in the classical part of the system. The *orientational polarization* comes from the translation of charges and rotations of dipoles in the system. If the simulation time is large enough to average out all rotational and translational motions, then the "global" effective dielectric constant of the whole system could be computed as a macroscopic statistical average. However, this is possible only for relatively simple systems. The dynamics of such complex systems as proteins embedded into the lipid membrane is so slow that true thermodynamic equilibrium is never reached in simulations (for example, the rotational degrees of freedom of a protein are almost never sampled in MD simulations). Despite this undersampling, the orientational polarization of the solvent in the vicinity of the active sites of proteins or other molecules of interest is still taken into account accurately. This can be done because the dynamics of the solvent is considered to be fast enough to produce reliable statistical averages.

3.3 Hydrogen Bonding Effects

Formation of intermolecular H-bonds by the probing dyes can be used for characterizing the H-bond donor and acceptor potential of the medium. These bonds can be typically formed by carbonyl groups as proton acceptors and hydroxyl groups as proton donors. When these bonds are formed, different changes of fluorescence emission can be observed: the changes in intensity (enhancement or quenching) and wavelength shifts. The general rule is that the formation of H-bonds with proton donors produces much stronger effects than with proton acceptors, which allows observing strong effects on the changes of solvent *proticity*. The other rule is that the H-bonding in the excited state is much stronger than in the ground state (due to increase of partial charges on correspondent groups), which, in some cases, may allow following the dynamics of formation of new bonds.

Hydrogen bonds of the dye carbonyl groups with protic partners can be recognized by strong spectral shifts and the changes in shape of fluorescence spectra at low protic component concentrations, at which there should be no significant polarity effects. There are interesting reported cases when in contrast to polarity effects that display the spectral shifts only, the essential transformations of spectra are observed and in this case they are so specific that can distinguish primarily formed strong H-bonds and weaker bonds formed at high protic cosolvent content [34]. In common solvatochromic dyes, such as Prodan, these effects occur in the same direction as the increase of polarity and, therefore, it is hard to distinguish them. They can be of the same magnitude as the spectral shifts over the full scale of polarity [35].

3.4 Empirical Correlations

Both scientists and users of scientific results like simplicity. The empirical methods offer such simplicity by describing the polarity with a single parameter. Providing often satisfactory description of polarity-related properties in many systems, they are very popular. They use the calibration on the basis of the correlation of spectroscopic changes with the indices characterizing solvent polarity. Several empirical solvent polarity scales were suggested [1, 35]. They are not totally equivalent and the scatter of experimental data against these scales is significant, even within the families of structurally related solvents. One of these scales is based on the wavelength shifting of the long-wavelength CT absorption band for *nonfluorescent betaine* dye 2,6-diphenyl-4-(2,4,6-triphenylpyridinium-1-yl)phenolate. The polarity index $E_T(30)$ is the spectral shift translated to kcal/mol, and its normalized version, E_T^N, describes relative variation of polarity between two extremes: highly polar water ($E_T^N = 1.0$) and low polar tetramethylsilane ($E_T^N = 0$) [1].

This scale is considered classical as the probing dye is soluble in a vast majority of solvents and solvent mixtures and its single parameter reflects both polarity and, to some extent, H-bonding interactions. In many investigations, the parameter E_T^N is used to check the response to polarity of different *fluorescent* dyes, which extends the range of application of this approach. Meantime, the more recently suggested dyes as polarity calibrators based on absorption or fluorescence are neither universal enough to cover the broadest range of solvents nor specific enough to investigate the mechanisms of deviations from simple regularities provided by specific features of every solvent. They allow, however, in a semiquantitative manner to characterize both polarity and H-bonding properties of the environment [36].

There were attempts to extend the single-parameter approach to distinguish several types of intermolecular interactions by using the comparison between the effects of similar dyes with different response to H-bonding as donors and acceptors [1]. For characterizing heterogeneous systems, this approach is of questionable value as the relative distribution of these dyes in studied systems can vary in an unpredictable manner.

Other empirical polarity scales are also known. The relative intensities of two vibronic bands of fluorescence spectrum of *pyrene* exhibit the polarity-dependent change, and this effect was also suggested for polarity probing under the name of Py scale [37]. Computational tools allow providing interpretation of this effect [38]. Development of the methods of covalent immobilization of pyrene on solid surfaces [39] opened many possibilities in the studies of interfaces.

The empirical correlations between fluorescence lifetime and polarity in some *polymethine dyes* were also suggested for determining this solvent parameter [40].

Concluding this section we can state that whereas polarity of a molecule can be easily calculated from distribution of its charge density, regarding the polarity in molecular ensembles and in condensed media we have to consider a composite function that involves different types of intermolecular interactions. These noncovalent interactions provide significant contribution to the energy of solute solvation (stabilization) in condensed medium. If the solute is a chromophore, then these interactions influence the energies of electronic transitions that are seen as the shifts in light absorption and fluorescence spectra. Physical modeling that does not consider specific interactions demonstrate that the interactions of dipoles and localized charges play the most important contribution to composite polarity function. This contribution can be most closely associated with the dielectric constant ε. Specific interactions such as the H-bonding can play an equally important role that makes the polarity function essentially multidimensional. In using empirical polarity scales, one strongly relies on the selection of reference solvents, and the derived polarity is "an equivalent polarity" based on comparison with these references.

4 Dynamics of Molecules and "Microfluidity"

Characterizing dynamics in condensed systems on the level of molecules and their groups of atoms is the great task. Here also, the mesoscopic approach is inevitable as any macroscopic properties such as diffusion rate and viscosity are not always descriptive and the observation of dynamics of selected atoms and of their groups does not allow characterizing the whole system. Macroscopic measurements of viscosity (as of the resistance to liquid flow) hide special features of dynamics on molecular level, which cannot be tolerated in the studies of nano-size objects. This required introduction of "microviscosity" or "*nanoviscosity*" as molecular-scale analogs of macroscopic property that are based on molecular properties of the probe and on the approximation of its environment as a continuous medium. Here again, we observe the division into empirical and model-based approaches.

4.1 Empirical and "Rotating Sphere" Methods

Empirical methods use various types of fluorescence response calibrated in standard solvents with the known "macroscopic" viscosity. This allows using the same *poise* units as for macroscopic viscosity. Selecting the dyes as viscosity sensors is also simple: the strong temperature dependence of response in high-viscosity liquids, such as glycerol and triacetin, can be used for that. Then, based on these data on dynamics in different liquid compositions, nanostructured systems and even the condensed media with poorly understood molecular properties, such as ionic liquids and supercritical liquids, can be characterized.

For sensing MD in condensed-phase systems such as liquids, polymers, and glasses, a fluorescent dye should respond to the change of its *translational or rotational diffusion* in this system by the change of any of its spectroscopic parameters. The oldest and simplest method introduced by Perrin to measure solvent viscosity is to apply the dye exhibiting molecular rotation during fluorescence lifetime with the detection of anisotropy [41]. The *rotational correlation time* φ in isotropic liquid medium can be determined from the measurement of anisotropy, r, and lifetime, τ_F, based on Perrin equation:

$$r = \frac{r_0}{1 + \tau_F/\varphi}.$$

This method in its present version needs the measurement of time-resolved anisotropy. Often these studies show deviations from the behavior of isotropic rotor due to nonspherical shapes of probing molecules, their specific interactions with the environment, and nonexponential decay of intensity. Structural anisotropy of the medium is also of great concern, but the dynamic component can be recognized in time-resolved experiment, so that:

$$r(t) = [(r_0 - r_\infty)\exp(-t/\varphi)] + r_\infty,$$

where φ is the apparent orientational relaxation time, r_0 is the fundamental anisotropy, and r_∞ is residual anisotropy.

The rotational correlation time can be related to viscosity η assuming that the medium is isotropic and that the rotating unit of volume V has a spherical shape (Debye–Stokes–Einstein model):

$$\varphi = \frac{\eta V}{kT}. \tag{2}$$

In order to account the deviation of probing molecule from spherical shape and dielectric friction during probe rotation, (2) is decorated with additional factors. By assuming a continuous solvent model, it can provide quantitative results only on the condition that the individual solvent molecules are smaller than the solute. Specific solvent–solute interactions produce additional problems [42]. Therefore, scientists often prefer to use empirical correlations between φ and macroscopic viscosity η that depend upon the probing dye and, of course, are different for H-bonding/polar and bond-free/nonpolar environments [43].

The time-resolved anisotropy can be combined with the observations of wavelength shifts. The benefit of that is the possibility of more detailed characterization of dye location sites in terms of polarity and rotational freedom. This is especially important in the studies of systems that cannot be easily characterized by other methods, such as silica gel-glass nanoporous materials [44].

Fluorescence quenching is another approach that can be used for evaluation of nanoscopic fluidity. There are a number of dyes with segmental mobility that provides channels for relaxation to the ground state without emission. In low-viscosity medium their segments are free to rotate, and such rotation induces the quenching [45]. In rigid media, such dynamics is frozen and a bright emission appears. Typical in this respect is the behavior of triphenylmethane dyes, such as Crystal Violet and Malachite Green. They possess the three-blade propeller-like phenyl rings joined by the central carbon atom. Their lifetimes τ_F depend strongly on the solvent viscosity and, due to the absence of groups forming specific non-covalent bonds, they practically do not depend on other solvent properties. Depending on viscosity, τ_F may change by as much as four orders of magnitude.

A class of organic dyes called *molecular rotors* responds to viscosity due to control by solvent dynamics of excited-state twisting [46]. In addition to segmental diffusion, their response can be due to formation of twisted intramolecular charge transfer (TICT) states. An example of such molecules is 4-tricyanovinyl-[*N*-(2-hydroxyethyl)-*N*-ethyl]aniline (TC1). Such molecules can be incorporated into nanoscale system without loss of sensitivity to fluidity [47]. Time-resolved microscopy allows applying molecular rotors for measuring viscosity inside the living cells [48].

One more approach is based on the possibility of two pyrene molecules to form intramolecular excimers. The excimer emission differs from that of pyrene monomer by strong wavelength shift, loss of vibrational structure, and decreased lifetime. The excimers are formed in a diffusion-controlled manner and when they are connected by flexible chain their response becomes concentration independent. Typically, 1,3-dipyrenylpropane is used for this purpose. This approach has found some application in the studies of solvent mixtures [49] and detergent micelles [50]. There were many attempts to apply this method to the studies of biological membranes, but this method is currently not in frequent use. The problem is in indefinite location of the probe in such media.

Isomerization dynamics can be coupled with the dynamics of formation/breaking of specific complexes with solvent molecules. Such effects were observed in asymmetric polymethine dyes [51]. Interplay between two forms, one emitting in solid (emitting at 700–710 nm) and the other in liquid environments (emitting at 760–770 nm), allows simple wavelength-ratiometric detection. Amazingly, neither the position of these bands nor the intensity ratio shows notable dependence on any solvent parameter except the viscosity [51]. Also, the styryl pyridinium dye showing the viscosity-dependent dissociation of ground-state complexes with water displays dual excitation peaks at 469 and 360 nm detected at emission wavelengths 500–650 nm. This property was useful for intracellular studies [52].

In the study of rotations of small molecules used as probes, it is the *dielectric friction* in addition to common viscous friction that determines the rotation rate [53]. The difference is that whereas viscous friction is determined by solute size, the dielectric friction is determined by solute dipole moment and senses the dynamics of its interaction with polar medium. It arises from longer range charge–dipole or dipole–dipole electrostatic interactions. When an ion or a dipolar solute moves in a polar solvent, there is an extra friction that arises from the fact that its polarization field induced in the surrounding solvent must readjust to its motion. Because this readjustment cannot occur infinitely rapidly, the solvent polarization will lag behind the solute and cause a systematic retarding force or friction on its motion [54]. In these dynamics, the rearrangements of intermolecular H-bonds can play an additional role.

The interaction of excited molecule with its dielectric medium can produce not only retardation but also acceleration effect. If the probe dye molecule is smaller than surrounding molecules, then its own rotation can be "induced" by electric field created in the medium of less mobile solvent molecules [55]. This *induced rotation* occurs with the rate of dielectric relaxations and can be recognized as an additional fast component in anisotropy decays that disappears at excitation red edge.

4.2 Spectroscopy of Molecular Relaxations

Solvation dynamics is the process of rearrangement of solvent dipoles around an instantaneously created or reorganized solute charge or dipole in the excited state. Electronic excitation results in instantaneous redistribution of electronic density generating the change in the dipole moment of the probe. Equilibrium in interaction with surrounding dipoles is disrupted and their relaxation to new equilibrium can be observed as the shift of fluorescence spectra in time. The ability to reorient under initial stress is an important characteristic of MD. The solvent relaxation is often approximated by the very simple Debye model that considers this relaxation mono-exponential. The *dielectric relaxation time* τ_R corresponds to that time obtained in dielectric measurements of bulk solvents, but it has the feature of molecular-scale resolution being used as the direct measure of MD in the dye environment. When we talk about "environment", we mean that the majority (about 85%) of response arises from the first solvation shell of a fluorescent probe [56]. This resolution is sufficient for studies of dynamics in first hydration layer in biological macromolecules [57].

The relaxation time is assumed to be equal to the rotational relaxation time of the single molecule τ_D. This time can be computed from the Stokes law in the hydrodynamic limit using well-known formulae for the coefficient of rotational diffusion

$$D_R = k_B T/(8\pi\eta R^3),$$

where η is the macroscopic solvent viscosity and R is the effective radius of the solute molecule. Using the Einstein relation for the reorientation time $\tau_D = 1/2D_R$, the Debye relaxation time is given by

$$\tau_D = 4\pi\eta R^3/k_B T.$$

The Debye treatment is very simplified. Particularly it does not account to the fact, that the local friction around the solute molecule is different from that in the bulk liquid. This means that the local viscosity is also different from the macroscopic bulk viscosity. Introducing such local viscosity at the phenomenological level leads to the so-called extended Debye treatment [58]. The relaxation time used in fluorescence measurements, τ_R, is in fact the reflection of relaxation of reactive field acting on fluorophore and the connection is:

$$\tau_R \approx [(n^2 + 2)/(\varepsilon + 1)]\tau_D.$$

In the case of highly polar media, $\tau_R \approx 0.1\ \tau_D$ [59]. Its determination in fluorescence experiments is based on observation of time-dependent spectral shifts in picosecond–nanosecond time range [60]. For describing these shifts, the correlation function is used:

$$C(t) = \frac{\nu(t) - \nu(\infty)}{\nu(0) - \nu(\infty)}.$$

Here, $\nu(t)$ is the position of spectrum in cm^{-1} as a function of time, and $\nu(0)$ and $\nu(\infty)$ are values extrapolated to time zero and infinity, respectively. A simplified method for determination of $C(t)$ based on single decay curve and steady-state spectrum was also described [61]. The application of $C(t)$ allows connecting the motion of spectrum to a full scale of this event and making comparisons of relaxation rates obtained with different dyes. Mazurenko and Bakhshiev [59] using a continuum model with a single Debye relaxation time have shown that $C(t)$ decays exponentially with time constant τ_R. In reality its change as a function of time is often nonexponential, and then the averaged $\langle\tau_R\rangle$ value is taken as dynamics variable [56]. This approach allowed establishing very interesting phenomenon – a decrease in mobility of water molecules by as much as 2–4 orders of magnitude in nano-size molecular ensembles, such as biopolymers and biomembranes [62]. Slow structural relaxation allows keeping preorganized electrostatic fields in reactive sites of enzymes [63], which is a strong contribution to their catalytic action.

Dielectric relaxation time and, therefore, τ_R can be connected directly, though nonlinearly, with solvent viscosity [58].

The prominent example of MD simulations, which complement experimental findings, comes from the studies of slow solvent dynamics of DNA molecules [64]. In this work, the calculated solvation responses for the dye molecule Hoechst 33258 bound to DNA were decomposed in terms of the components present in the system: water, DNA, and the ions. It was shown that the longest time scale of the solvent response observed experimentally (19 ns) is caused by the DNA itself. An ability to decompose the response into the components coming from different parts of the system is unique for MD simulations and allows interpreting experimental results properly. It is expectable that such combined studies will become increasingly popular.

Many results obtained either by "fluidity-based" or "relaxation-based" approaches demonstrate that the dynamics on nanoscopic scale can be very different from that obtained by bulk viscosity measurements. The experiments show a dramatically decreased fluidity that cannot be understood simply by intuitive extrapolation of bulk properties [65]. Friction and lubrication between nanoscale materials that determine their macroscopic properties should be understood on atomic and molecular level [66].

5 Inhomogeneous Broadening and Red-Edge Effects

In order to increase the information content of fluorescence method, scientists are trying to study the complex functions of fluorescence parameters, such as the time-resolved anisotropy and the spectral dependence of lifetime and anisotropy over excitation and emission bands. In this content, the shapes of absorption and emission bands and wavelength-selective excitation with recording of spectral dependence of parameters of emission may contain both static and dynamic information about the fluorophore environment.

5.1 Static Red-Edge Effects

Usually the electronic transitions generating absorption and emission spectra are presented as two-dimensional functions of vibrational and solvation coordinates (see [9]). The main difference between these coordinates is the intrinsically over-damped nature of solvation modes and their very low frequencies. This behavior is in contrast to vibrational modes. The latter are quantized, achieved in a very fast Franck–Condon process and, according to Kasha rule, relax rapidly to a lowest energy level. This allows treating solvation coordinate as a classical coordinate with continuous availability of electronic states. Thus, at any finite temperature a Boltzmann distribution in the population of different solvent configurations is responsible for the inhomogeneous broadening in the steady-state spectra [26]. Such broadening arises from the solute–solvent distribution in excitation energy that reflects the distribution in energy of dye interactions with its dielectric environment. The stronger will be these interactions, the broader the distribution. Laser hole-burning experiments in cryogenic media allow obtaining the inhomogeneous site energy distribution function [67], which provides important insight into the systems with low level of molecular order and can be extremely important for characterizing the environments of interfacially located dyes. But because of frozen mobility in the system, it brings only static information.

For obtaining both site photoselection and dynamic information, a different approach can be applied that allows operating in a broad variety of experimental conditions, including room and elevated temperatures. Within the inhomogeneously broadened spectrum, one may select the species interacting stronger than the mean of this distribution with the environment by selective excitation at the long-wavelength edge of the spectrum. The positions of their fluorescence spectra, their lifetimes, and anisotropies will then be different from the mean. This is the essence of *Red-Edge effect* [68]. In structurally disordered or low-ordered systems, the strongest are the dipole–dipole interactions with polar environments. Therefore, the width of distribution and the extent of Red-Edge effects should correlate with solvent polarity.

The Red-Edge effects depend strongly on the properties of a fluorophore. The dyes with a strong increase of the dipole moment in the excited state will exhibit a broad excited-state distribution on interaction energies, and the dyes interacting stronger in the ground state – the broader ground state distributions. These features will determine their properties in spectral, anisotropy, and time domains. Meantime, if the dyes in heterogeneous system are distributed between the sites of different polarity, the spectra may also exhibit variation in these properties. In our work [69], we tried to introduce the criteria that distinguish the effects of inhomogeneous broadening from those that originate from positional (ground-state) heterogeneity. In the case of dye location at the surface or interface, the positional heterogeneity can arise from imprecise location and orientation of the dye, even on atomic scale of distances. The increased distribution on interaction energies brings new site-selection effects that overlap those resulting from common inhomogeneous broadening.

The distinction between these cases can be observed when the excitation wavelength dependencies of fluorescence spectra are analyzed (Fig. 5). The Red-Edge effects demonstrate a very characteristic shape with the absence of the shifts of fluorescence spectra as a function of excitation wavelength at the maxima and short-wavelength shoulders of excitation bands and unlimited increase of effect at the red excitation edge [70, 71]. In contrast, the ground-state heterogeneity comes from superposition of absorption (excitation) spectra of the dyes residing

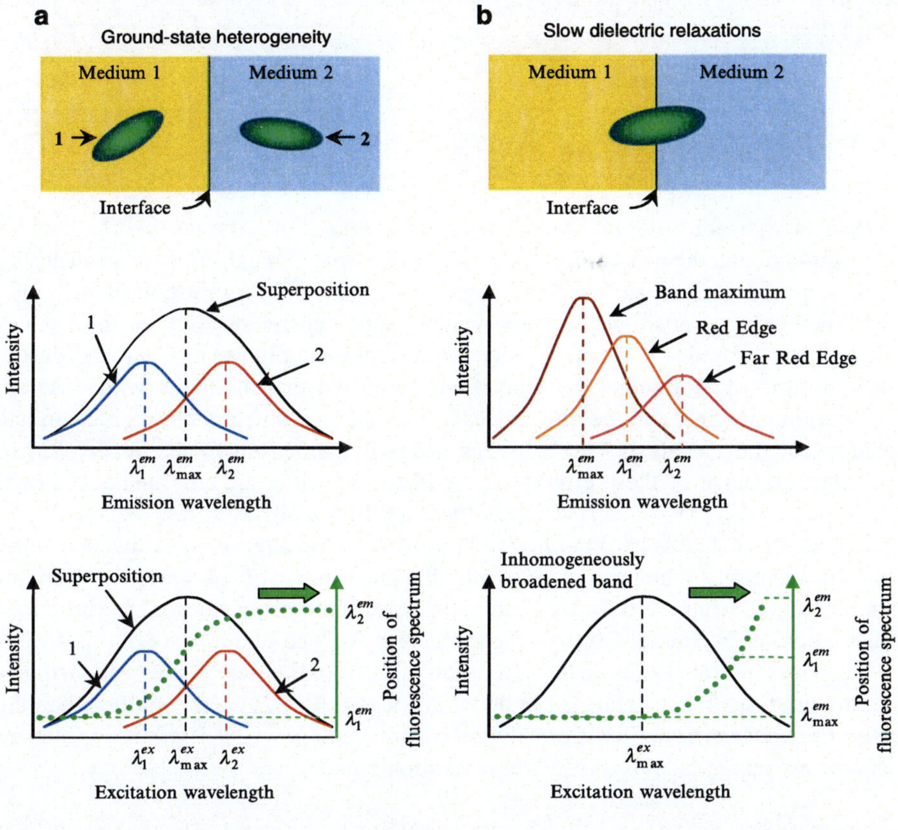

Fig. 5 Schematic representation of spectroscopic effects of (**a**) ground-state heterogeneity (e.g., the dye is distributed between two contacting media) and (**b**) slow dielectric relaxations (e.g., a single dye located at the interface and exhibiting inhomogeneous broadening) leading to Red-Edge effects. In the case (**a**) variation of excitation wavelength leads to photoselection between the species possessing difference in excitation spectra, so that the dye excited at shorter wavelengths (λ_1^{ex}) exhibits blue emission at (λ_1^{em}) and, correspondingly, excited at (λ_2^{ex}) exhibits red-shifted emission at (λ_2^{em}). The shifting of excitation wavelength leads to change of relative contributions of two emissions. In the case (**b**) there is a single ground-state form but the shift of excitation wavelength from band maximum, λ_{max}^{ex}, to the red edge of excitation leads to progressive shift of emission spectra (two of them, λ_1^{em} and λ_2^{em} are shown) to longer wavelengths

in different locations. Because the quantum yield, anisotropy, and lifetime of a dye in these locations can differ, the shapes of measured excitation-wavelength dependences can be variable, even sigmoid [24]. The dyes with low dependence for absorption spectra on solvent polarity and a very strong dependence for emission spectra could be ideal for this application. Good in this respect is tryptophan, which is the constituent of many proteins, and therefore we observe many applications of this method in protein research [68, 70]. Regarding different fluorescence dyes used for probing, the situation is variable. For instance, Nile Red with its strong depending on the environment polarity variation of both excitation and emission spectra can display both the Red-Edge effects and the ground-state heterogeneity [72], and the analysis of these effects is complicated by its strong variation of lifetime.

5.2 *Dynamic Red-Edge Effects*

When the dye molecules in the studied system are distributed between different sites (the ground-state heterogeneity), this distribution may not exhibit strong sensitivity to external conditions, such as temperature and pressure. In contrast, the distribution and therefore the positions of fluorescence bands depend strongly on the rates of dielectric relaxations in the cases when the environment dipoles move (relax) during the lifetime of emission. The relaxation mixes the environments with different excitation energies, and the differences in emission parameters between mean and site-selected dyes vanish. This is the essence of dynamic Red-Edge effects [69, 73].

The full range of these effects can be observed when the environment is rigid ($\tau_R > \tau_F$). When ($\tau_R < \tau_F$), which is the condition realized in low-viscous liquid solvents, this effect is hard to observe because of rapid averaging of species within the distribution. When $\tau_R \approx \tau_F$, with the knowledge of τ_F we get a tool for estimating τ_R from the shifts of steady-state spectra by using external perturbations that modulate τ_R (temperature or pressure) or τ_F (dynamic quenchers).

The method for obtaining the relaxation rates from the steady-state data using τ_F as the time marker is described in detail elsewhere [70, 71]. Essentially, when the Red-Edge effect in the limit of slow relaxations $\left(\nu_{t=0} - \nu_{t=0}^{\text{edge}}\right)$ is known, then τ_R can be estimated from this effect in relaxation zone:

$$\nu - \nu^{\text{edge}} = (\nu_{t=0} - \nu_{t=0}^{\text{edge}})\tau_R/(\tau_R + \tau_F).$$

Time-resolved spectroscopy allows observing the dynamics of molecular relaxations of Red-Edge selected species and comparing them with the relaxations of mean in ensemble [74, 75].

These experiments showed the appearance of the Red-Edge effect in polar solvent with the increase of solvent viscosity by lowering the temperature. The problem here is that the high and wavelength-selective temporal resolution is needed in the conditions when the dielectric relaxation rates match the measured time scale of fluorescence lifetimes. For low-viscosity liquids relaxing on a scale of several picoseconds, this scale is inconvenient for routine measurements. However, these methods can be efficiently used in highly viscous media (in which the relaxations are nanosecond or longer) and allow characterizing nanoscale systems, such as biomembranes and protein molecules [69]. In fluid aqueous media, even when the macroscopic viscosity of studied systems is low, these objects of tens or even several nanometers in dimension behave as *nanoscopic solids*.

There is an attractive but still rarely explored possibility to amplify the Red-Edge effects by observing the site-photoselection-dependent excited-state reactions, such as intramolecular electron transfer [76, 77], intramolecular proton transfer [78], and intermolecular resonance energy transfer [77, 79].

Selected can be the dyes, in which the difference in interactions with the environment from the mean of distribution results in variation of relative intensities of fluorescence bands belonging to reactant and reaction product species. Coupled with dielectric relaxations, these reactions will produce the spectra that will signal on the polarity and dynamics of dielectric relaxations on nanoscopic scale. Site photoselection at the red edge is commonly observed for ICT states, which decreases the probability of excited-state intramolecular proton transfer and intermolecular resonance energy transfer. Coupled with dielectric relaxations, these reactions produce new spectral bands that signal on the polarity and dynamics of dielectric relaxations on nanoscopic scale.

6 Variations of Solvation Shell Composition

The fact that it is the local interactions but not the average solvent effects that dominate in the response of fluorescent dyes can be derived from the results obtained in solvent mixtures, ionic liquids, and supercritical systems.

6.1 Preferential Solvation: Statics and Dynamics

For a long time it has been known that the macroscopic properties of solvent mixtures (density, viscosity, etc.) can deviate substantially from their linear behavior in proportion to concentrations of components. The origin of such abnormalities can be analyzed on nanoscopic level using fluorescence probes. Ideal solvation should be observed when any property obeys a simple regularity:

$$P_{\mathrm{mix}} = P_1 X_1 + P_2 X_2, \tag{3}$$

where P_{mix} is the property of interest in binary mixture, P_1 and P_2 are the measures of this property in neat solvents 1 and 2, and X_1 and X_2 are the mole fractions of the solvents in their mixture. If we insert any molecule or nanoparticle into the solution composed of the mixture of molecules differing in charge, polarity, or ability to form specific bonding, then the local concentrations of solvent components at or close to its surface can differ from their bulk values. Its environment will be enriched with the component possessing higher affinity – polar if the object is polar itself or unpolar if, correspondingly, the object will be unpolar. Such deviation from macroscopic solvent composition is called *preferential solvation* [80].

Elucidation of spectroscopic data in equilibrium and time-resolved kinetic experiments helps comprehending different phenomena related to preferential solvation. In steady-state experiments, it can be observed that the experimental points deviate from linearity expressed by (3). Curvature of the plot indicates that the solute environment is preferentially enriched with one of the solvent components [81]. Such enrichment can be an indication of clustering of solvent molecules already existing without the probe, so that one component of the binary mixture preferentially binds to molecules of the same type in its neighborhood, forming clusters. Preferential solvation can be considered as the solute inclusion into these clusters. Meantime, the role of probing dye can also be active in a sense that this dye forms its environment by saturating its intermolecular interactions. There can be "dielectric enrichment", so that the polar dye is preferentially solvated by polar component. It can also be enrichment by forming specific interactions, particularly, by H-bonds [82].

MD simulations allow studying the preferential solvation in atomic details. The typical setup of such simulations includes the binary solvent with highly polar and miscible lower polar components (such as methanol and water, or DMSO and water). The solutes may range from single ions to the realistic chromophores [83]. Coumarin 153 is used as a prototype in the studies of solvation dynamics. The charge distributions in the ground and the first excited state of this dye were computed using the quantum chemistry methods and used in subsequent classical all-atom MD simulations. The MD trajectories provide solvation responses, characteristic times, and Stokes shifts of the dye, which could be compared with experimental results directly. The dye was equilibrated in the ground state. Then the charge distribution was changed to match the first excited state and the relaxation of the solvent molecules was monitored. It was shown that in dimethylsulfoxide (DMSO)–water mixtures there exists a strong preferential solvation by DMSO contributing most significantly to the solvent effects. The contribution of water molecules is small and their relaxation is only slightly coupled with the charge transfer in the coumarin dye [84]. Similar studies in benzene–acetonitrile and benzene–methanol mixtures allow reproducing quite well the experimental time-dependent Stokes shifts. The motion of benzene molecules and rotational diffusion of acetonitrile in the vicinity of the dye were studied in atomic details. In the case of methanol, the complex dynamics of the hydrogen bonding network around the dye was observed

[85]. These examples show that classical MD simulations could provide significant insight into the behavior of the mixed solvents and give atomistic description of preferential solvation, which correlates with the experimental data.

Clustering in binary solvents (such as methanol–chloroform mixtures) is easily revealed in MD simulations by analyzing the radial distribution functions of the mixtures at different concentrations. It was also demonstrated that in such a mixture the hydrophilic and hydrophobic parts of amphiphilic solutes are preferentially solvated by methanol and chloroform, respectively [86].

Energetics of solute enrichment in the dye solvation shell is determined by interaction energy of the solute dye balanced by the decrease in the entropy of mixing. It may contain dielectric and H-bonding components, which should depend strongly on the properties of solute dye [87]. In these cases, this "disturbance" is in fact so great that the solute molecule acts no longer as a "probe" of the properties of the solvent mixture, but actually forms its own local environment, which can be quite different from the bulk of the mixture [82]. Therefore, fluorescence response cannot be taken as the measure of macroscopic "polarity" but is essentially a nanoscopic property. This property, if we can estimate it, is more important than the bulk "polarity" for understanding the solubility and reaction kinetics in mixed solvents.

Fluorescent dyes allow detecting preferential solvation as the deviations of local (solvation shell) polarity or proticity values from linearity (3). Meantime, these dyes allow doing more – providing more detailed characterization of solvation sites, particularly, separating the polarity and H-bonding effects or electronic and orientational polarizability. Spectacular in this respect are the studies of 3HC dye in toluene–acetonitrile mixtures [88] (Fig. 6).

Toluene is an aprotic solvent with high electronic polarizability ($f(n) = 0.226$) and low polarity ($f(\varepsilon) = 0.239$), while acetonitrile is an aprotic solvent with low polarizability ($f(n) = 0.175$) and high polarity ($f(\varepsilon) = 0.480$). The study of their

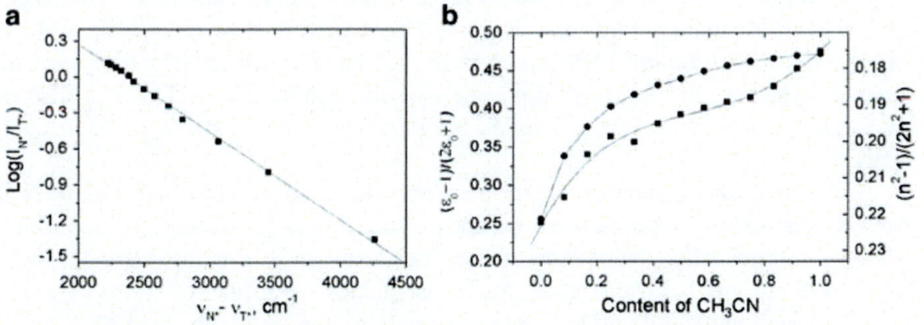

Fig. 6 Results of wavelength ratiometric response of 3HC dye 4′-diethylamino-3-hydroxyflavone on the titration of toluene with acetonitrile. (**a**) Logarithm of two band intensity ratio I_{N*}/I_{T*} versus band separation $\nu_{N*} - \nu_{T*}$. (**b**) Estimated values of polarity $f(\varepsilon)$ (*filled circle*) and polarizability $f(n)$ (*filled square*) as a function of volume content of acetonitrile in toluene [88]

mixtures with the 3HC dye possessing multiparametric response allowed separating these effects on spectra. First, we observe that for intensity ratio of two emissive bands, N* and T*, $\log(I_{N*}/I_{T*})$ shows perfect linear correlation with the band separation for the mixtures of different content (Fig. 6a) and such linear fit is close to that obtained for extended number of solvents, which means that proposed algorithm of multiparametric analysis can be applied in this case. Since this system is aprotic, it is possible to estimate the polarity function $f(\varepsilon)$ from the obtained I_{N*}/I_{T*} ratios. Then the solvent polarizability function $f(n)$ can be found as described in [18]. The obtained values of $f(\varepsilon)$ and $f(n)$ as a function of acetonitrile volume content are presented in Fig. 6b. We observe that the variation of $f(\varepsilon)$ is not a linear function of polar solvent concentration; it exhibits a steep initial rising part and, at high acetonitrile content, tends to saturation. Meanwhile, the electronic polarizability function $f(n)$ does not decrease in the same manner. It demonstrates a more complex dependence on concentration of polar component. At low acetonitrile concentrations, $f(n)$ changes steeply but slower than $f(\varepsilon)$. This can be explained by the fact that the dye structure is composed of highly polar and low polar interaction sites. Therefore, it can be assumed that in the solvation shell, the polar sites are occupied first by the polar component with stronger effect on $f(\varepsilon)$ than on $f(n)$. In the middle of titration curve, both functions follow the same trend. At the final step of titration, acetonitrile molecules substitute the remaining toluene molecules in low polar sites with stronger influence on $f(n)$ than on $f(\varepsilon)$. Thus, because of the distribution of polar and apolar interaction sites on the dye molecule, toluene can contribute to the solvation even when its concentration in acetonitrile is low. Thus, a multiparametric approach in the analysis of spectroscopic information allows studying new features of the theoretically and practically important phenomenon of preferential solvation.

Dynamic component in preferential solvation appears if it is triggered by an increase of solvation energy by one of the components due to increase of the probe charge redistribution in the excited state [89]. This sudden change of solute properties starts the diffusion-controlled exchange of solvent molecules between the bulk and the solvation sphere. This process leads to a time-dependent shift of emission spectrum in the nanosecond time domain. The relatively slow relaxation is attributed to the *translational* diffusion of polar solvent molecules, which replace the nonpolar ones in the first solvation shell of the solute. This effect can also be accounted when the spectral kinetics is analyzed in the systems of nanoscopic heterogeneity.

Both static and dynamic solvation effects of coumarin 153 and 102 dyes in hexane–methanol mixtures were recently studied experimentally. The nonlinear solvatochromic shifts dependent on the methanol mole fraction were detected. These effects are believed to be the consequence of formation of a hydrogen bond between methanol and the carbonyl group of the coumarin dye [90].

A special case of such dynamic solvation is seen with ionic liquids and the mixtures of ionic and common liquids. In ionic liquids, the relaxations occur by translational diffusion of solvent components to equilibrate the excited-state charge redistribution in the solute dye [91].

The cations and small anions possess different diffusion rates, which create the temporal effects. Two different, fast and slow, components of the solvation are detected originating due to local motion and diffusional motions of the ions [92]. Gradual addition of water decreases substantially both the solvation time and the rotational relaxation time.

Thus, the results discussed above show that the preferential solvation makes the solute–solvent and solvent–solvent interactions more complex than that observed in pure solvents. Two phases of the solvent relaxation around the dye were detected. The first phase corresponds to local reorientation of the solvent in the vicinity of the dye, while the second slower phase corresponds to the diffusional motions of the ions. Preferential solvation is often observed even in the relatively simple binary mixtures, like the 1-propanol–water. The effect of preferential solvation is observed as the strong dependence of the rotational diffusion and solvent relaxation times on the water content detected using the coumarin 153 probe [93].

6.2 *Supercritical and Gas-Expanded Liquids*

Local compositions in supercritical, near-critical, and gas-expanded fluids may differ substantially from bulk compositions [94], and such differences have important effects on spectroscopic observations, phase equilibria, and chemical kinetics [95]. *Supercritical fluid* is a unique state of condensed matter, in which such properties as density, viscosity, and the diffusion rates can be tuned continuously in broad ranges by simply adjusting temperature and pressure values without changing the solvent composition. Strong pressure-dependent density fluctuations in supercritical fluids result in specific effects in fluorescence dyes that cannot be accounted when empirical methods are formally applied. The clustering of solvent molecules may result in local solvent density around a solute to be higher than the actual bulk value [96] and fluorescence lifetimes can vary in extremely broad ranges due to intensive molecular collisions. This can explain a dramatic pressure-dependent variation in fluorescence response of wavelength-ratiometric 3-hydroxychromone dye exhibiting the ESIPT reaction in supercritical CO_2 [97], which cannot be interpreted as the change of polarity, as it is usually done in common solvents (Fig. 7).

The MD simulations of the gas-expanded liquids are of great interest because they reveal the properties of the cybotactic regions (the regions where the solute alters the solvent properties) with atomic details. Particularly, MD simulations of the gas-expanded methanol$-CO_2$ and acetone$-CO_2$ mixtures demonstrated local density enhancements similar to those seen in supercritical fluids, although less dramatic [98].

The effects of preferential solvation of the fluorescent dyes are observed in the gas-expanded liquids as well. Particularly, the coumarin 153 dye is solvated preferentially by organic species in the CO_2-expanded methanol and acetone [95]. The dynamics of the solvent molecules in the gas-expanded liquids is often

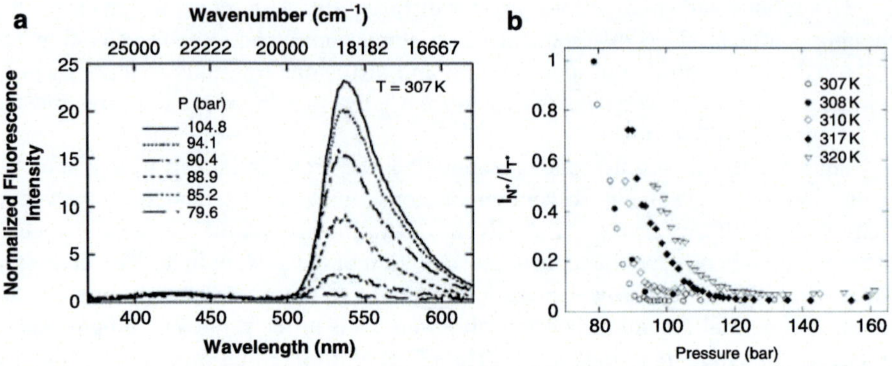

Fig. 7 The results of fluorescence studies in supercritical CO_2 at different pressures and *T* 307 K using a wavelength-ratiometric dimethylamino-3-hydroxyflavone dye. (**a**) The spectra normalized to the maximum of the normal band (428 nm) to emphasize the change in the relative intensities of the normal and tautomer bands. (**b**) The pressure dependence of ratiometric response at different temperatures. Excitation wavelength was 350 nm. Reproduced with permission from [97]

nontrivial and involves the phase of fast rotational diffusion in the cybotactic region followed by the slower redistribution of the solvent between the cybotactic region and the bulk solvent. Such behavior was recently observed both experimentally and in MD simulations for solvatochromic dye *trans*-4-dimethylamino-4′-cyanostilbene (DCS) solvated by the CO_2-expanded acetonitrile [99]. It was also shown that the cybotactic region surrounding DCS molecule is enriched in acetonitrile, and the extent of this enrichment is greater in the excited state. 10-*bis*(phenylethynyl)anthracene (PEA) and coumarin 153 dyes also exhibit preferential solvation and strong solvatochromism in CO_2-expanded acetonitrile and methanol. It was shown that the spectral shifts of PEA depend nonlinearly on the local compositions in these mixtures, which suggests substantial preferential solvation of both solutes by the liquid components of the mixtures [100].

The results of this section witness that even in macroscopically homogeneous systems the nanoscopic heterogeneity may exist that is responsible for variation of many fluorescence parameters. The solute dye does not simply "probe" the environment; it can form this environment preferentially interacting with different solvent components. If such interactions change in the excited state, then the spectral kinetics can be easily observed.

7 Remarks on the Properties of Fluorescent Dyes

As it was stated elsewhere [101], fluorescent dyes can be classified into two broad categories: no responsive (used for labeling) and responsive (used as probes and reporters in molecular sensing). The first perform based just on their presence in particular medium or at particular site. Ideally, their response should be *directly proportional* to dye concentration and *independent* of any factors that influence fluorescence parameters (quenching or enhancing of emission, wavelength shifting). Second, in contrast, should respond to different stimuli required in sensing and probing by the *most significant change* of these or some other parameters and, ideally, this response should not depend on their concentration. If the electronic density is delocalized over the whole fluorophore, which is characteristic of resonant or mesomeric dyes, such as fluoresceins, rhodamines, and cyanines, the spectral sensitivity to environment is small [102]. In contrast, the dyes exhibiting ICT from the initially excited normal (N*) electronic state are characterized by strong excited-state dipole moment, which allows them to interact differently with the environment of different polarity, electronic polarizability, and proticity, making them efficient molecular probes [26]. Below we review in short the properties of responsive dyes on view of their optimal parameters for their applications in fluorescence sensing and imaging (Fig. 8).

The dyes for *intensity sensing*. There is a large variety of organic dyes that perform in this way [103]. Commonly, they respond by the mechanism of photoinduced electron transfer (PET) that results in total quenching [9]. Dye-doped nanoparticles [104, 105] and conjugated polymers [106] can exhibit "superquenching", in which the quenching of one among many coupled emitters quenches the emission of the whole ensemble. The response in intensity is not internally calibrated, which creates many problems in its analysis [7, 107].

The dyes for *anisotropy sensing*. The response in anisotropy is internally calibrated [7, 107] but its dynamic range depends strongly on fitting of fluorescence lifetime to rotational correlation time of sensor units. Therefore, the dyes with long lifetimes offer more possibilities.

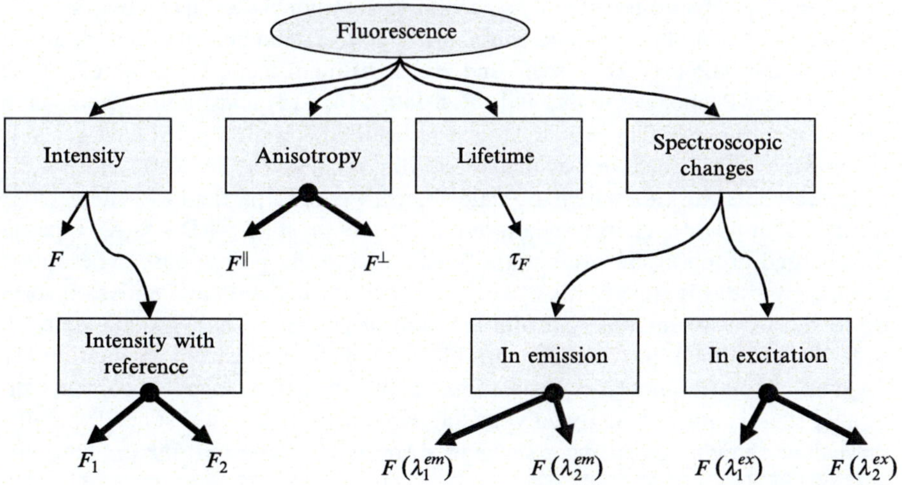

Fig. 8 Parameters used in fluorescence sensing and imaging. *Thick arrows* indicate the techniques, in which the self-referencing is provided by simultaneous recording of two parameters

The dyes for *lifetime sensing*. These dyes should exist in two switchable emissive forms with different lifetimes. Such dyes are rare, though their application is efficient including microwell formats used in high-throughput screening [108]. Recently reported squaraine dyes are attractive for different applications including immunoassays [109] (see also [110]). The dyes that gradually change their lifetimes find their use as oxygen and viscosity sensors based on quenching effects. Variation of lifetimes of FRET donors is another possibility frequently used in sensing.

The dyes for *wavelength ratiometry*. Different reporter mechanisms and selection of correspondent dyes are extensively discussed in [107]. There are many possibilities to use coupled dyes exhibiting excimer formation (pyrene), Főrster resonance energy transfer (FRET), or using molecular reference. Meantime, the most attractive is obtaining the ratiometric reporter signal from a single dye. This can be achieved with the dyes exhibiting wavelength shifts or generation of new bands.

The dyes exhibiting wavelength shifts are very popular despite the fact that many of them exhibit strong solvent-dependent variation of intensity, so that substantial quenching is typically observed in strongly polar and protic solvents [18]. Their general property is the distributed charge along their π-electronic system. These dyes possess the excited-state ICT states that are characterized by strong dipoles [9]. Among them, the strongest polarity-dependent effect is observed for the dye called fluoroprobe [111]. Next in efficiency are ketocyanines [86], dialkylaminonaphthalenes (such as Prodan [112]), 4-(*N*,*N*-dimethylamino)-4′-nitrostilbene (DANS) or -cyanostylbene [113], and other dyes, the typical feature of which is the presence of electron donor and acceptor groups on different sides of their molecules. Prodan has a single broad emission spectrum with a dramatic solvent-dependent shift (130 nm), ranging from 401 nm in cyclohexane to 531 nm in water. The charge-transfer dyes containing carbonyls as acceptors, such as Prodan, exhibit the fluorescence shifts, also due to H-bonding, that are almost as strong as the whole polarity scale and are undistinguishable from the effects of polarity [35]. Phenoxazone dye Nile Red [80, 114, 115] and fluorene analogs of Prodan with twice higher brightness and red-shifted emission [116] have recently become the most attractive wavelength-shifting probes.

Researchers working with nanoscale objects would be particularly interested in the dyes that possess charge separation in the ground state (of D^+–π–A^- type) as the charged group(s) are preferentially solvated in the polar phase. These dyes exhibit inverse solvatochromism, i.e., the shifts of excitation and emission spectra to shorter wavelengths with the increase of polarity. Their ground-state CT state is transformed into almost neutral D–π–A state on electronic excitation [1]. As their dipole moment decreases upon photoexcitation, in polar solvents the ground state is more strongly solvent stabilized than the excited state. Typically, these dyes show much stronger shifts in absorption (excitation) than in emission spectra [117].

It is a challenge to separate the effects of polarity and H-bonding in response of fluorescent dyes. One possibility realized in 3-hydroxychromone derivative is to block sterically the carbonyl group from the access to protic partner [18] so that the dye can respond to polarity only. The other possibility is to obtain both characteristics of the environment separately by using two or more parameters of fluorescence emission that could provide differential response. Thus, it was suggested to use dramatic quenching effect observed in some dyes in protic media (e.g., in Nile Red) so that this quenching could be taken as the measure of hydrogen bonding potential whereas the spectral shift could provide a response to the change of polarity [115]. However, this type of quenching is difficult to quantify and distinguish from other quenching effects. Also, the two parameters can be strongly connected: H-bonding also induces the spectral shifts whereas variations of polarity producing the shifts may influence fluorescence intensity.

It could be more productive to get benefit from the fact that the excitation and emission spectra can exhibit different sensitivity to different types of interactions. The change in ground-state energy results in shifts of both excitation and emission spectra and the change of energy of the relaxed excited state – only in emission spectra. The transitions between different ground- and excited-state forms can amplify this effect. Thus, the response of the 10-hydroxybenzo[h]quinoline molecule (HBQ) allows observation of the changes in absorption spectra of the normal tautomer together with the changes of fluorescence emission of the proton-transfer (ESIPT) tautomer [36]. The wavelength shifts of the absorption band depend essentially on polarity with a minor contribution of H-bond donation acidity, contrary to the fluorescence band shifts, which exhibit a greater dependence on H-bond acidity than on polarity.

Much stronger effects can be observed when one of the ground-state forms exhibits and the other form does not exhibit excited-state reaction generating new strongly wavelength-shifted fluorescence band. Thus, switching in the ground-state equilibrium can bring orthogonal information to that derived from the excited-state equilibrium. The ground-state equilibrium can be established between the species with and without intermolecular H-bonds characterizing *proticity* (H-bond donor potential) of the environment. This idea was realized with 3-hydroxychromone (3HC) dyes [118]. The ESIPT reaction in these dyes is observed only for the H-bond free form, and the distribution of intensity between their initially excited and strongly wavelength-shifted ESIPT bands is governed by *polarity* of the environment [18]. When these dyes are applied to test the unknown properties of their environment, one can observe three partially overlapped emission bands, and their excited-wavelength-dependent deconvolution allows obtaining independently the polarity and hydrogen bonding potential of this environment [119]. The validity of this deconvolution is increased when an important principle based on the mechanism of this reaction is applied. The N* and T* bands generate from the same ground state and therefore the ratio of intensities F_{N*}/F_{T*} should not depend on excitation wavelength, whereas the H-bonded complex formed in the ground state can be recognized by a different excitation spectrum.

The advantages of polarity-sensitive dyes exhibiting excited-state equilibrium between ICT and ESIPT states deserve special remark. The ICT response to the change of polarity is observed as a common spectral shift, but the change in interaction energy behind it produces the change of relative intensities of well-separated ICT and ESIPT bands. The ratiometric signal becomes robust to temperature and solvent dynamic quenching and to variations of all instrumental factors since under their influence the two bands change proportionally their intensities without affecting the ratio [107]. Dramatic amplification in sensitivity to polarity in 3HCs is observed when the ratio of intensities F_{N*}/F_{T*} is compared with the shift of N* band [18]. A complete switching between N* and T* emissions is observed within narrow polarity range. A series of different dyes with the variation of the N* state dipole moment had to be developed for determination of polarity in its narrow ranges [18]. These dyes found numerous applications that are mentioned in different sections of this chapter.

Regarding sensitivity to electric fields, styryl dyes and particularly 4-dialkyl-aminostyrylpyridinium derivatives with electron-donor and electron-acceptor substituents at the opposite ends of their rod-shaped aromatic conjugated moieties are the most popular [32]. They respond by the shifts of their excitation spectra that allow ratiometric detection, so that the most convenient for observation shifts of emission spectra are less prominent. The change of dipole moment in such dyes on electronic excitation is one of the most dramatic. It is determined by a partial positive charge that is reversed in the excited state by electronic ICT from dialkylamino group to aromatic amine. This electronic polarization along the extended π-electron system can be modulated in broad ranges by external electric field.

Advantages of the family of 3HCs as electrochromic dyes are in easy detection of two-band ratiometric signal in fluorescence emission [18]. Since the parent fluorophore is low polar but allows substitutions of different charge and polarity, various constructions are possible based on these dyes. This allows the possibility of targeted insertion of chromophore to a desired location in heterogeneous systems.

8 Location of Fluorophores with Subnanometer Precision

In the study of nanoscale structures and interfaces with fluorescence reporter dyes, there appear many problems. The gradients of noncovalent interactions and their dynamics in these cases are so steep even on the scale of fluorophore dimension that it makes meaningless many studies of "polarity" or "viscosity" performed without clear indication of fluorophore location. Such results can be used only for the observation of formation or disruption of structures (such as protein denaturation or formation of detergent micelles) but not for obtaining characteristics of these structures or their interfaces.

It can be argued that if the nanoscale object exhibits cooperative structural transition, this transition being rather extended in space should always be sensed by attached fluorescence reporter. There are a number of cases that show the opposite. A popular dye, ANS, when bound to the surface of phospholipid membrane, does not sense its transition between gel and liquid crystalline phases, and the dye label attached to the surface of protein molecule does not sense protein denaturation as its local environment does not change. Moreover, interfacial ordering and induction of structural anisotropy may influence substantially the fluorescence spectra as it was shown for incorporated ICT dyes in nematic liquid crystals [120]. Therefore, locating the dye in particular site and in specified orientation is extremely important. MD simulations may help to specify this location and provide the reference that is needed for quantitative analysis of probe effect.

8.1 Available Methods to Localize the Dyes

Fluorescent probes sense their molecular environment at the distances comparable with dimensions of their molecules. There are several possibilities for increasing precision in location of fluorescence probes.

8.1.1 Covalent Labeling to a Desired Molecule, Interface or Nanostructure

These methods are well developed, especially those involving modification of SH– and amino groups [121]. With their use, the labeling of proteins has become a routine procedure. In order to provide site-specific labeling, the Cys residues can be incorporated into protein structures in desired positions allowing exposure of their SH– groups. In biomembrane studies very popular is the labeling of phospholipid heads of phosphatidylethanolamine. Meantime, caution is needed when the target for labeling is a flexible molecule. Thus, covalent attachment of the relatively polar NBD group to flexible acyl chains of lipids does not allow its localization close to the bilayer center [122], because the chains loaded with the groups of higher polarity loop to the polar interface [123].

In the studies of polymer–glass interfacial region, fluorescent dye can be covalently attached to organosilane-modified glass surface before the deposition of polymer [124]. Immobilization with the aid of spacer groups allows achieving the distance dependence. The minimum size by which the spacers can be varied is nominally a single CH_2 group, which is about 2.5 Å.

8.1.2 Using Complementarity in Multipoint Noncovalent Interactions

With designing of docking, complementarity of charges, and proper hydrophobicity–hydrophilicity balance, one can achieve the precise location of the dye. Small noncovalent interactions have a tendency to provide collective effects that allow manipulating with several substituents to achieve high affinity to a desired site. Figure 9 illustrates one of the possibilities of using the dyes of similar structure and spectroscopy but differing in polarity. A number of sophisticated dyes with a

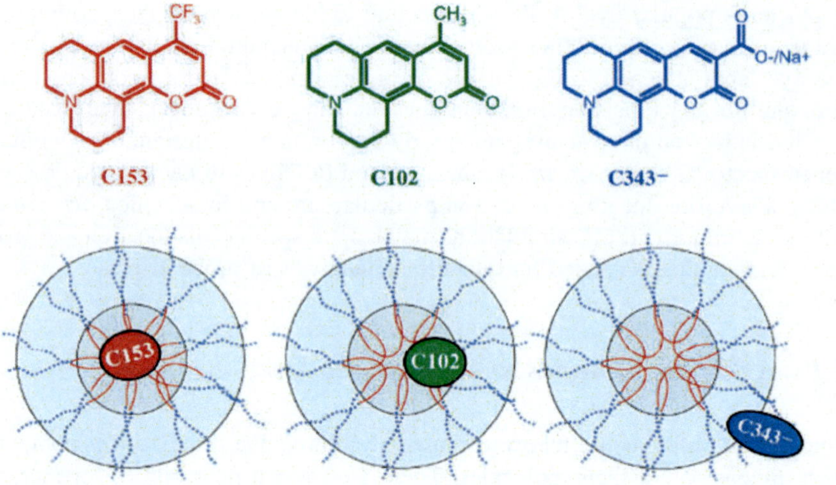

Fig. 9 The three 7-aminocoumarin probe molecules C153, C102, and C343 (*above*) share very similar molecular geometries, volumes, and spectroscopic characteristics but they have greatly varying hydrophobicities. By selecting coumarin fluorescence probe molecules with different solubilities, their different locations can be achieved in the F88 triblock copolymer micelles, which allows reporting on different polarities [126]

variety of low-polar, polar, and charged substituents are developed for the study of biological membranes. Their structures, locations, and properties will be discussed in this volume [125].

8.1.3 Achieving Selectivity with the Dye Optical Properties

There are a number of dyes with severely quenched emission in protic environments, particularly, in water [103]. Among them are 1,8-ANS, Nile Red, dialkylamino derivatives of 3HC, and many other solvatochromic dyes. The mechanisms in the background of such behavior are the relaxations along intermolecular vibrational modes, the electron transfer to solvent traps, and formation of nonfluorescent aggregates. The dyes exhibiting opposite behavior are also known: they exhibit strongly enhanced emission in protic media [127]. Application of quenchers with selective solubility in particular phase or in particular location is an additional tool in research. The dyes that exhibit strong polarity-dependent shifts in absorption spectra can be selectively excited in polar or unpolar medium by variation of excitation wavelength. For Nile Red, the bathochromic shift with increasing solvent polarity is as much as 108 nm from *n*-hexane to water. If such dyes are located in two types of sites differing in polarity, selective excitation allows obtaining fluorescence spectra from one type of sites [72].

8.1.4 Application of Location-Selective Spectroscopic Techniques

Among the optical methods that are both probe and locus selective, the resonance-enhanced second-harmonic generation (SHG) is an attractive tool, especially in the studies of interfacial phenomena [128]. SHG is a second-order nonlinear optical spectroscopic method that allows for the selective measurement of an effective excitation spectrum of the dye at asymmetrical surface or interface. The intensity of SHG light reflects the nonlinear susceptibility and orientation of the interfacial substances. When the light frequency corresponds to electronic transition in the dye, we observe the resonance-enhanced SHG, and from the excitation spectrum recorded with the aid of a wavelength tunable laser one can extract information on polarity of the dye local environment. In a typical SHG experiment, a single laser beam is focused on the interface under study, and a nonlinear polarization with double frequency is detected. The latter intensity is proportional to the square of the second-order susceptibility that is essential only in anisotropic environments. Liquids themselves are isotropic so that the dye orientations are averaged producing no signal, but when they are immobilized at an interface they generate detected SHG signal. Therefore, this technique is intrinsically adapted to discriminate between anisotropically located (e.g., surface-bound) and free reporter molecules. This method can be used in total internal reflection conditions and with resolution in anisotropy and lifetime. Its drawback is the level of recorded signal that is by several orders of magnitude lower than in common measurements of emission.

Thus, it is very hard to achieve a subnanometer precision that is needed for operation on molecular length scales.

8.2 Control for the Probe Location

8.2.1 Imaging

The limit of spatial resolution of common fluorescence microscope is 200–500 nm. It exists for all types of microscopy (confocal, two-photonic, or total internal reflection) and is due to diffraction limit. This resolution is not sufficient to see individual molecules at the sites of their location. Stimulated-emission depletion (STED) microscopy and related methods [129] demonstrated overcoming this limit, but it still did not reach resolution sufficient for locating small molecules. Reference [130] describes how to use electron microscopy for locating a particular type of dye molecules after additional staining.

8.2.2 Application of Localized Fluorescence Quenchers

When the location of fluorescence quencher molecules or ions is known, then the quenching effect can be used for determining the location of fluorescence dye.

Quenching requires direct contact between the dye and the quencher, so their proximity can be determined with relatively high precision. Dynamic quenching requires formation of this contact by diffusion of partners during fluorescence lifetime, which makes location less certain. Location of dye at liquid–solid interface can be determined by application of quenchers dissolved in liquid phase. Very popular in this respect is the determination of Trp residues on protein surface by quenching effect of heavy ions (cesium and iodine) [70]. Nitroxide compounds are dissolved in different media and can quench fluorescence with the rate that approach their diffusional limit [26]. For determining the depth of location of fluorescent probes in the membranes, the method based on quenching by spin-labeled lipids was developed [131, 132]. Small nitroxide label can be attached at different positions of lipid molecule, so the areas in the membrane can be formed in which the quenching occurs with high probability. Still, this method did not reach subnanometer precision.

8.2.3 X-Ray Crystallography

Publications with the application of this method are very rare and are limited to structural analysis of crystallized dye–protein complexes [133]. Meantime, they can bring interesting information. Thus, the structure of ribonuclease A covalently labeled with *N*-[[(iodoacetyl)-amino]ethyl]-5-naphthylamine-1-sulfonic acid (1,5-IAENS) demonstrates the location of the dye on protein surface, whereas fluorescence spectrum can be interpreted in terms of "low polarity" of dye environment [134]. Probably, the slow rate of dielectric relaxations of contacting protein groups and hydration water are responsible for this effect.

8.2.4 Additional Experiments

There are other methods that help to establish fluorophore location, orientation, and dynamics. One can mention NMR methods [135], cyclic and alternating current voltammetry [136] and, among fluorescence methods, anisotropy in steady-state and time-resolved domains and the observation of Red-Edge effects [68]. The latter two methods can be especially useful for the studies of liquid–solid interfaces and of biological membranes. The dye immobilization at the interface can be recognized by highly polarized emission and by the appearance of these effects.

8.2.5 MD Simulation

The MD simulations can be used to obtain detailed atomic-scale information on the location and orientation of the fluorescent probes in studied media as well as the information about their interactions and dynamics. The MD simulations have only started to be applied to this task. These simulations for fluorescent probes DPH [137, 138] and pyrene [139] in lipid membranes were reported (Fig. 10). Although

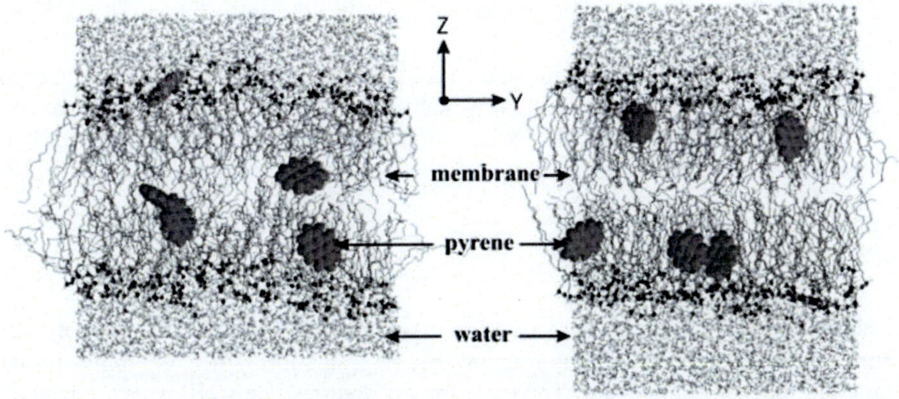

Fig. 10 Redistribution of pyrene molecules in the lipid bilayer in the course of 20 ns MD simulations from their initial placement (*left*) to final positions (*right*) [139]

surprisingly few in number, these works showed the suitability of MD for calculation of a variety of properties of fluorescence probes in organized ensembles.

Concluding this section, we state that the dyes used for probing are commonly larger than the solvent molecules and the groups of atoms composing the nanoparticles and interfaces. Solvation around such large solutes necessarily reflects an average of the environment sampled by the solute. Their response is sensitive to collective effects of the environment. So, possessing very high resolution in time and in interaction energy, they cannot offer subnanometer structural resolution. This resolution is also limited by our ability to control exact location, orientation, and distribution of them in an ensemble. Whereas accurate profiling is not possible, a lot can be done with spectroscopic and different complementary methods. Strong efforts should be provided for designing the probes with more specified location and orientation in the studied media, and a caution is always needed in interpretation of obtained data.

9 The Study of Surfaces and Interfaces

In many areas of science and technology, it is important to characterize the structure, dynamics, and interactions at liquid–liquid, liquid–solid, and solid–solid interfaces and at the liquid and solid surfaces exposed to air. Besides the broken symmetry, liquid–liquid, liquid–solid, and biological (e.g., cell membrane) surfaces are able to exhibit the interactions that are not present in composing bulk media. One of the major difficulties in studying interfacial phenomena is distinguishing the physical properties of the interface from that of the bulk components. Fluorescence probing can be selective to interface if the probe is covalently bound to the surface of one of the partners before forming an interface or if the probe has specific affinity to interfacial region.

The *small-size organic dye* molecules are commonly used to study interfaces to characterize interfacial polarity, fluidity, solvation dynamics, statics, and dynamics of orientation of solute molecules. We observe domination of empirical approach in determining these characteristics.

9.1 Air–Liquid Interface

The polar liquid surfaces contacting with air or other gasses are probably the simplest cases in which the fluorescence probes can successfully combine surfactant and reporting properties. The experiments demonstrate well-expected results: they often show the properties intermediate between vacuum and bulk solvent conforming to "half hydration" concept [140]. Meantime, it is obvious that obtained estimates of interfacial polarity depend strongly on the probing dye and on its interactions at the interface.

The air–water interface is probably ideal for the studies with time-resolved SHG spectroscopy [141], electronic sum frequency generation [142], and other surface-selective nonlinear methods. The application of the latter technique with the aid of Malachite Green as a viscosity indicator has shown higher local viscosity at the interface than in bulk water [143]. A slow component appears also in solvation dynamics.

The coupling of probe response to such properties as surface tension and surface diffusion is probably more important than determination of "polarity". The rotational dynamics of the probes at the air/water interface is influenced by the surface-active agents, especially such strongly charged species as sodium dodecyl sulfate (SDS). It was shown experimentally that the rotational dynamics of the coumarin 314 probe is three times slower at the presence of SDS, although the dependence of this effect on the surfactant charge density is rather weak [144]. The same system was studied in MD simulations, which revealed atomic details of the interface organization and the localization of the probe at different surfactant coverage. Two well-defined solvation environments for the probe were identified. In the first one, the probe lies adjacent to two-dimensional cluster of the surfactant molecules. In the second type of solvation, the coumarin is embedded within the compact surfactant domain. There is a gradual transition from parallel to perpendicular dipolar alignment of the probe with respect to the interface as the concentration of surfactant increases [145].

The influence of the other surfactant – stearic acid on this probe was also studied by means of the femtosecond pump−probe spectroscopy. It was shown that different solvation dynamics at the surfactant modified interface is attributed to rearrangement of the hydrogen bonding network of water molecules near the interface due to interactions with the hydrophilic carboxyl group of the surfactant [141].

The role of the hydrogen bonds in the dynamic behavior of the vapor–water and the metal–water interfaces could be revealed by MD simulations, which aids in understanding the atomistic details of the interfacial dynamics. Particularly, for both types of interfaces, the relaxation of hydrogen bonds is slower than in bulk water. However, the dynamics of the hydrogen bonds formed between the molecules at the interface and in its adjacent layer is faster than that in the bulk liquid [146]. It was shown that the accounting for electronic polarization effect in water is vital in such MD studies and the polarizable water models should be used instead of the standard fixed-charge models in order to obtain correct dynamic properties of the hydrogen bonds [147].

The MD studies of the liquid–vapor system are not limited to pure water. The studies of the trimethylamine-*N*-oxide [148], nitrobenzene [149], and ammonia [150] mixed with the water-vapor systems were reported recently. In the case of nitrobenzene, the surface rotation is much faster than in the bulk liquid, but slows down and approaches the bulk behavior as the solute becomes more polar. The reorientation dynamics of the nitrobenzene is quite anisotropic, with out-of-plane rotation of nitrobenzene faster than in-plane rotation [149].

9.2 Liquid–Liquid Interfaces

These interfaces are formed between contacting liquids that are immiscible and differ in many properties, such as polarity, electronic polarizability, and association via H-bonds. The strongly *amphiphilic* dye molecules (containing both apolar and polar sites) can act as surfactants concentrating at the interface between these liquids and reporting about the properties of the interface [151]. Many questions can be addressed in the studies with such fluorescent dyes. Are these interfaces sharp or there is a continuous transition from one phase to another? If such transition area exists, what is its length (the so-called dipole width)? How this width depends on the ability of contacting solvents to associate? If one of the solvents is a strong H-bond former (e.g., water), what happens with its unsaturated H-bonds at the interface? And, finally, is the continuous description of liquids useful for characterizing interfacial environments?

Currently, there are no full and definitive answers to these questions. Distribution between two different types of solvation sites was suggested based on the data obtained with popular fluorescence probes ANS [152] and coumarin 343 [145]. In contrast, sulforhodamine dye located at the interface demonstrates intermediate polarity between two phases as it is witnessed by time-resolved spectroscopy with total internal reflection observation [153]. This discrepancy probably exists because of nonsufficient precision of the dye location and of spatial resolution of response. Discrimination of response from the bulk solvents for collecting emission only from interface is not sufficient, and precise tuning of its location is needed across the interface. Moreover, the effects described for small solvatochromic dyes in phospholipid membranes by one of the present authors many years ago [55, 154] can take place in all cases of heterogeneous dye location (see also Sect. 4.1). Strong increase of their dipole moment on excitation can induce translational and rotational relaxation in the excited state along the polarity gradient that can induce additional heterogeneity to solvent response leading to false interpretation as the broadening of interface zone.

Enhancing selectivity of response to interface and decreasing ambiguity in interpretation can be achieved by both probe design and new spectroscopic techniques, such as surface-enhanced SHG (see Sect. 8.1.4). In order to optimize probe location, a special technique of "molecular rulers" to measure the dipolar width of weakly and strongly associating liquid–liquid interfaces was developed [155]. These dye derivatives with surfactant properties consist of anionic sulfate groups attached to hydrophobic, solvatochromic probes by different length alkyl spacers (Fig. 11). The probes themselves are based on *p*-nitroanisole, an aromatic solute, whose bulk solution excitation wavelength (λ_{max}^{ex}) monotonically shifts by more than 20 nm as solvent polarity changes from that of cyclohexane to that of water.

Comparison of resonance-enhanced SPG spectra for these rulers suggests that the water–cyclohexane interface possesses a molecularly sharp dipolar width (within 0.9 nm). A larger difference is observed at the water–octanol interface suggesting the presence of transition zone with intermixed components and polarity lower than that of bulk octanol or bulk water.

The applications of "molecular rulers" suggest but still do not prove the existence of transition zone between contacting liquids. Meantime, several experiments and theoretical studies suggested that the liquid/liquid interfaces are molecularly sharp and therefore can be modeled by a simple ideal plane. The interfacial cavity incorporating the dye can be viewed in the following way (see Fig. 12). The solute inserts a cavity formed by interlocking spheres and the cavity crosses the planar interface. The cavity surface and the plane are then partitioned into small elements.

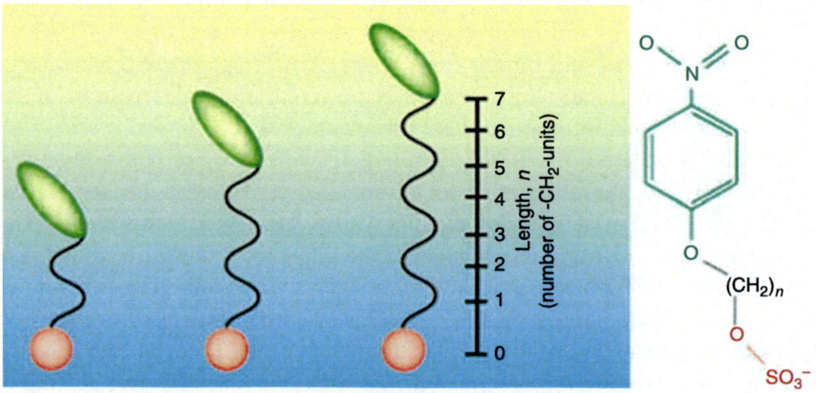

Fig. 11 Schematic representation of molecular ruler surfactants adsorbed to a liquid–liquid interface, and their general structure. As the alkyl spacer between the headgroup (*red circle*) and solvatochromic chromophore (*green ellipse*) lengthens, the hydrophobic probe can extend further into the organic phase. Surfactants are referred to as C_n rulers, where *n* corresponds to the number of methylene (–CH_2–) groups (2, 4, 6, 8) in the alkyl spacer [155]

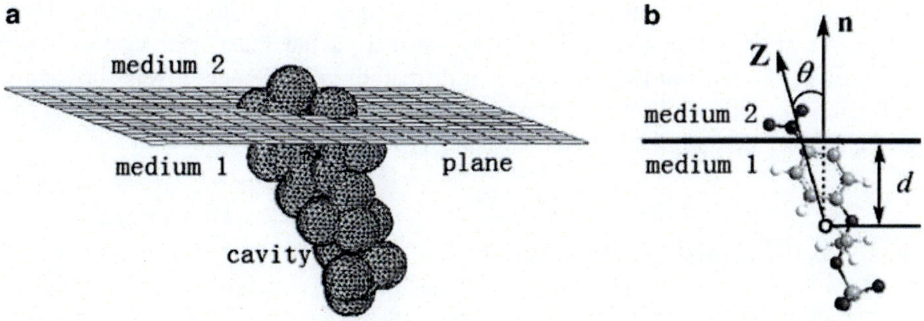

Fig. 12 Schematic depiction of molecular ruler surfactant at interface (**a**) and the orientation of the inertial principal axis *Z* and the embedding depth *d* of the molecular mass center into the water phase (**b**) [15]

The polarizable continuum model (PCM) was applied to calculate the solvation energy accounting for both electrostatic and nonelectrostatic solute–solvent interactions. In this way it was suggested to obtain the information about the position and orientation of the interfacial molecule [15].

There is a special interest to interfaces formed by water. The interface weakens the interactions between adjacent water molecules that cannot be substituted by bonding with organic molecules and this may result in significant interfacial structuring and molecular orientation on both sides of the interface. Upon solute addition the H-bonding between neighboring water molecules must change to adjust to its new solid, liquid, or gaseous neighbor.

In recent years, researchers have turned to molecular simulations to provide an insight into the physicochemical properties of the liquid–liquid interfaces [149]. The setups of such simulations are usually quite similar. The solute of interest is placed into the interface of two liquids and then into the bulk phase of each liquid. The solvent dynamic response is then computed in each position. The interfaces of different organic solvents such as 1-octanol, 1,2-dichloroethane, *n*-nonane, and carbon tetrachloride with water were studied in this way [156]. It was shown that the solvent dynamic response at the interface is more complex than in the bulk liquids. The relaxation involves multiple time scales corresponding to contributions from both solvents and from the unique structural and dynamic properties of the interface. It was shown that MD simulations could generate vibrational sum frequency (VSF) spectral profiles of water across the interfaces with several organic solvents such as carbon tetrachloride, chloroform, and dichloromethane. These spectral profiles are calculated as functions of both frequency and interfacial depth, providing a visual description of interfacial water structure that can be difficult to elucidate in experiments [157].

We may conclude that despite of many efforts the properties of liquid–liquid interface are not fully understood. It is still not known if the broad distribution in local polarity values observed with fluorescence probing results from surface roughness or it occurs due to probe distribution around a sharp interface. More definitive should also be the results of estimating the interfacial viscosity that determines the rates of two-dimensional diffusion. For instance, it was shown that diffusion of lipids at oil–water interface depends on mismatch of viscosities of two phases rather than on their absolute values [158].

9.3 Solid–Liquid Interfaces

In *solid–liquid interfaces*, some interactions that are saturated in bulk liquid appear unsaturated with some new interactions formed. Stability of suspensions of nanoparticles formed by these solids is determined by these interactions as well as the efficiency of catalyzed reactions at an interface. Meantime, the researcher should be extremely cautious in application of fluorescence probes, since being immobilized on solid surface the dyes may change dramatically their photophysical properties, up to generation of new bands [159].

The method of resonance-enhanced SHG could be used with great success to study solvation of such molecules as *p*-nitroanisole or indoline at different solid/liquid interfaces [160]. The SHG spectra of these compounds adsorbed to silica and solvated by organic solvents interfaces allow to probe either interfacial polarity or interfacial hydrogen bond donating/accepting abilities. The SHG results show that interfacial polarity probed by *p*-nitroanisole depends strongly on solvent structure, while the hydrogen bonding interactions probed by indoline are insensitive to solvent identity but depend on the hydrogen bond donating properties of the polar silica substrate. Complementary quantum ab initio simulations in the bulk solvent were also performed, which allowed to create a comprehensive picture of solvation mechanisms at different polar solid surfaces [160].

MD simulations of the solid–liquid interfaces are usually straightforward in the case of explicit water models. However, hybrid QM/MM simulations and implicit water simulations of solid–liquid interfaces become very nontrivial because of the lack of adequate theory. Recently, sophisticated theory of QM/MM simulations at the solid–liquid interface in the presence of implicit solvent was developed. It allows simulating the solute adsorbed at the solid surface, which extends periodically in two dimensions, at the quantum level and surrounded by continuous dielectric medium. This simulation technique was employed successfully to investigate the acid–base equilibrium at the TiO_2–water interface [161].

9.4 Solid–Solid Interfaces

Intermolecular interactions at solid–solid interfaces are critical to the function of different composite materials, since the interfacial regions determine many properties such as their strength, thermal stability, and durability. Their importance is well recognized, but the possibilities for their studies are still very limited and are frequently reduced to determination of the glass transition temperature (T_g). Due to the low weight fraction of the interface region, the conventional techniques of thermal analysis are not adequate to characterize such thermal transitions. Meantime in fluorescence studies it was frequently reported that at polymer interfaces T_g changes and its range broadens indicating the change of molecular organization and dynamics [162]. T_g is one of the fundamental properties of polymers. Above T_g, the material behaves like a supercooled liquid. Lowering the temperature brings rigidity to the polymer matrix. A number of experimental and theoretical studies describe a spatial heterogeneity at the glass transition that can be significantly perturbed by the contact with odd surface.

Technologically important glass–polymer composites allow chemical grafting a fluorescent probe onto an organosilane-modified glass surface with subsequent formation of polymer on this surface [124]. This suggests an efficient way to monitor the chemical and physical properties of the buried polymer/glass interfacial region by obtaining the signal only from interfacially localized dye. Sometimes the third component is used that serves for providing adhesion between interacting surfaces, and it can be also used for introduction of fluorescence sensing dyes. Meantime, the level of application and analysis of spectroscopic data is rather poor, even with the use of solvatochromic dyes, and commonly does not go beyond empirical "polarity" and "viscosity" correlations.

Ultrathin polymer films of several nanometers in the form of flat layers or as monolayer and multilayer shells of nanoparticles have found a variety of applications. Fluorescence studies have shown that in these films incorporated into fibers the surface-induced perturbations can propagate away from a fiber surface to induce the formation of a three-dimensional zone with properties different from those of the bulk polymer [163]. Particularly, concentration gradients can arise if one of the reacting components of the resin mixture has a higher affinity for the fiber surface than the other components. Therefore, the challenge lies in proper description of perturbations at surfaces and interfaces of polymer materials. The properties of these films strongly depend on interaction with substrates [10]. These findings indicate that the dynamics of nanoscale layers can be greatly influenced by adjacent domains even when the two polymers form a narrow interface.

9.5 The Surfaces of Solids

With its relatively low structural resolution, fluorescence probing cannot compete with other physical methods addressing the structures and properties of solid surfaces. Meantime, many researchers try to formally derive the empirical "polarity" using the calibration made in liquids (e.g., [164] and citations therein), thus ignoring the differences in molecular relaxations between these media that can be very strong. The sensitivity in response of adsorbed probe to its weak intermolecular interactions can be used here in a different way – to characterize the surface by a parameter related to inhomogeneous broadening (see Sect. 5) that could be a direct measure of distribution in the strengths of these interactions. This approach still needs its exploration.

Concluding this section we must stress the following. Surfaces and interfaces present the strongest molecular-scale gradients of many physical properties. Located at the interface, organic dye molecules experience strong anisotropy in these interactions, and being structurally anisotropic themselves, they adopt preferred location and orientation together with some distribution around these preferred values. This structural anisotropy makes the physical modeling based on reactive field of spherical symmetry nonproductive or even meaningless. No better are empirical approaches that intuitively assume the tested medium as isotropic and continuous. Even at very selective location, the probes may not catch very important features just because the change of molecular order and interactions at the interface appear to be very sharp extending to bulk phases only to the length of several atoms. The probe sizes are simply not commeasurable with the sizes of molecules and groups of atoms forming these gradients.

10 Guest–Host Composites and Nanocavities

Macrocyclic hosts such as cyclodextrins and calixarenes open a wide range of opportunities for new supramolecular chemistry and materials. They can include organic dyes with a dramatic change of spectroscopic response, and such complexes have found versatile use in sensing technologies [7]. Likewise chelating and catalytic properties of nanoporous materials are in the frontier of research and development. Complexation reactions with them serve as excellent models for understanding the general behavior of dyes in structurally confined environments. We have to see how the analysis of their fluorescence response could be used for the description of this behavior.

10.1 Cyclodextrins

Cyclodextrins (CDs) are cyclic oligosaccharides consisting of six or more glucopyranose units attached by α-1,4-linkages. They are water-soluble natural products, inexpensive, commercially available, nontoxic, and readily functionalized. With hydroxylic groups facing the outer space, these molecules are highly soluble in water, whereas the inner space can accommodate different low-polar molecules, including fluorescent dyes. Their binding affinity is determined by cavity size. It increases from α-cyclodextrin (α-CD) to β-cyclodextrin (β-CD) and to γ-cyclodextrin (γ-CD) [165]. Comparative studies on the binding of different neutral, anionic, and cationic dyes [166] demonstrated that this binding is mainly hydrophobic and is only insignificantly affected by the charge of guest molecule. In all studied cases the tight complexes are formed, which is manifested in the time-resolved studies of anisotropy that detects rotation of the whole complex. Ultrafast dielectric relaxations are manifested, they were detected by time-resolved methods [167]. These systems exhibit also an ultraslow ($\geq$1 ns) component of solvation dynamics [168] suggesting that confinement of water and dynamic exchange of bound and free water are the main sources of such slow motions [169].

It is interesting to note that steric restriction in CD cavities can provide more significant effects than the change of polarity estimated from empirical correlations based on reference spectra in liquid solvents. This is clearly seen for the dyes with their response based on ICT [92] and ESIPT [170].

Solvation of coumarin 153 dye in the hydrophobic cavities of di- and trimethylated-cyclodextrins (CD) in aqueous solutions was recently studied using MD simulations [171]. The partial atomic charges of the dye were computed in ab initio calculations in the ground and excited states and used in classical MD simulations. Both equilibrium and dynamical characteristics could be obtained in such studies. MD simulations provide direct structural information about location of the probe inside the CD cavities. It was shown that two alternative orientations of the dye are possible. In the so-called slanted geometry, the dye lies mostly parallel to one of the glucose units of the CD, but in the upright geometry it is oriented parallel to the axis of CD cavity (Fig. 13). The rotational dynamics of the trapped coumarin includes diffusive motions within a local restrictive environment coupled to the overall

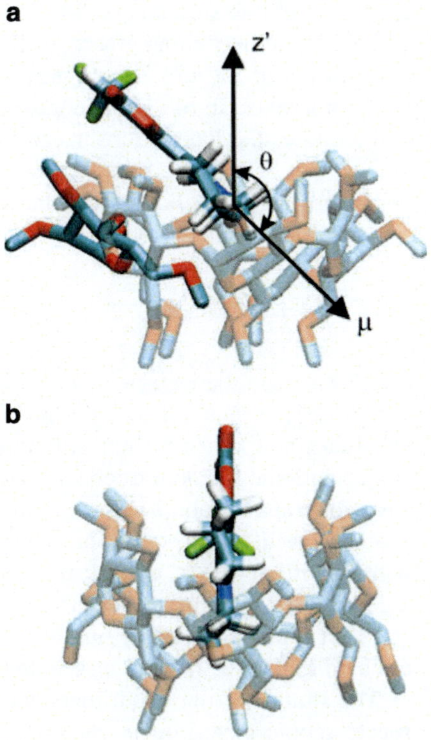

Fig. 13 Alternative configurations of the coumarin 153-cyclodextrin complexes in aqueous solutions (**a**) slanted geometry, where the dye is inclined relative to the axis of CD cavity and lies parallel to one of the glucose units of the CD (shown in *bright colors*); (**b**) upright geometry, where the dye lies parallel to the axis of CD cavity. Picture from [171]

rotational motion of the surrounding CD. There are two distinct components of the solvation response, which correspond to water relaxation within 2–3 ps time scale and to fast subpicosecond relaxation that involves intramolecular motions in CDs, respectively. Slow *gauche-trans* interconversions in primary hydroxyl chains of the CD, which take place in the nanosecond time scale, are unlikely to be connected to the electronic excitation of the probe.

10.2 Calixarenes

Calixarenes are a class of polycyclic compounds that can allow (by proper chemical substitutions) achieving high-affinity binding of many small molecules. Calixarenes are cyclic oligomers (usually 4-mers, calix[4]arenes) obtained by phenol–formaldehyde condensation. They exist in a "cup-like" shape with a defined upper and lower rim and a central annulus. Their rigid conformation of skeleton forming a cavity together with some flexibility of side groups enables them to act as host molecules. By functionally modifying the upper and/or the lower rims it is possible to prepare various derivatives with differing selectivity for various guest molecules, which makes them good chelators and sensors for small molecules and ions. A variety of environment-sensitive dyes were used for attachment, particularly, to an upper rim of their structures [144]. In addition, calix[4]arenes chelate different dyes with dramatic quenching effect [172], so that sensing response can be achieved on their displacement with the targets. This, probably, makes difficult the studies of their complexes. In our knowledge, no systematic studies on fluorescence probing of intermolecular interactions in calixarene-dye complexes have been performed by now. It can be expected that the probing dyes will exhibit "interfacial behavior" being surrounded by both low-polar groups in the cavity and the polar solvent.

10.3 Dendrimers

Dendrimers are the well-defined, symmetric tree-like macromolecules that possess a high degree of order and the possibility to contain selected chemical units in predetermined sites of their structure. Due to the presence of some degree of conformational freedom of chain segments, they possess structural flexibility sufficient for trapping and holding small molecules including fluorescent dyes [173]. Probing the dendrimer interior was provided by different polarity-sensitive dyes with the reference to empirical scales of polarity [9, 174]. This description is definitely not conclusive because of solvent molecules also present in dendrimer interior (in nonaccountable number and location) and of unknown relaxation rates of chain segments that can also influence the recorded spectra.

The studies of the dendrimers show remarkable complementarity in results of spectroscopic experiments and MD simulations. The instantaneous structure of dansyl terminated dendrimer of propylene amine dissolved in CH_2Cl_2 was investigated in MD simulations. Subsequent incorporation of several eosin Y dyes into the dendrimer was also studied. The guest molecules are distributed at two main distances from the center of the dendrimer and their surroundings are very dynamic. Eosins move inside the hyperbranched molecule in a way similar to what the solvent molecules do and sometimes aggregate [175]. More recently, the time-resolved fluorescence spectroscopy and MD simulations were applied simultaneously to study internal dynamics and excitation energy transfer from dansyls to eosin in these dendrimers. Combination of results from fluorescence anisotropy and the MD simulation suggested three time domains for the internal dynamics of the dendrimers: about 60 ps for motions of the outer-sphere dansyls, 500–1,000 ps for restricted motions of back-folded dansyls, and 1,500–2,600 ps for the overall rotation. Solvation relaxation around a single dansyl of a dendrimer could be analyzed using the time-dependent spectral shifts of fluorescence. The characteristic time of relaxation was estimated to be 20–30 ps. Dendrimer-independent excitation energy transfer from dansyls to eosins was also observed [176].

10.4 Sol-Gel Derived Materials

Sol-gel derived materials are the materials that can be obtained with very high nanoscale porosity and can incorporate a large amount of guest together with the solvent molecules. Functional properties of these materials depend strongly on the composition of gel-forming precursors, the type of solvent, the concentrations of impurities, etc. Fluorescent dyes can be incorporated into this system on the step of gel formation, and the information that they provide can be very valuable for understanding their nanoscale structure and dynamics [177]. In these media, the presence of voids of different sizes and shapes, the chemical disorder of the matrix, and the presence of trapped solvent molecules result in diversity of local effects that are to be expected when the response of probing dye is examined.

Thus, with the aid of environmentally sensitive dye Nile Red, it was found that the peak fluorescence emission could be tuned by addition of solvents with different dielectric constants indicating the dye solvation within the matrix pore structure [178]. The lower limit of this wavelength range was mainly determined by interaction with the host, while the upper limit was related to emission from an added solvent mixed with water. Specificity in response to the presence of polar solvent can be provided by the probe exhibiting ESIPT reaction, as it was demonstrated with benzoxazole dye [38]. Switching of fluorescence emission between normal and ESIPT forms was observed as a function of the ratio of inorganic and organic components in nanocomposites [179].

The study of translational and orientational motion by diffusional quenching and by time-resolved anisotropy seems to be straightforward methods when applied to nanoporous systems [180]. But there were no attempts to separate viscous diffusion, dielectric friction, and dielectric induced rotations, which as stated above (Sect. 4.1) can dramatically modulate probe response.

The dielectric response of the polar fluid trapped in the nanocavities was recently investigated in the series of MD simulations. Although this study was not focused on the real solvent and the simple model fluid consisting of spherical dipolar particles was used, it provides some insight into dielectric properties of such confined systems. Particularly, dynamic relaxation of the total dipole moment of the liquid depends strongly on dielectric constant. At the same time, the density profile and local orientational order parameter of the dipoles turn are rather insensitive to dielectric constant of the medium. Such studies demonstrate the ability of MD simulations to capture important properties of confined solvents [181].

11 Conclusions and Prospects

The power of molecular probing method is the ability to operate with the systems of quite different complexity. Meantime, this complexity cannot be seen directly in fluorescence probe response. The probe responds in some integrated way to intermolecular interactions and dynamics in its environment only, so by observing these collective molecular effects we are able to derive only a small number of output variables. Therefore, intuitively or explicitly we operate with mesoscopic models. Light absorption and emission of fluorophore obeys quantum mechanical laws but the description of collective effects in these interactions can be described in classical terms. This classical description allows averaging and building the bridge to quasi-continuous approximation.

Such approach proved to be productive when the probing system is structurally homogeneous. In the case of nanoscale heterogeneity, the local "polarity" and the local "fluidity" within or between organized assemblies are appreciably different from that in the bulk medium. They include structural anisotropy and anisotropic mobility in a way that cannot be accounted and analyzed with empirical approach based on the references derived from the study of homogeneous materials. The benefit of simplicity turns out to be the loss of information that is needed for understanding different molecular events, such as adsorption, self-assembly, and catalysis.

The methods are needed that allow molecular-scale description of interactions and of their dynamics. To achieve this goal, several steps have been made already. The techniques of introduction of organic dyes by covalent labeling are nicely complemented by rapidly developing methodology based on affinity labeling and molecular recognition. Meantime, this trend has almost reached the limit in its perfection and it does not help to resolve the gradients of physical parameters that are steeper than the probe dimensions. Since these problems refer to the systems studied and to spatial sensitivity of response, they cannot be resolved in full by the improvement of detection methods. What remains is to change the paradigm. The factors complicating the analysis should be transformed to the source of new information. The progress can be seen with the development of means to characterize the ensemble- and time-correlated distribution of dye interaction energies with their local environment. This could be an important characteristic of interaction energies across the interface zone (the analogs of the gradients of polarity), which could be achieved by extension in the analysis of inhomogeneous broadening effects. The gradients of interaction energies can be seen as an additional factor broadening the spectra. The rates of reorganization within the ensemble of states forming this distribution (relaxation phenomena) will then appear as the variables characterizing the dynamics (the analogs of the gradients of viscosity).

Fundamental understanding of dynamics and interactions at the molecular level cannot be achieved without the aid of molecular simulation based on as quantum-mechanics and classical MD. Atomic resolution of these methods is beyond the possibilities of any spectroscopic techniques, which provides extremely precise structural information about the studied systems. The interactions between the molecule of interest and its surrounding could be decomposed into components in MD simulations, which allow studying the contribution of various molecules, functional groups, or even individual atoms. Such simulations are the especially valuable tools in the studies of systems with nanoscale structural and dynamic heterogeneity where the description based on reduction to single or several parameters are insufficient or even inappropriate.

Like any other technique of research, fluorescence probing has its merits and limitations. In this chapter, we tried to outline these limitations and show that some of them are of fundamental character and the others can be reduced or even eliminated with the improvement in techniques and analyses of data. This opens a broader way for applications in different areas, including the studies in systems with strong nanoscale heterogeneity. With these prospects, fluorescence probing has a bright future.

References

1. Reichardt C (1990) Solvents and solvent effects in organic chemistry, 2nd edn. VC, Germany
2. Berendsen HJC (1996) Bio-molecular dynamics comes of age. Science 271:954 955
3. van Gunsteren WF, Bakowies D, Baron R, Chandrasekhar I, Christen I, Daura X, Gee P, Geerke DP, Glattli A, Hunenberger PH, Kastenholz MA, Oostenbrink C, Schenk M, Trzesniak D, van der Vegt NFA, Yu HB (2006) Biomolecular modeling: goals, problems, perspectives. Angew Chem Int Ed 45:4064–4092
4. Leontiadou H, Mark AE, Marrink S-J (2007) Ion transport across transmembrane pores. Biophys J 92:4209–4215
5. Singh UC, Kollman PA (1986) A combined ab initio quantum mechanical and molecular mechanical method for carrying out simulations on complex molecular systems: applications to the $CH_3Cl + Cl-$ exchange reaction and gas phase protonation of polyethers. Journal of Computational Chemistry 7(6):718–730
6. Ding L, Kang J, Lü F, Gao L, Yin X, Fang Y (2006) Fluorescence behaviors of 5-dimethylamino-1-naphthalene-sulfonyl-functionalized self-assembled monolayer on glass wafer surface and its sensing properties for nitrobenzene. Thin Solid Films 515:3112–3119

7. Demchenko AP, Yesylevskyy SO (2009) Nanoscopic description of biomembrane electrostatics: results of molecular dynamics simulations and fluorescence probing. Chem Phys Lipids 160:63–84
8. van der Spoel D, Lindahl E, Hess B, Groenhof G, Mark AE, Berendsen HJC (2005) GROMACS: fast, flexible and free. J Comp Chem 26:1701–1718
9. Case DA, Darden TA, Cheatham TE, Simmerling CL, Wang J, Duke RE, Luo R, Walker RC, Zhang W, Merz KM, Roberts BP, Wang B, Hayik S, Roitberg A, Seabra G, Kolossváry I, Wong KF, Paesani F, Vanicek J, Wu X, Brozell SR, Steinbrecher T, Gohlke H, Cai Q, Ye X, Wang J, Hsieh M-J, Cui G, Roe DR, Mathews DH, Seetin MG, Sagui C, Babin V, Luchko T, Gusarov S, Kovalenko A, Kollman PA (2010) AMBER 11. University of California, San Francisco
10. Sulpizi M, Rohrig UF, Hutter J, Röthlisberger U (2005) Optical properties of molecules in solution via hybrid TDDFT/MM simulations. Int J Quantum Chem 101:671–682
11. Ando K (2008) Ligand-to-metal charge-transfer dynamics in a blue copper protein plastocyanin: a molecular dynamics study. J Phys Chem B 112:250–256
12. Dal Peraro M, Villa AJ, Carloni P, Klein ML (2007) Role of zinc content and the catalytic efficiency of B1 metallo β-Lactamases. JACS 129:2808–2816
13. Car R, Parrinello M (1985) Unified approach for molecular dynamics and density-functional theory. Phys Rev Lett 55:2471–2474
14. Kina D, Arora P, Nakayama A, Noro T, Gordon MS, Taketsugu T (2009) Ab initio QM/MM excited-state molecular dynamics study of coumarin 151 in water solution. Int J Quantum Chem 109:2308–2318
15. Ma JY, Wang JB, Li XY, Huang Y, Zhu Q, Fu KX (2008) A study on orientation and absorption spectrum of interfacial molecules by using continuum model. J Comput Chem 29:198–210
16. Feig M, Brooks CLI (2004) Recent advances in the development and application of implicit solvent models in biomolecule simulations. Curr Opin Struct Biol 14:217–224
17. Klamt A, Schüürmann G (1993) COSMO: a new approach to dielectric screening in solvents with explicit expressions for the screening energy and its gradient. J Chem Soc Perkin Trans 2:799–805
18. Yesylevskyy SO, Klymchenko AS, Demchenko AP (2005) Semi-empirical study of two-color fluorescent dyes based on 3-hydroxychromone. J Mol Struct THEOCHEM 755:229–239
19. Chipman DM (2009) Vertical electronic excitation with a dielectric continuum model of solvation including volume polarization I. Theory. J Chem Phys 131:014103
20. Chipman DM (2009) Vertical electronic excitation with a dielectric continuum model of solvation including volume polarization II. Implementation and applications. J Chem Phys 131:014104
21. Mennucci B (2006) Time dependent solvation: a new frontier for quantum mechanical continuum models. Theor Chem Acc 116:31–42
22. Suppan P, Ghoneim N (1997) Solvatochromism. Royal Society of Chemistry, Cambridge, UK
23. Mataga N, Kubota T (1970) Molecular interactions and electronic spectra. Marcel Dekker, New York
24. Sharma VK, Sahare PD, Rastogi RC, Ghoshal SK, Mohan D (2003) Excited state characteristics of acridine dyes: acriflavine and acridine orange. Spectrochim Acta A 59:1799–1804
25. Umadevi M, Vanelle P, Terme T, Ramakrishnan V (2006) Spectral investigations on 2, 3-bis (chloromethyl)-1, 4- anthraquinone: solvent effects and host-guest interactions. J Fluoresc 16:569–579
26. Nemkovich NA, Rubinov AN, Tomin VI (1991) In: Lakowicz JR (ed) Topics in fluorescence spectroscopy, principles. Plenum, New York, pp 367–428
27. Bakhshiev NG (1972) Spectroscopy of intermolecular interactions. Nauka, Leningrad

28. Hunt J, Hunt KLC (2003) Non-local dielectric functions on the nanoscale: electronic polarization and fluctuations. J Mol Struct THEOCHEM 633:145–155
29. Bublitz GU, Boxer SG (1997) Stark spectroscopy: applications in chemistry, biology, and materials science. Annu Rev Phys Chem 48:213–242
30. Liptay W (1969) Electrochromism and solvatochromism. Angew Chem Int Ed 8:177–188
31. Tsuchida M, Tsujita W, Majima Y, Iwamoto M (2001) Interfacial electrostatic phenomena in phthalocyanine Langmuir-Blodgett films under photoillumination. Jpn J Appl Phys 40:1315–1321
32. Clarke RJ (2010) Electric field sensitive dyes. In: Demchenko AP (ed) Advanced fluorescence reporters in chemistry and biology I: Fundamentals and molecular design. Springer Series on Fluorescence 8. Springer, Heidelberg, pp 331–344
33. Demchenko AP, Mely Y, Duportail G, Klymchenko AS (2009) Biophysical properties of lipid membranes by environment-sensitive fluorescent probes. Biophys J 96:3461–3470
34. Pivovarenko VG, Klueva AV, Doroshenko AO, Demchenko AP (2000) Bands separation in fluorescence spectra of ketocyanine dyes: evidence for their complex formation with monohydric alcohols. Chem Phys Lett 325:389–398
35. Catalan J, Perez P, Laynez J, Garcia-Blanco F (1991) Analysis of the solvent effect on the photophysics properties of 6-propionyl-2-(dimethylamino)naphthalene (PRODAN). J Fluoresc 4:215–223
36. del Valle JC, Catalan J (2001) Understanding the solvatochromism of 10-hydroxybenzo[h] quinoline. An appraisal of a polarity calibrator. Chem Phys 270:1–12
37. Zhang L, Wang L, Kao Y-T, Qiu W, Yang Y, Okobiah O, Zhong D (2007) Mapping hydration dynamics around a protein surface. PNAS 104:18461–18466
38. Aschi M, Fontana A, Di Meo EM, Zazza C, Amadei A (2010) Characterization of electronic properties in complex molecular systems: modeling of a micropolarity probe. J Phys Chem 114:1915–1924
39. Mazur M, Blanchard GJ (2005) Probing intermolecular communication with surface-attached pyrene. J Phys Chem B 109:4076–4083
40. Berezin MY, Lee H, Akers W, Achilefu S (2007) Near infrared dyes as lifetime solvatochromic probes for micropolarity measurements of biological systems. Biophys J 93:2892–2899
41. Lakowicz JR (2007) Principles of fluorescence spectroscopy. Springer, New York
42. Dela Cruz JL, Blanchard GJ (2002) The influence of chromophore structure on intermolecular interactions. A study of selected rhodamines in polar protic and aprotic solvents. J Phys Chem A 106:10718–10724
43. Horng M-L, Gardecki JA, Maroncelli M (1997) Rotational dynamics of Coumarin 153: time-dependent friction, dielectric friction, and other nonhydrodynamic effects. J Phys Chem A 101:1030–1047
44. Ferrer ML, del Monte F, Levy D (2001) Microviscosities at the porous cage of silica gel-glasses and Ormosils through fluorescence anisotropy. J Phys Chem B 105:11076–11080
45. Baptista MS, Indig GL (1998) Effect of BSA binding on photophysical and photochemical properties of triarylmethane dyes. J Phys Chem 102:4678–4688
46. Haidekker MA, Nipper M, Mustafic A, Lichlyter D, Marianna Dakanali M, Theodorakis EA (2010) Dyes with segmental mobility: molecular rotors. In: Demchenko AP (ed) Advanced fluorescence reporters in chemistry and biology I: Fundamentals and molecular design. Springer Series on Fluorescence 8. Springer, Heidelberg, pp 267–308
47. Haidekker MA, Theodorakis EA (2007) Molecular rotors–fluorescent biosensors for viscosity and flow. Org Biomol Chem 5:1669–1678
48. Kuimova MK, Yahioglu G, Levitt JA, Suhling K (2008) Molecular rotor measures viscosity of live cells via fluorescence lifetime imaging. J Am Chem Soc 130:6672–6673
49. Pandey S, Fletcher KA, Baker SN, Baker GA (2004) Correlation between the fluorescent response of microfluidity probes and the water content and viscosity of ionic liquid and water mixtures. Analyst 129:569–573

50. Zana R (1999) Microviscosity of aqueous surfactant micelles: effect of various parameters. J Phys Chem B 103:9117–9125
51. Volovik Z, Demchenko A, Skursky S (1994) Solvent-dependent photophysics of non-symmetric polymethine dyes as fluorescence probes: dual emission and inhomogeneous broadening. Proc SPIE Int Soc Opt Eng 2137:600–607
52. Wandelt B, Cywinski P, Darling GD, Stranix BR (2005) Single cell measurement of micro-viscosity by ratio imaging of fluorescence of styrylpyridinium probe. Biosens Bioelectron 20:1728–1736
53. Dutt GB, Raman S (2001) Rotational dynamics of coumarins: an experimental test of dielectric friction theories. J Chem Phys 114:6702
54. Papazyan A, Maroncelli M (1995) Rotational dielectric friction and dipole solvation: tests of theory based on simulations of simple model solutions. J Chem Phys 102:2888–2919
55. Gakamsky DM, Nemkovich NA, Rubinov AN (1992) Wavelength-dependent rotation of dye molecules in a polar solution. J Fluoresc 2:81–92
56. Maroncelli M, Castner EW, Bagchi B, Fleming GR (1988) Dipolar solvation dynamics. Faraday Discuss Chem Soc 85:199–210
57. Bhattacharyya K (2008) Nature of biological water: a femtosecond study. Chem Commun (Camb):2848–2857
58. Magee MD (1974) Dielectric relaxation time, a non-linear function of solvent viscosity. J Chem Soc Faraday Trans 70:929–938
59. Mazurenko YT, Bakhshiev NG (1970) The influence of orientational dipolar relaxation on spectral, temporal and polarizational properties of luminescence in solutions. Optika Spektr 28:905–913
60. Bagchi B, Fleming GR (1990) Dynamics of activationless reactions in solution. J Phys Chem 94:9–20
61. Gardecki JA, Maroncelli M (1999) Comparison of the single-wavelength and spectral-reconstruction methods for determining the solvation-response function. J Phys Chem A 103:1187–1197
62. Bhattacharyya K (2003) Solvation dynamics and proton transfer in supramolecular assemblies. Acc Chem Res 36:95–101
63. Childs W, Boxer SG (2010) Solvation response along the reaction coordinate in the active site of ketosteroid isomerase. J Am Chem Soc 132:6474–6480
64. Furse KE, Corcelli SA (2010) Molecular dynamics simulations of DNA solvation dynamics. J Phys Chem Lett 1:1813–1820
65. Granick S (1991) Motions and relaxations of confined liquids. Science 253:1374–1379
66. Butt H, Graf K, Kappl M (2003) Physics and chemistry of interfaces. Wiley-VCH, Weinheim
67. Renge I (2006) Influence of temperature and pressure on shape and shift of impurity optical bands in polymer glasses. J Phys Chem A 110:3533–3545
68. Demchenko AP (2002) The red-edge effects: 30 years of exploration. Luminescence 17:19–42
69. Demchenko AP (2008) Site-selective Red-Edge effects, Chap. 4. Methods Enzymol 450:59–78
70. Demchenko AP (1986) Ultraviolet spectroscopy of proteins. Springer, Berlin
71. Demchenko AP (1988) Red-edge-excitation fluorescence spectroscopy of single-tryptophan proteins. Eur Biophys J 16:121–129
72. Datta A, Mandal D, Pal SK, Bhattacharyya K (1997) Intramolecular charge transfer processes in confined systems Nile Red in reverse micelles. J Phys Chem B 101:10221–10225
73. Rubinov AN, Tomin VI, Bushuk BA (2008) Kinetic spectroscopy of orientational states of solvated dye molecules in polar solutions. J Luminesc 26:377–391
74. Vincent M, Gallay J, Demchenko AP (1995) Solvent relaxation around the excited-state of indole – analysis of fluorescence lifetime distributions and time-dependence spectral shifts. J Phys Chem 99:14931–14941
75. Rubinov AN, Tomin VI, Bushuk BA (1982) Kinetic spectroscopy of orientational states of solvated dye molecules in polar solutions. J Luminesc 26:377–391

76. Demchenko AP, Sytnik AI (1991) Solvent reorganizational red-edge effect in intramolecular electron transfer. Proc Natl Acad Sci USA 88:9311–9314
77. Demchenko AP, Sytnik AI (1991) Site-selectivity in excited-state reactions in solutions. J Phys Chem 95:10518–10524
78. Demchenko AP, Ercelen S, Klymchenko AS (2002) Site-selective Red-Edge spectroscopy of disordered materials and microheterogeneous systems: polymers, phospholipid membranes and proteins. Proc SPIE 4938:294–304
79. Nemkovich NA, Gulis IM, Tomin VI (1982) Excitation-frequency dependence of the efficiency of directed nonradiative energy transfer in two-component solid solutions of organic compounds. Opt Spectr 53:140–143
80. Ghoneim N (2000) Photophysics of Nile red in solution: Steady state spectroscopy. Spectrochim Acta A Mol Biomol Spectr 56:1003–1010
81. Giusti LA, Marini VG, Machado VG (2009) Solvatochromic behavior of 1-(p-dimethylaminophenyl)-2-nitroethylene in 24 binary solvent mixtures composed of amides and hydroxylic solvents. J Mol Liq 150:9–15
82. Jozefowicz M (2008) Spectroscopic determination of solvation shell composition of fluorenone and 4-hydroxyfluorenone in binary solvent mixtures. Spectrochim Acta A Mol Biomol Spectrosc 71:537–542
83. Samios J, Durov VA (2004) Novel approaches to the structure and dynamics of liquids: experiments, theories and simulations. Kluwer Academic, Dordrecht
84. Martins LR, Tamashiro A, Laria D, Skaf MS (2003) Solvation dynamics of coumarin 153 in dimethylsulfoxide–water mixtures: molecular dynamics simulations. J Chem Phys 118:5955–5963
85. Kozinski M, Jarzeba W, Vuilleumier R (2004) Solvation dynamics of coumarin 153 in benzene-acetonitrile and benzene-methanol mixtures: a Molecular Dynamics study. In: Martin MM, Hynes JT (eds) Femtochemistry and femtobiology: ultrafast events in molecular science. Maison de la Chimie, France, Paris
86. Gratias R, Kessler H (1998) Molecular dynamics study on microheterogeneity and preferential solvation in methanol/chloroform mixtures. J Phys Chem 102:2027–2031
87. Lerf C, Suppan P (1992) Hydrogen bonding and dielectric effects in solvatochromic shifts. J Chem Soc Faraday Trans 88:963–969
88. Klymchenko AS, Demchenko AP (2003) Multiparametric probing of intermolecular interactions with fluorescent dye exhibiting excited state intramolecular proton transfer. Phys Chem Chem Phys 5:461–468
89. Wetzler DE, Chesta C, Fernández-Prini R, Aramendía PF (2002) Dynamic solvation of aminophthalimides in solvent mixtures. J Phys Chem A 106:2390–2400
90. Molotsky T, Huppert D (2003) Site specific solvation statics and dynamics of coumarin dyes in hexane−methanol mixture. J Phys Chem 107:2769–2780
91. Samanta A (2010) Fluorescence probing of the physicochemical characteristics of the room temperature ionic liquids. In: Demchenko AP (ed) Advanced fluorescence reporters in chemistry and biology III: Applications. Springer Series on Fluorescence 10. Springer, Heidelberg, pp 65–90
92. Chakrabartya D, Chakrabortya A, Setha D, Hazraa P, Sarkar N (2004) Dynamics of solvation and rotational relaxation of Coumarin 153 in 1-butyl-3-methylimidazolium hexafluorophosphate [bmim][PF6]-water mixtures. Chem Phys Lett 397:469–474
93. Shirota H, Castner EW (2000) Solvation in highly nonideal solutions: a study of aqueous 1-propanol using the coumarin 153 probe. J Chem Phys 112:2367
94. Tucker SC (1999) Solvent density inhomogeneities in supercritical fluids. Chem Rev 99:391–418
95. Gohres JL, Kitchens CL, Hallett JP, Popov AV, Hernandez R, Liotta CL, Eckert CA (2008) A spectroscopic and computational exploration of the cybotactic region of gas-expanded liquids: methanol and acetone. J Phys Chem B 112:4666–4673

96. Abbott AP, Hope EG, Palmer DJ (2007) Probing solute clustering in supercritical solutions using solvatochromic parameters. J Phys Chem B 111:8119–8125
97. Barroso M, Chattopadhyay N, Klymchenko AS, Demchenko AP, Arnaut LG, Formosinho SJ (2006) Dramatic pressure-dependent quenching effects in supercritical CO_2 assessed by the fluorescence of 4′-dimethylamino-3-hydroxyflavone. Thermodynamic versus kinetics control of excited-state intramolecular proton transfer. J Phys Chem A 110:13419–13424
98. Shukla CL, Hallett JP, Popov AV, Hernandez R, Liotta CL, Eckert CA (2006) Molecular dynamics simulation of the cybotactic region in gas-expanded methanol−carbon dioxide and acetone−carbon dioxide mixtures. J Phys Chem 110:24101–24111
99. Swalina C, Arzhantsev S, Li H, Maroncelli M (2008) Solvation and solvatochromism in CO_2-expanded liquids. 3. The dynamics of nonspecific preferential solvation. J Phys Chem 112:14959–14970
100. Li H, Arzhantsev S, Maroncelli M (2007) Solvation and solvatochromism in CO_2-expanded liquids. 2. Experiment−simulation comparisons of preferential solvation in three prototypical mixtures. J Chem Phys 111:3208–3221
101. Demchenko AP (2010) Comparative analysis of fluorescence reporter signals based on intensity, anisotropy, time-resolution, and wavelength-ratiometry. In: Demchenko AP (ed) Advanced fluorescence reporters in chemistry and biology I: Fundamentals and molecular design. Springer Series on Fluorescence 8. Springer, Heidelberg, pp 3–24
102. Resch-Genger U, Grabolle M, Nitschke R, Nann T (2010) Nanocrystals and nanoparticles versus molecular fuorescent labels as reporters for bioanalysis and the life sciences: a critical comparison. In Demchenko AP (ed) Advanced fluorescence reporters in chemistry and biology II: Molecular constructions, polymers and nanoparticles. Springer Series on Fluorescence 9. Springer, Heidelberg, pp 3–40
103. de Silva AP, Gunaratne HQN, Gunnaugsson T, Huxley AJM, McRoy CP, Rademacher JT, Rice TE (1997) Signaling recognition events with fluorescent sensors and switches. Chem Rev 97:1515–1566
104. Borisov SM, Mayr T, Mistlberger G, Klimant I (2010) Dye-doped polymeric particles for sensing and imaging. In: Demchenko AP (ed) Advanced fluorescence reporters in chemistry and biology II: Molecular constructions, polymers and nanoparticles. Springer Series on Fluorescence 9. Springer, Heidelberg, pp 193–228
105. Liang S, John CL, Xu S, Chen J, Jin Y, Yuan Q, Tan W, Zhao JX (2010) Silica-based nanoparticles: design and properties. In: Demchenko AP (ed) Advanced fluorescence reporters in chemistry and biology II: Molecular constructions, polymers and nanoparticles. Springer Series on Fluorescence 9. Springer, Heidelberg, pp 229–252
106. Reppy MA (2010) Structure, emissive properties, and reporting abilities of conjugated polymers. In: Demchenko AP (ed) Advanced fluorescence reporters in chemistry and biology II: Molecular constructions, polymers and nanoparticles. Springer Series on Fluorescence 9. Springer, Heidelberg, pp 357–388
107. Demchenko AP (2010) The concept of lambda-ratiometry in fluorescence sensing and imaging. J Fluoresc 20:1099–1128
108. Hoefelschweiger BK, Pfeifer L, Wolfbeis OS (2005) Screening scheme based on measurement of fluorescence lifetime in the nanosecond domain. J Biomol Screen 10:687–694
109. Povrozin YA, Kolosova OS, Obukhova OM, Tatarets AL, Sidorov VI, Terpetschnig EA, Patsenker LD (2009) Seta-633 – a NIR fluorescence lifetime label for low-molecular-weight analytes. Bioconjug Chem 20:1807–1812
110. Patsenker LD, Tatarets AL, Terpetschnig EA (2010) Long-wavelength probes and labels based on cyanines and squaraines. In: Demchenko AP (ed) Advanced fluorescence reporters in chemistry and biology I: Fundamentals and molecular design. Springer Series on Fluorescence 8. Springer, Heidelberg, pp 65–104
111. Hermant RM, Bakker NAC, Scherer T, Krijnen B, Verhoeven JW (1990) Systematic studies of a series of highly shaped donor-acceptor systems. J Am Chem Soc 112:1214–1221

112. Adhikary R, Barnes CA, Petrich JW (2009) Solvation dynamics of the fluorescent probe PRODAN in heterogeneous environments: contributions from the locally excited and charge-transferred states. J Phys Chem B 113:11999–12004
113. Arzhantsev S, Zachariasse KA, Maroncelli M (2006) Photophysics of trans-4-(dimethylamino)-4′-cyanostilbene and its use as a solvation probe. J Phys Chem A 110:3454–3470
114. Kucherak OA, Oncul S, Darwich Z, Yushchenko DA, Arntz Y, Didier P, Mely Y, Klymchenko AS (2010) Switchable Nile Red-based probe for cholesterol and lipid order at the outer leaflet of biomembranes. J Am Chem Soc 132:4907–4916
115. Cser A, Nagy K, Biczók L (2002) Fluorescence lifetime of Nile Red as a probe for the hydrogen bonding strength with its microenvironment. Chem Phys Lett 360:473–478
116. Kucherak OA, Didier P, Mely Y, Klymchenko AS (2010) Fluorene analogues of Prodan with superior fluorescence brightness and solvatochromism. J Phys Chem Lett 1:616–620
117. Terenziani F, Painelli A, Girlando A (2004) From solution to Langmuir–Blodgett films: Spectroscopic study of a zwitterionic dye. J Phys Chem B 108:10743–10750
118. Shynkar VV, Klymchenko AS, Piemont E, Demchenko AP, Mely Y (2004) Dynamics of intermolecular hydrogen bonds in the excited states of 4′-dialkylamino-3-hydroxyflavones. On the pathway to an ideal fluorescent hydrogen bonding sensor. J Phys Chem A 108:8151–8159
119. Caarls W, Celej MS, Demchenko AP, Jovin TM (2009) Characterization of coupled ground state and excited state equilibria by fluorescence spectral deconvolution. J Fluoresc 20:181–190
120. Tajalli H, Gilani AG, Zakerhamidi MS, Tajalli P (2008) The photophysical properties of Nile red and Nile blue in ordered anisotropic media. Dyes Pigments 78:15–24
121. Hermanson GT (1995) Bioconjugation techniques. Academic, San Diego
122. Huster D, Muller P, Arnold K, Herrmann A (2001) Dynamics of membrane penetration of the fluorescent 7-nitrobenz-2-oxa-1, 3-diazol-4-yl (NBD) group attached to an acyl chain of phosphatidylcholine. Biophys J 80:822–831
123. Loura LM, Ramalho JP (2007) Location and dynamics of acyl chain NBD-labeled phosphatidylcholine (NBD-PC) in DPPC bilayers. A molecular dynamics and time-resolved fluorescence anisotropy study. Biochim Biophys Acta 1768:467–478
124. Lenhart JL, van Zanten JH, Dunkers JP, Zimba CG, James CA, Pollack SK, Parnas RS (2000) Immobilizing a fluorescent dye offers potential to investigate the glass/resin interface. J Colloid Interf Sci 221:75–86
125. Yesylevskyy SO, Demchenko AP (2011) Fluorescence probing in structurally anisotropic materials. From liquid crystals to macromolecules, micelles and lipid bilayers. In: Demchenko AP (ed) Advanced fluorescence reporters in chemistry and biology III: Applications. Springer Series on Fluorescence. Springer, Heidelberg, pp 119–158
126. Grant CD, Steege KE, Bunagan MR, Castner EW Jr (2005) Microviscosity in multiple regions of complex aqueous solutions of poly(ethylene oxide)-poly(propylene oxide)-poly (ethylene oxide). J Phys Chem B 109:22273–22284
127. Uchiyama S, Takehira K, Yoshihara T, Tobita S, Ohwada T (2006) Environment-sensitive fluorophore emitting in protic environments. Org Lett 8:5869–5872
128. Gengeliczki Z, Rosenfeld DE, Fayer MD (2010) Theory of interfacial orientational relaxation spectroscopic observables. J Chem Phys 132:244703
129. Hell SW (2009) Microscopy and its focal switch. Nat Methods 6:24–32
130. Spagnuolo C, Joselevich M, Leskow FC, Jares-Erijman EA (2010) Tetracysteine and bipartite tags for biarsenical organic fluorophores. In: Demchenko AP (ed) Advanced fluorescence reporters in chemistry and biology III: Applications. Springer Series on Fluorescence 10. Springer, Heidelberg, pp 263–296
131. Asuncion-Punzalan E, London E (1995) Control of the depth of molecules within membranes by polar groups: determination of the location of anthracene-labeled probes in model membranes by parallax analysis of nitroxide-labeled phospholipid induced fluorescence quenching. Biochemistry 34:11460–11466

132. Kachel K, Asuncion-Punzalan E, London E (1998) The location of fluorescence probes with charged groups in model membranes. Biochim Biophys Acta 1374:63–76
133. Ory JJ, Banaszak LJ (1999) Studies of the ligand binding reaction of adipocyte lipid binding protein using the fluorescent probe 1, 8-anilinonaphthalene-8-sulfonate. Biophys J 77:1107–1116
134. Baudet-Nessler S, Jullien M, Crosio MP, Janin J (1993) Crystal structure of a fluorescent derivative of RNase A. Biochemistry 32:8457–8464
135. Bammel BP, Hamilton DD, Haugland RP, Hopkins HP, Schuette J, Szalecki W, Smith JC (1990) NMR, calorimetric, spin-label, and optical studies on a trifluoromethyl-substituted styryl molecular probe in dimyristoylphosphatidylcholine vesicles and multilamellar suspensions: a model for location of optical probes. Biochim Biophys Acta 1024:61–81
136. Osakai T, Sawada J, Nagatani H (2009) Potential-modulated fluorescence spectroscopy of the membrane potential-sensitive dye di-4-ANEPPS at the 1, 2-dichloroethane/water interface. Anal Bioanal Chem 395:1055–1061
137. Repáková J, Čapková P, Holopainen JM, Vattulainen I (2004) Distribution, orientation, and dynamics of DPH probes in DPPC bilayer. J Phys Chem B 108:13438–13448
138. Repakova J, Holopainen JM, Morrow MR, McDonald MC, Capkova P, Vattulainen I (2005) Influence of DPH on the structure and dynamics of a DPPC bilayer. Biophys J 88:3398–3410
139. Hoff B, Strandberg E, Ulrich AS, Tieleman DP, Posten C (2005) 2H-NMR study and molecular dynamics simulation of the location, alignment, and mobility of pyrene in POPC bilayers. Biophys J 88:1818–1827
140. Watanabe H, Yamaguchi S, Sen S, Morita A, Tahara T (2010) "Half-hydration" at the air/water interface revealed by heterodyne-detected electronic sum frequency generation spectroscopy, polarization second harmonic generation, and molecular dynamics simulation. J Chem Phys 132:144701
141. Benderskii AV, Eisenthal KB (2000) Effect of organic surfactant on femtosecond solvation dynamics at the air-water interface. J Phys Chem 104:11723–11728
142. Yamaguchi S, Tahara T (2006) Determining electronic spectra at interfaces by electronic sum frequency generation: one- and two-photon double resonant oxazine 750 at the air/water interface. J Chem Phys 125:194711
143. Pal N, Verma SD, Sen S (2010) Probe position dependence of DNA dynamics: comparison of the time-resolved stokes shift of groove-bound to base-stacked probes. JACS 132:9277–9279
144. Nguyen KT, Shang X, Eisenthal KB (2006) Molecular rotation at negatively charged surfactant/aqueous interfaces. J Phys Chem 110:19788–19792
145. Pantano DA, Sonoda MT, Skaf MS, Laria D (2005) Solvation of Coumarin 314 at water/air interfaces containing anionic surfactants. I. Low coverage. J Phys Chem 109:7365–7372
146. De Paula R, Machado AED, de Miranda JA (2004) 3-benzoxazol-2-yl-7-(N, N-diethylamino)-chromen-2-one as a fluorescence probe for the investigation of micellar microenvironments. J Photochem Photobiol A Chem 165:109–114
147. Liu P, Harder E, Berne BJ (2005) Hydrogen-bond dynamics in the air−water interface. J Phys Chem 109:2949–2955
148. Paul S (2009) Liquid–vapour interfaces of aqueous trimethylamine-N-oxide solutions: a molecular dynamics simulation study. Chem Phys 368:7–13
149. Johnson ML, Rodriguez C, Benjamin I (2009) Rotational dynamics of strongly adsorbed solute at the water surface. J Phys Chem 113:2086–2091
150. Paul S, Chandra A (2005) Liquid-vapor interfacial properties of water-ammonia mixtures: dependence on ammonia concentration. J Chem Phys 123:174712
151. Perera JM, Stevens GW (2009) Spectroscopic studies of molecular interaction at the liquid–liquid interface. Anal Bioanal Chem 395:1019–1032
152. Bessho K, Uchida T, Yamauchi A, Shioya T, Teramae N (1997) Microenvironments of 8-anilino-1-naphthalenesulfonate at the heptane–water interface: time-resolved total internal reflection fluorescence spectroscopy. Chem Phys Lett 264:381–386

153. Ishizaka S, Kim HB, Kitamura N (2001) Time-resolved total internal reflection fluorometry study on polarity at a liquid/liquid interface. Anal Chem 73:2421–2428
154. Demchenko AP, Shcherbatska NV (1985) Nanosecond dynamics of charged fluorescent-probes at the polar interface of a membrane phospholipid-bilayer. Biophys Chem 22:131–143
155. Steel WH, Beildeck CL, Walker RA (2004) Solvent polarity across strongly associating interfaces. J Phys Chem 108:16107–16116
156. Michael D, Benjamin I (2001) Molecular dynamics computer simulations of solvation dynamics at liquid/liquid interfaces. J Chem Phys 114:2817
157. Walker DS, Richmond GL (2008) Interfacial depth profiling of the orientation and bonding of water molecules across liquid–liquid interfaces. J Phys Chem 112:201–209
158. Negishi M, Sakaue T, Yoshikawa K (2010) Mismatch of bulk viscosity reduces interfacial diffusivity at an aqueous-oil system. Phys Rev E Stat Nonlin Soft Matter Phys 81:020901
159. El-Rayyes AA, Al-Betar A, Htun T, Klein UKA (2005) Fluorescence emission from rhodamine-B lactone adsorbed at solid catalysts. Chem Phys Lett 414:287–291
160. Brindza MR, Walker RA (2009) Differentiating solvation mechanisms at polar solid/liquid interfaces. JACS 131:6207–6214
161. Sánchez VM, Sued M, Scherlis DA (2009) First-principles molecular dynamics simulations at solid–liquid interfaces with a continuum solvent. J Chem Phys 131:174108
162. Lenhart JL, van Zanten JH, Dunkers JP, Parnas RS (2001) Studying the buried interfacial region with an immobilized fluorescence probe. Macromolecules 34:2225–2231
163. Ellison CJ, Torkelson JM (2002) Sensing the glass transition in thin and ultrathin polymer films via fluorescence probes and labels. J Polym Sci B Polym Phys 40:2745–2758
164. Fischer K, Prause S, Spange S, Cichos C, von Borczyskowski C (2003) Surface polarity of cellulose derivates observed by Coumarin 151 and 153 as solvatochromic and fluorochromic probes. J Polym Sci B Polym Phys 41:1210–1218
165. Szejtli J (1998) Introduction and general overview of cyclodextrin chemistry. Chem Rev 98:1743–1753
166. Balabai N, Linton B, Napper A, Priyadarshy S, Sukharevsky AP, Waldeck DH (1998) Orientational dynamics of beta-cyclodextrin inclusion complexes. J Phys Chem B 102: 9617–9624
167. Douhal A (2004) Ultrafast guest dynamics in cyclodextrin nanocavities. Chem Rev 104:1955–1976
168. Sen P, Roy D, Mondal SK, Sahu K, Ghosh S, Bhattacharyya K (2005) Fluorescence anisotropy decay and solvation dynamics in a nano-cavity: Coumarin 153 in methyl beta-cyclodextrins. J Phys Chem A 109:9716–9722
169. Nandi N, Bhattacharyya K, Bagchi B (2000) Dielectric relaxation and solvation dynamics of water in complex chemical and biological systems. Chem Rev 100:2013–2046
170. Organero JA, Tormo L, Sanz M, Roshal A, Douhal A (2007) Complexation effect of gamma-cyclodextrin on a hydroxyflavone derivative: formation of excluded and included anions. J Photochem Photobiol A Chem 188:74–82
171. Rodriguez J, Martí J, Guardia E, Laria D (2008) Exploring the picosecond time domain of the solvation dynamics of Coumarin 153 within β-cyclodextrins. J Phys Chem B 112:8990–8998
172. Kubinyi M, Vidoczy T, Varga O, Nagy K, Bitter I (2005) Absorption and fluorescence spectroscopic study on complexation of oxazine 1 dye by calix 8 arenesulfonate. Appl Spectrosc 59:134–139
173. Bergamini G, Marchi E, Ceroni P (2010) Luminescent dendrimers as ligands and sensors of metal ions. In: Demchenko AP (ed) Advanced fluorescence reporters in chemistry and biology II: Molecular constructions, polymers and nanoparticles. Springer Series on Fluorescence 9. Springer, Heidelberg, pp 253–284
174. Richter-Egger DL, Tesfai A, Tucker SA (2001) Spectroscopic investigations of poly(propyleneimine)dendrimers using the solvatochromic probe phenol blue and comparisons to poly (amidoamine) dendrimers. Anal Chem 73:5743–5751

175. Teobaldi G, Zerbetto F (2003) Molecular dynamics of a dendrimer-dye guest-host system. JACS 125:7388–7393
176. Aumanen J, Kesti T, Sundström V, Teobaldi G, Zerbetto F, Werner N, Richardt G, van Heyst J, Vögtle F, Korppi-Tommola J (2010) Internal dynamics and energy transfer in dansylated POPAM dendrimers and their eosin complexes. J Phys Chem 114:1548–1558
177. Dunn B, Zink JI (1997) Probes of pore environment and molecule−matrix interactions in sol−gel materials. Chem Mater 9:2280–2291
178. Hungerford G, Ferreira JA (2001) The effect of the nature of retained solvent on the fluorescence of Nile Red incorporated in sol–gel-derived matrices. J Luminesc 93:155–165
179. Grando SR, Pessoa CM, Gallas MR, Costa TM, Rodembusch FS, Benvenutti EV (2009) Modulation of the ESIPT emission of benzothiazole type dye incorporated in silica-based hybrid materials. Langmuir 25:13219–13223
180. Baumann R, Ferrante C, Kneuper E, Deeg F-W, Bräuchle C (2003) Influence of confinement on the solvation and rotational dynamics of Coumarin 153 in ethanol. J Phys Chem A 107:2422–2430
181. Blaak R, Hansen JP (2006) Dielectric response of a polar fluid trapped in a spherical nanocavity. J Chem Phys 124:144714

Part II
Probing Condensed Media

Fluorescence Probing of the Physicochemical Characteristics of the Room Temperature Ionic Liquids

Anunay Samanta

Abstract Room temperature ionic liquids are salts, which in their pure state are liquid at ambient conditions. These substances have attracted a great deal of attention in recent years for the various opportunities they offer. Several physicochemical characteristics of these novel materials can be understood from fluorescence studies on carefully chosen probe molecules in these media. In this chapter, we address some of the properties of these complex media that have been obtained from these fluorescence studies. Specifically, we discuss about the polarity of these liquids, solvent relaxation dynamics in these media and properties that arise as a consequence of slow solvation and diffusion in these viscous environments. The microheterogeneous nature of these solvents, which is indicated in some of the fluorescence studies, is also highlighted. The steady state and time-resolved fluorescence response of the molecular systems presented here is expected to provide insight into the various structural and dynamical aspects of these complex media and will also initiate further studies to throw light into some of the properties that are not yet fully understood.

Keywords Dynamic fluorescence Stokes shift · Excitation wavelength-dependent fluorescence · Fluorescence recovery after photobleaching · Ionic liquids · Red-edge effect · Solvation dynamics

Contents

A. Samanta
School of Chemistry, University of Hyderabad, Hyderabad 500046, India
e-mail: assc@uohyd.ernet.in

A.P. Demchenko (ed.), *Advanced Fluorescence Reporters in Chemistry and Biology III: Applications in Sensing and Imaging*, Springer Ser Fluoresc (2011) 10: 65–90,
DOI 10.1007/978-3-642-18035-4_2,

1 Introduction

A greater awareness and concern of the pollution caused by the volatile organic compounds, which are used in large quantities for various purposes, has prompted significant effort in recent years directed towards finding environment-friendly replacement for these harmful chemicals. This quest for green chemistry [1] has led to the realization of the importance of solvent-free synthesis [2] and use of water [3], supercritical carbon dioxide [4], and room temperature ionic liquids (RTILs) as possible alternative reaction media [5–19].

The RTILs, which are currently the focus of worldwide attention, are salts consisting entirely of ions. However, unlike the ionic solids, these are viscous liquids at ambient conditions. The most commonly used RTILs are the ones based on nonsymmetrically substituted *N,N′*-dialkylimidazolium cations and non-coordinating bulky inorganic anions. 1-Butyl-3-methylimidazolium ion $[\text{bmim}]^+$ and 1-ethyl-3-methylimidazolium ion $[\text{emim}]^+$ are the most common cationic components and PF_6^-, BF_4^-, and $[(CF_3SO_2)_2N]^-$ are the anionic counterparts. The chemical formulae of some common RTILs are shown in Fig. 1.

The low melting point of these salts is believed to be the result of weak electrostatic attraction between the ionic partners and poor packing of the crystal. The former is due to the bulkiness of the constituent ions and charge delocalization in them. The poor packing is attributed to the lack of symmetry in one or both the ionic partners. In addition to possessing the essential solvent properties, the features

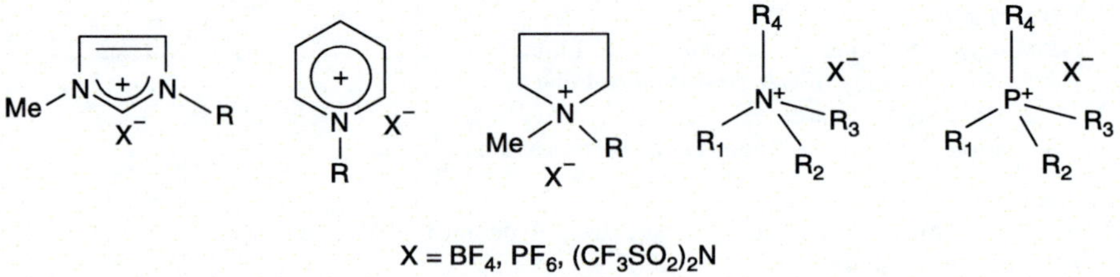

Fig. 1 Chemical formulas of some frequently used RTILs

that these liquids do not evaporate, do not burn, and can be recycled have made them quite appealing as possible environment friendly solvents for various fundamental studies and applications [5, 7, 10, 13, 19, 20]. Even though extensive studies during the past decade have brought into light a number of drawbacks pointing to some of the limitations of the RTILs in replacing the volatile organic compounds [17, 21–24], the fact that these studies have also revealed several distinct features of these substances useful as specialty chemicals [25–27] explains the continued interest in these substances. A large number of papers and review articles have appeared already describing various physiochemical properties and applications of the RTILs [28–81].

A number of properties of these substances have been understood from fluorescence studies in these media despite the fact that the imidazolium ionic liquids absorb in the near UV region and exhibit weak emission with the fluorescence behavior dependent on the excitation wavelength [75, 82, 83]. The optical properties of the RTILs do pose some constraints to the fluorescence studies in these media, particularly those involving substances that absorb in the UV region and are weakly fluorescent. However, the studies carried out so far clearly suggest that these constraints are not problematic in most cases and a wealth of information concerning the structure and dynamics of these substances can be obtained from fluorescence studies on the probe molecules in RTILs [74–77].

2 Polarity of the Room Temperature Ionic Liquids

2.1 Methodology of Determination of Polarity

Solvent polarity is an important parameter that characterizes how a solvent interacts with the solutes. Dielectric constant, refractive index, or functions involving these two quantities are frequently used as macroscopic solvent polarity parameters. However, as the polarity of a solvent is indicative of all possible interactions at the microscopic level, the solvent polarity is most often expressed in terms of empirical parameters, which take into account all possible specific and nonspecific interactions between the solute and the solvent [84]. The $E_T(30)$ scale, which is based on the transition energy of a zwitterionic betaine dye (Fig. 2) that exhibits exceptional negative solvatochromism, is the most common one [84–87]. The $E_T(30)$ value of a solvent is defined as the molar transition energy of this dye (in kcal mol^{-1}) at room temperature (25°C) and normal pressure (1 bar) and is related to the wavelength corresponding to the maximum of the long wavelength absorption band of the dye as

$$E_T(30)/\text{kcal mol}^{-1} = 28{,}591/(\lambda_{max}/\text{nm}). \quad (1)$$

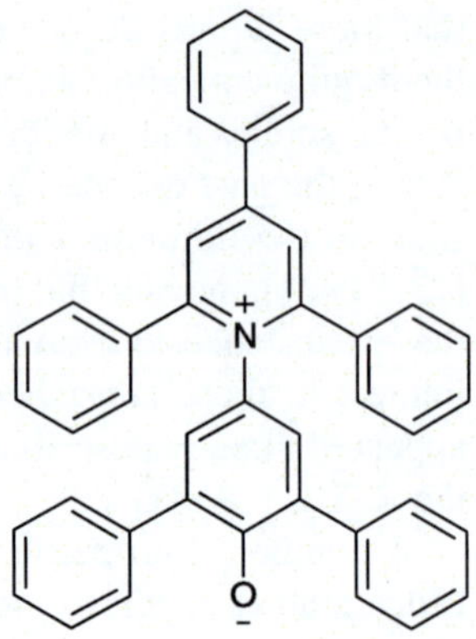

Fig. 2 Molecular structure of the betaine dye used in the $E_T(30)$ scale of solvent polarity

Even though the $E_T(30)$ scale of polarity is defined with respect to the betaine dye, the $E_T(30)$ values are often estimated using other substances because of convenience and sometimes because of necessities such as the poor solubility of this dye in the medium. Apart from the most commonly used $E_T(30)$ scale, several other polarity parameters are also used, which can be found elsewhere [84, 88].

One of the most common approaches to the estimation of the polarity of an unknown medium using fluorescence technique involves the use of dipolar fluorescent systems [88]. As the excited state of this type of molecules is usually more polar than the ground state, the excited state is stabilized to a greater extent compared to the ground state in a polar medium leading to a Stokes shift of the emission maximum (Fig. 3). As the magnitude of the Stokes shift or the wavelength/energy corresponding to the fluorescence maximum of the dipolar molecule in a solvent is a measure of the various interactions between the two, the dipolar molecules often serve as polarity probes for the study of complex systems.

2.2 *Polarity Estimates*

Several studies have been carried out for quantification of the polarity of the RTILs from the measured frequency of the fluorescence maximum of various dipolar molecular systems [29, 40, 41, 48, 89] such as coumarin 153 (C-153), 4-aminophthalimide (AP), 6-propionyl-2-dimethylaminonaphthalene (PRODAN), etc. (Fig. 4). In this method, one first obtains a calibration line by plotting the measured wavenumbers corresponding to the fluorescence maximum ($\bar{\upsilon}_{max}^{flu}$) of the probe molecule in several conventional solvents of known $E_T(30)$ values. The unknown $E_T(30)$ value of an RTIL is then obtained from the measured $\bar{\upsilon}_{max}^{flu}$ value of the probe in the given RTIL. These studies have indicated that most of the RTILs are more polar than acetonitrile, but less polar than methanol [29, 89]. In fact, the $E_T(30)$ values of most of the ionic liquids have been found comparable to those of the

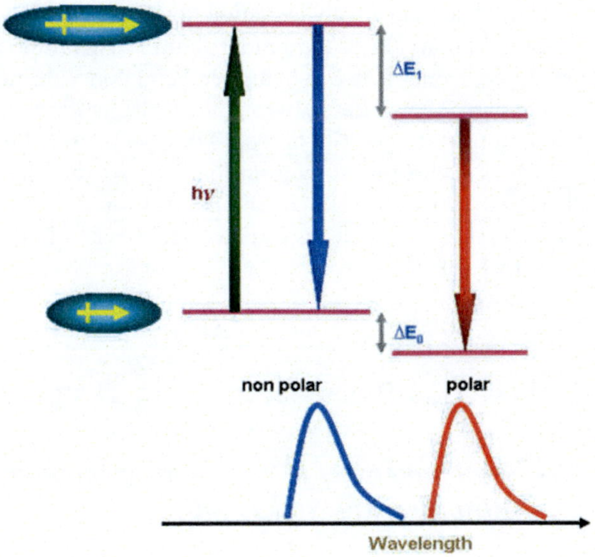

Fig. 3 Typical energy level diagram of a dipolar molecule in nonpolar and polar media illustrating the solvent stabilization of the ground and excited states and consequent shift of the emission frequencies

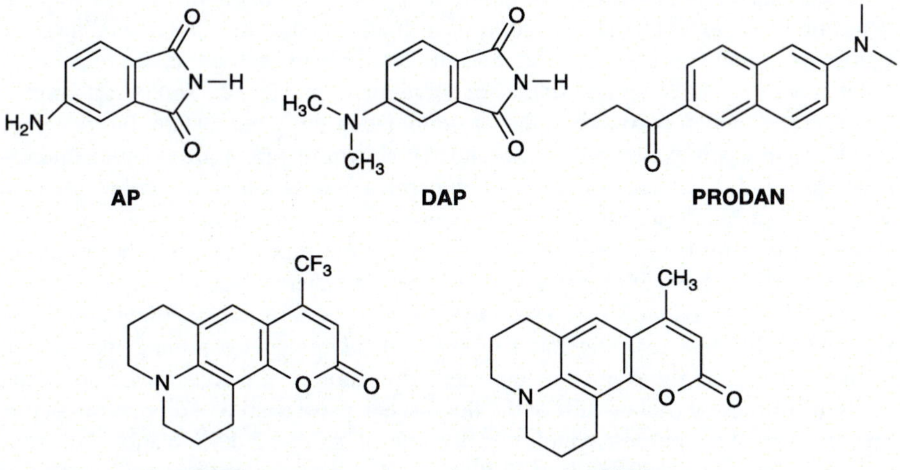

Fig. 4 Chemical formula of some dipolar fluorescence probes used in the measurement of solvent polarity and solvation dynamics

short-chain aliphatic alcohols [29, 30]. In this context, one cannot ignore, however, the fact that despite a large variety of structures of constituent ions, the polarity of most of the RTILs comprising different imidazolium ionic liquids, as estimated from the microenvironment sensed by the probe molecules, is very similar to that of the short-chain alcohols.

This observation can be rationalized taking into consideration the hydrogen bonding interaction of the imidazolium ion with the probe molecules [30, 89], in which the former acts as a hydrogen bond donor through the acidic hydrogen of the C–H moiety, which lies between the two ring nitrogen atoms (see Fig. 1). A detailed discussion on the polarity of the RTILs can be found in a recent review of Reichardt [89]. It is interesting to note in this context that the dielectric constant, which is also used as an indicator of the polarity of a medium, pictures the imidazolium ionic liquids as much less polar media with dielectric constant values around 11.4–15.1 [78, 79].

3 Solvation Dynamics in Room Temperature Ionic Liquids

3.1 *Fundamentals of Solvation Dynamics and Time-Dependent Fluorescence Stokes Shift*

Solvation dynamics commonly refers to reorientation of the solvent molecules around an instantly created solute dipole in a polar solvent. While static solvation determines the extent of stabilization of a dipolar solute in a polar medium, the dynamics of solvation controls the rates of electron and proton transfer reaction in polar solvent. In fluorescence studies of solvation dynamics, the solute is a fluorescence probe, which is weakly polar or nonpolar in the ground state, but becomes highly polar upon excitation. C-153 and AP shown in Fig. 4 are popular fluorescence probes for the study of solvation dynamics. Excitation of the probe molecule with an ultrashort laser pulse instantly separates charge creating a dipole, which disturbs the equilibrium arrangement of the solvent dipoles around the probe. In response, the solvent molecules start rearranging themselves around the newly created charge-separated species to form a new equilibrium configuration. The time taken for this rearrangement of the solvent molecules around the excited probe molecule is referred to as the relaxation time of the solvent. As solvation leads to substantial Stokes shift of the fluorescence spectrum of a dipolar molecule, one of the most commonly exploited procedures of studying the solvation dynamics is to follow the time-dependent shift of the fluorescence spectrum of a dipolar probe molecule, termed as dynamic Stokes shift [90–97], following its electronic excitation using a short pulse of light (Fig. 5).

The dynamic Stokes shifts are measured from the time-resolved emission spectra (TRES), which are usually constructed by an indirect procedure starting with the measurement of a series of fluorescence decay profiles monitoring several wavelengths across the entire steady state emission spectrum [98]. A wavelength-dependent fluorescence intensity decay profile is observed as a general case. The wavenumbers corresponding to the emission maxima at different times ($\overline{\nu}(t)$) are subsequently obtained by fitting the time-resolved emission spectra to a line-shape function. The solvent reorganization dynamics is usually followed by constructing a

Fig. 5 Dynamic Stokes shift of a dipolar molecule in a polar environment

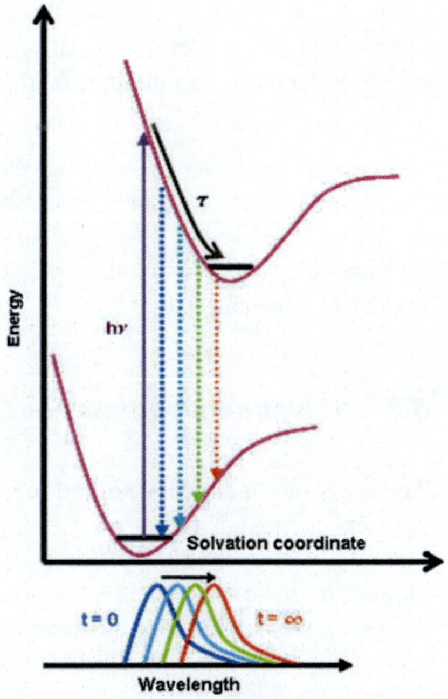

dimensionless spectral shift correlation function, $C(t)$, the value of which varies between 1 and 0.

$$C(t) = \frac{\overline{\nu}(t) - \overline{\nu}(\infty)}{\overline{\nu}(0) - \overline{\nu}(\infty)}, \tag{2}$$

where $\overline{\nu}(t)$, $\overline{\nu}(0)$, and $\overline{\nu}(\infty)$ are the emission peak frequencies (in cm^{-1}) at times, t, $t = 0$ and $t = \infty$ following laser excitation of the solute probe molecule. The longitudinal relaxation time of the solvent, τ_L, is an exponential decay function of $C(t)$ with time, so that $C(t) = \exp(-t/\tau_L)$. However, experimentally found $C(t)$ functions are often multiexponential and hence, an average solvation time $\langle\tau_L\rangle$, defined as

$$\langle\tau_L\rangle = \sum_{i=1}^{n} a_i \tau_i \tag{3}$$

is usually estimated. The simple continuum theory equates the longitudinal relaxation time, τ_L, with much slower Debye relaxation time, τ_D according to the following equation [97]:

$$\tau_L = \frac{2\varepsilon_\infty + \varepsilon_c}{2\varepsilon_0 + \varepsilon_c}\tau_D, \tag{4}$$

where ε_∞ and ε_0 ($\varepsilon_\infty < \varepsilon_0$) are the infinite-frequency and low-frequency dielectric constant of the solvent. ε_c is the dielectric constant of the cavity containing the probe molecule. The equation is often simplified as

$$\tau_L \cong \frac{2n^2 + 1}{2\varepsilon + 1} \tau_D \cong \frac{n^2}{\varepsilon} \tau_D. \quad (5)$$

Assuming, $\varepsilon_\infty = n^2$, where n is the refractive index and replacing ε_0 by static dielectric constant, ε.

3.2 Solvation Dynamics in Conventional Solvents

The study of dynamics of solvation is an active area of interest from both theoretical and experimental point of view [90–97]. Water and alcohols are the most commonly used media for this purpose, though solvation dynamics in organized assemblies have also recently become an attractive topic for investigation [99–105].

The solvent relaxation process in conventional solvents is usually very fast at room temperature. Using ultrafast time resolution of the order of few femtoseconds it has been found that in major class of solvents, where specific interaction like hydrogen bonding is unimportant, the solvation times are in the subpicosecond time scale at room temperature. The linear correlation between experimentally observed average relaxation time and longitudinal relaxation time, τ_L as predicted by simple continuum theory, depends on the nature of solute–solvent interaction and temperature of the medium. Deviation from the above, when observed, can be accounted for considering molecular nature of the solvent, translational contribution to the solvent relaxation and specific hydrogen bonding ability of the protic solvent [84]. This relaxation time obviously depends on the viscosity of the medium, molecular structure of the solvent and temperature of the system. Obviously, the relaxation times are much slower in viscous solvents and in membranes or proteins [100–105].

3.3 Timescale and Mechanism of Solvation Dynamics in Room Temperature Ionic Liquids

As the RTILs are sufficiently polar, both static solvation and dynamics of solvation are important processes in these media. The time-dependent fluorescence behavior of dipolar solutes in these media is therefore expected to yield information that is useful for an understanding of the time-scale and mechanism of solvation dynamics [74, 75, 77]. Early studies revealed that common dipolar molecules such as C-153, AP, and PRODAN exhibit wavelength-dependent fluorescence decay profiles (Fig. 6) and time-dependent fluorescence Stokes shift in the ps–ns timescale [39–41] (Fig. 7).

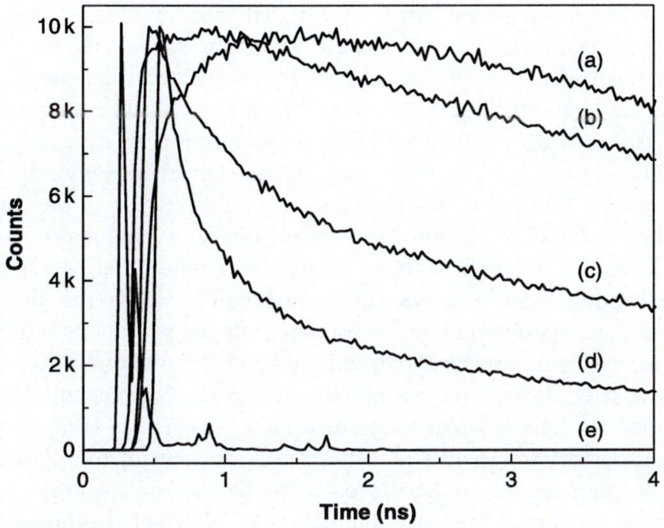

Fig. 6 Time-resolved fluorescence behavior of C153 in [bmim][BF_4] at (*a*) 580, (*b*) 530, (*c*) 500, and (*d*) 480 nm; $\lambda_{exc} = 375$ nm. The excitation profile is shown in (*e*). Reproduced with permission from [39]

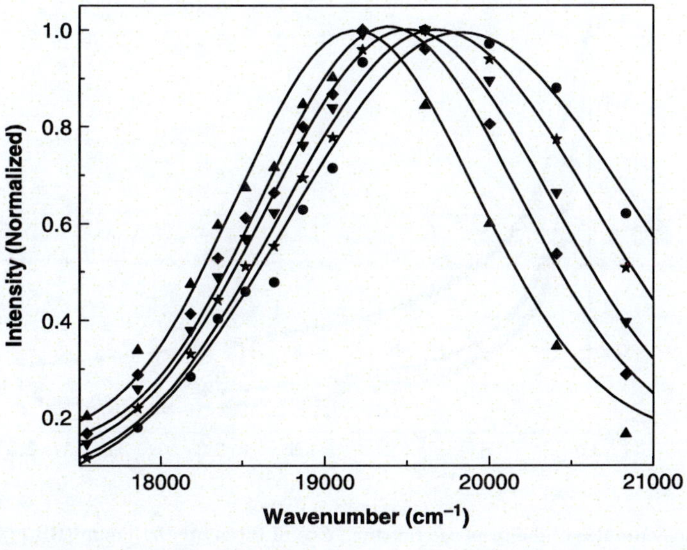

Fig. 7 Time-resolved normalized emission spectra of C153 in [bmim][BF_4] at 0 (*filled circle*), 100 (*asterisk*), 250 (*filled inverted triangle*), 500 (*filled diamond*), and 2,000 ps (*filled triangle*). Reproduced with permission from [39]

This observation suggested that solvation dynamics in RTILs is slow compared to the common molecular solvents such as acetonitrile, methanol, or water. These early studies not only showed that slow solvation dynamics is a general feature in viscous RTILs, but also indicated that the time-resolvable component of the dynamics, as obtained from the fits to *C*(*t*) versus *t* plots (Fig. 8), was biexponential in nature. The average solvation time was found to be dependent on the viscosity of the media and on the probe molecule though no correlation of the time with the probe molecule could be established. Interestingly, even though the solvation dynamics in RTILs was found to be slow, it was noted that nearly 50% of the spectral relaxation dynamics was quite fast compared to the time-resolution (~25 ps) of the experimental time-correlated single photon counting (TCSPC) setup and hence, could not be captured in most of the measurements. The studies carried out at later stages have corroborated the early findings and they also have been extended to cover a wider range of RTILs [42–47, 51, 106]. Attempts have also been made to time-resolve the ultrafast component of the dynamics, which is missed in these studies and believed to be faster than the time resolution of the instrumental setup. Using a combination of Kerr-gated emission and time-correlated single photon counting techniques, the complete solvation response of a dipolar probe in several imidazolium ILs and one pyrrolidinium IL has recently been captured by following the fluorescence Stokes shift dynamics [50]. These measurements have revealed a rapid subpicosecond component (10–20% of the total relaxation) and a much longer component that relaxes in the ps–ns time scale.

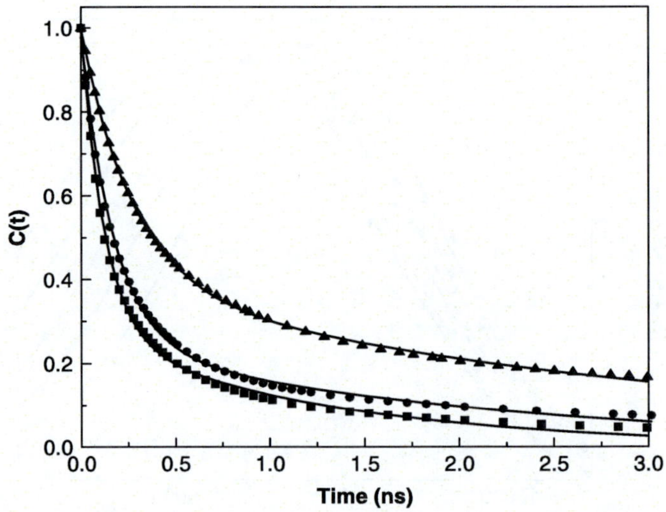

Fig. 8 Decay profiles of the response function, *C*(*t*) of PRODAN in [emim][BF_4] (*filled circle*), PRODAN in [bmim][BF_4] (*filled triangle*) and C153 in [emim][BF_4] (*filled square*). The *points* denote the actual values of *C*(*t*), while the *solid line* represents the biexponential fit. Reproduced with permission from [41]

3.4 *Physical Origin of the Solvation Components and Mechanism of Solvation*

The mechanism of solvation in RTILs is fundamentally different from that in conventional solvents. In the later case, the solvent response to enhanced charge separation in the probe molecule consists of reorientation of the dipolar solvent molecules. However, in RTILs, where the ion–dipole interaction between the solvent and solute dominates, the solvent response is expected to be the translational motion of the constituent ions and other collective motions associated with this. This difference is pictorially illustrated in Fig. 9.

However, considering the large size and polarizable nature of the constituent ions, the dipole–dipole interaction may not be insignificant and hence, the orientational relaxation of these free ions also cannot be ruled out. As some of the ions of the RTILs exist in associated forms as ion pairs or higher ion aggregates [107, 108], the translational and rotational motion of these species cannot be ruled out as response to the charge separation process in the probe molecule. Therefore, even when additional complexities such as the heterogeneity of the RTILs [69, 73, 109, 110] are kept out of the consideration, the mechanism of solvent reorganization in these media is a quite complex process and the assignment of the various time constants to specific motions of the constituents is a difficult exercise [74, 75].

The early interpretations of the solvation dynamics data of RTILs were based on the literature data on high temperature solvation dynamics in the molten salts [111] and molecular dynamics simulation studies [112, 113]. The short and long components of the dynamics were assigned to the translational motion of the individual

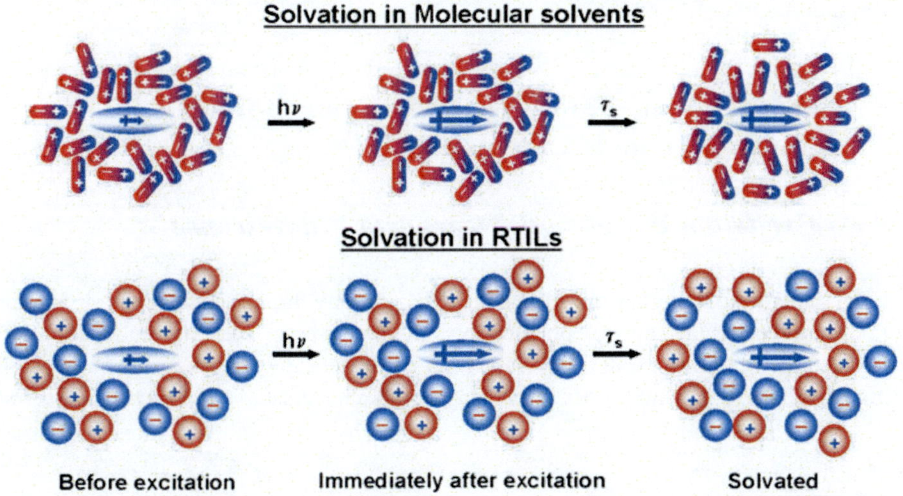

Fig. 9 An oversimplified diagram highlighting the fundamental difference of the mechanism of solvation of a dipolar probe in conventional solvents and in RTILs. Reproduced with permission from [75]

ions and collective motion of the constituent ions, respectively [40, 41]. A different school of thought however considered the relaxation as a nonexponential process and a stretched exponential fit to the data was found more appropriate [43, 51]. Nevertheless, irrespective of the nature of the fit, it is the average solvation time whose dependence on the various parameters (viscosity of the medium, probe molecule) is generally studied [75–77] as it is felt that the separate values of the two or more components may not mean much in the absence of a clear understanding of the mechanism of solvation dynamics in these media.

Despite the complex nature of the mechanism of solvation dynamics in RTILs, which is presented above and reviewed from time to time, some inferences can however be made. It is obvious that the ultrafast component, which is often missed in most of the studies, is due to some small amplitude motion of the ions, which lie in the close proximity of the probe molecule. The nature of the physical motion that contributes to this component, however, is a matter of speculation and debate [50, 75, 77]. As stated already, the early interpretations of solvation dynamics in RTILs were based primarily on the results of simulation studies and literature information available at that time [112, 113]. Even though the ultrafast component was previously thought to be associated with the small amplitude motions of the planar and polarizable imidazolium cations in the vicinity of the probe molecule, this was found not to be the case in later studies [48]. Based on the recent experimental findings and simulation studies [71], the molecular picture currently accepted is that the subpicosecond component of the dynamics, which constitutes ~10–20% of the total solvent response, arises from the immediate response of the ions. The slow component, which spans the ps–ns time domain and is viscosity dependent, arises from complex dynamical processes that resemble the dynamics occurring in supercooled liquids and involve structural reorganization such as cage formation and deformation.

4 Unconventional Fluorescence Behavior in Room Temperature Ionic Liquids

4.1 Excitation Wavelength-Dependent Fluorescence

Even though most dipolar molecules in RTILs display fluorescence behavior similar to that expected in short-chain alcohols, a few systems such as ANF (Fig. 10) interestingly exhibits excitation wavelength-dependent fluorescence spectra in these media, a behavior, which is not in accordance with the Kasha's rule and not observed

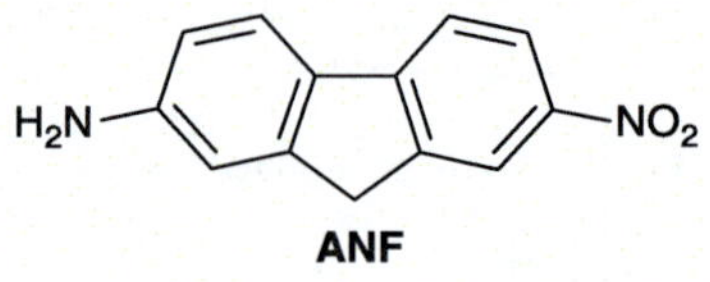

Fig. 10 Chemical formula of 2-amino-7-nitrofluorene (ANF)

in short-chain alcohols [110]. A typical excitation wavelength-dependent fluorescence behavior is highlighted in Fig. 11. A plot of the emission peak wavelength versus the excitation wavelength is illustrated in Fig. 12. The extent of the excitation

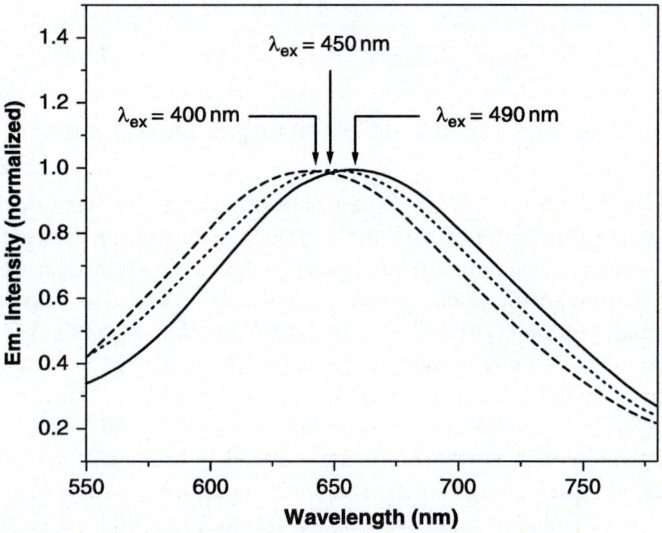

Fig. 11 Normalized fluorescence spectra of ANF in [bmim][BF_4] at room temperature as a function of excitation wavelength, λ_{ex}: 400 nm (*dash*), λ_{ex}: 450 nm (*dot*) and λ_{ex}: 490 nm (*solid*). The fluorescence spectra have been corrected for the instrumental response. Reproduced with permission from [110]

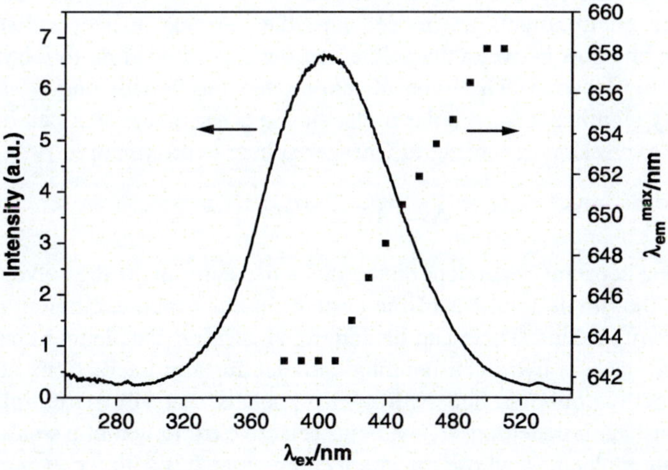

Fig. 12 Wavelength corresponding to the fluorescence maximum (λ_{em}^{max}) versus excitation wavelength (λ_{ex}) plot of ANF in [bmim][BF_4] (*filled square*) superimposed on the absorption spectrum. Reproduced with permission from [110]

wavelength-dependent shift of the emission maximum is found to be dependent on the viscosity of RTILs. That the observed behavior is not due to the ionic nature of the constituents of the RTILs, but is due to their highly viscous nature is evident from the fact that ANF exhibits similar excitation wavelength dependence in highly viscous conventional solvent, glycerol.

4.2 Origin of the Excitation Wavelength Dependence

Even though excitation wavelength-dependent fluorescence behavior of the kind described above is uncommon for simple molecular systems in conventional media of low viscosity, a behavior of this nature is however often observed in highly viscous or frozen media such as low temperature glasses, or polymer matrices and organized assemblies such as membranes, proteins, etc. [99, 114–119]. This phenomenon is commonly termed as "red-edge effect" (REE) [114, 116], although terminologies such as "edge excitation shift" (EES) [117], "edge excitation red shift" (EERS) [115] and "red edge excitation shift (REES)" [88, 119] are also used. Since this phenomenon is frequently encountered in biological systems, REE has been exploited quite extensively for studies in biological systems and an excellent review article by Demchenko is available on this topic [99]. The first chapter of this book also contains a discussion on REE [87].

In order to observe REE two conditions must be satisfied. The first requirement is the inhomogeneous broadening of the absorption spectrum of the molecular system (Fig. 13), which, in the case of dipolar systems, can arise from the distribution of the solute–solvent interaction energies. As each molecule in a condensed medium does not necessarily experience an identical environment and since a large number of different environments are indeed possible, one can expect a distribution of interaction energies between the solute and the solvent leading to a broadening of the absorption spectra. This type of inhomogeneous broadening is shown to be particularly significant for dipolar molecules in polar media as it can be described by simple expression according to Onsager sphere approximation [99]

$$\Delta \nu = A \Delta \mu a^{-3/2} (kT)^{1/2} \tag{6}$$

where A is a constant dependent on the dielectric constant of the solvent, $\Delta\mu$ is the change in the dipole moment of the system on electronic excitation and a is the Onsager cavity radius. There can be additional spectral broadening due to specific interactions such as hydrogen bonding and electrostatic interactions or due to the presence of hydrophobic and hydrophilic pockets that allow multiple solvation sites. Additional broadening, if any, helps selective excitation of a small population of the solute molecules, whose transition energy matches with the excitation energy [99]. However, the presence of an ensemble of energetically different molecules, which allows photoselection of the individual species, *alone* cannot give rise to the excitation wavelength-dependent fluorescence behavior of a system as rapid

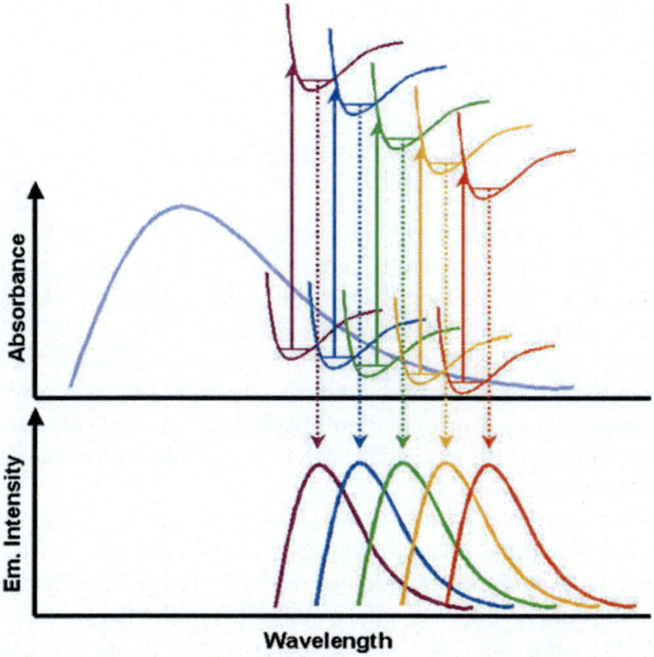

Fig. 13 A pictorial illustration of the Red-Edge Effect, which arises from the presence of an ensemble of energetically different molecules in the ground state and inefficient relaxation in the excited state. Reproduced with permission from [75]

relaxation in the excited state results in emission from the lowest excited state irrespective of the excitation [99, 118].

The second condition that must be satisfied for the occurrence of REE is that the excited state relaxation of the fluorescent species, which can be solvation of the fluorescent state or energy transfer from the fluorescent state to a low lying energy state, must be slower or comparable to the fluorescence lifetime of the species [99, 118].

Taking into consideration the two essential conditions described above it is possible to rationalize why ANF and a few other molecular systems exhibit excitation wavelength-dependent fluorescence behavior in RTILs [110]. For ANF, the inhomogeneous broadening of the absorption spectrum, which is necessary for photoselection of energetically different species and is determined by (6), is quite high due to its large (25 D) $\Delta\mu$ value. Moreover, the fluorescence lifetime of ANF in RTILs is not only very short (~100 ps) compared to other dipolar probe molecules but also much shorter than the solvent reorganization time in RTILs ($\langle T \rangle_{solv} \geq 1$ ns). Consequently, it is not possible to observe fluorescence from the fully equilibriated excited state of the molecule. Instead, fluorescence from a nonrelaxed or partially relaxed state, whose position is largely determined by the excitation wavelength, is observed. REE resulting from inhomogeneous broadening caused by the electrostatic and hydrogen bonding interactions and microheterogeneity of the RTILs is also reported in these media.

5 Structural Heterogeneity of the RTILs

Even though the RTILs appear as homogeneous fluid through naked eye, a number of experimental and simulation studies have revealed that these are microheterogeneous systems comprising long-lived hydrophobic and hydrophilic domains such as micelles, reverse micelles, proteins, membranes, etc. [73, 109, 110, 120, 121]. Despite the fact that no direct evidence of this heterogeneity has so far been obtained from the fluorescence measurements, there are however several instances where the microheterogeneous nature of the RTILs has been indicated in fluorescence studies [53–56, 110, 122, 123]. A few representative studies are presented below to illustrate this point.

In molecular solvents, 4-*N,N*-dimethylaminobenzonitrile (DMABN) exhibits dual fluorescence and the relative intensities of the two (LE and ICT) emission bands are independent of the excitation wavelength [124, 125]. However, in RTILs, as can be seen from Fig. 14, with an increase in the excitation wavelength, the ICT emission intensity increases relative to the LE emission [122]. If it is assumed that the molecular geometry of DMABN in RTILs is not very different from that in conventional solvents, the excitation wavelength dependence can only be explained considering the microheterogeneous nature of these media wherein DMABN

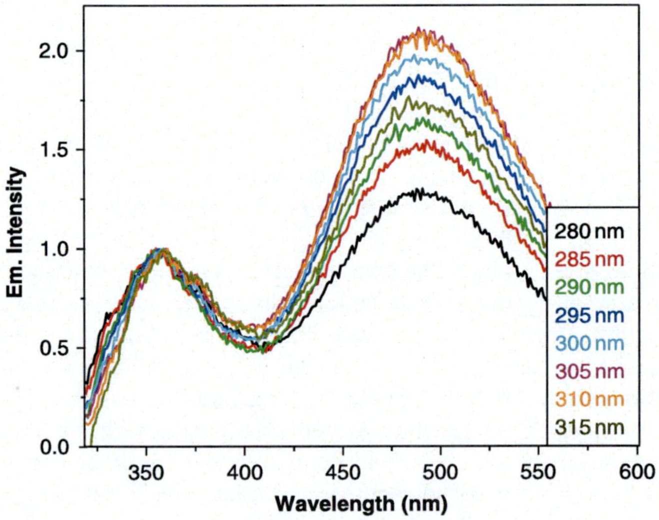

Fig. 14 Fluorescence spectra of DMABN in [bmim][PF_6] for excitation wavelengths of 280–315 nm. The spectra have been normalized at the LE emission peak position and corrected for the fluorescence of [bmim][PF_6]. Reproduced with permission from [122]

molecules occupy various possible environments having different polarities. With increase in the excitation wavelength, DMABN molecules, which lie in more polar regions, are preferentially excited. As the LE $\rightarrow$ ICT barrier depends on the polarity of the medium and is lower for molecules that lie in a more polar environment, it is understandable why an increase of the excitation wavelength results in an enhancement of the ICT emission.

Photoinduced electron transfer (PET) reaction between pyrene and *N,N*-dimethylaniline (DMA) has been extensively studied in conventional media by monitoring the fluorescence quenching of pyrene upon addition of DMA [126]. As this PET reaction is diffusion controlled, one expects considerable slowing down of the rate of this process in RTILs due to the viscous nature of the media. While this is indeed the case, what appeared quite surprising was the finding that the estimated rates of fluorescence quenching due to PET, as obtained from the steady state and time-resolved fluorescence measurements, were higher than the diffusion controlled rates (estimated using bulk viscosities of the media) by a factor of 2–4 [55]. Results of this kind can only be explained if the medium is heterogeneous comprising domains whose microviscosities are different from the bulk viscosities of the media.

Indirect evidence of the microheterogeneous nature of [bmim][PF_6], one of the extensively studied RTILs, has been obtained recently from a study based on multiphoton confocal fluorescence microscopy technique [122]. In this work, a few microlitres of a dilute solution of 4-(*N,N*-dimethylamino)benzonitrile (DMABN) in [bmim][PF_6] is placed between a clean glass slide and a coverslip and used for photobleaching and analysis. A square region 43 μm × 43 μm of interest is scanned under the microscope and a circular region of interest of radius 4.3 μm is photobleached by multiphoton excitation. The diffusion characteristics of DMABN in [bmim][PF_6] are obtained by acquiring sequence of images (Fig. 15) of the fluorescence recovery after photobleaching (FRAP). The diffusion coefficient of DMABN in [bmim][PF_6] estimated from this experiment is found to be different from that given by the Stokes–Einstein relation. This observation can only be rationalized considering the microheterogeneous nature of the RTIL.

Fluorescence technique is frequently used for the measurement of microviscosity of a variety of organized assemblies, which are microheterogeneous in nature. One can use for this purpose molecular rotors, which are molecular systems whose fluorescence quantum yield is highly sensitive to the viscosity of the medium. Malonitrile derivatives are well-known examples of molecular rotors or fluorescent microviscosity probes. 9-(dicyanovinyl)julolidine (DCVJ, Fig. 16) is one of the extensively used systems for the characterization of a variety of microheterogeneous chemical and biological systems [125]. DCVJ is rather weakly fluorescent (quantum yield, $\Phi_f \approx 10^{-3}$) in conventional less viscous solvents (<2 cP), whereas in highly viscous glycerol, Φ_f is as high as 0.1. Studies on DCVJ have also shown that the microviscosity, as experienced by this probe molecule and estimated from its fluorescence efficiency in RTILs, can be significantly different from the respective bulk viscosities indicating the microheterogeneous nature of these media [56].

In addition to the above examples, excitation wavelength-dependent fluorescence behavior of molecular systems of the kind observed in the case of ANF and described

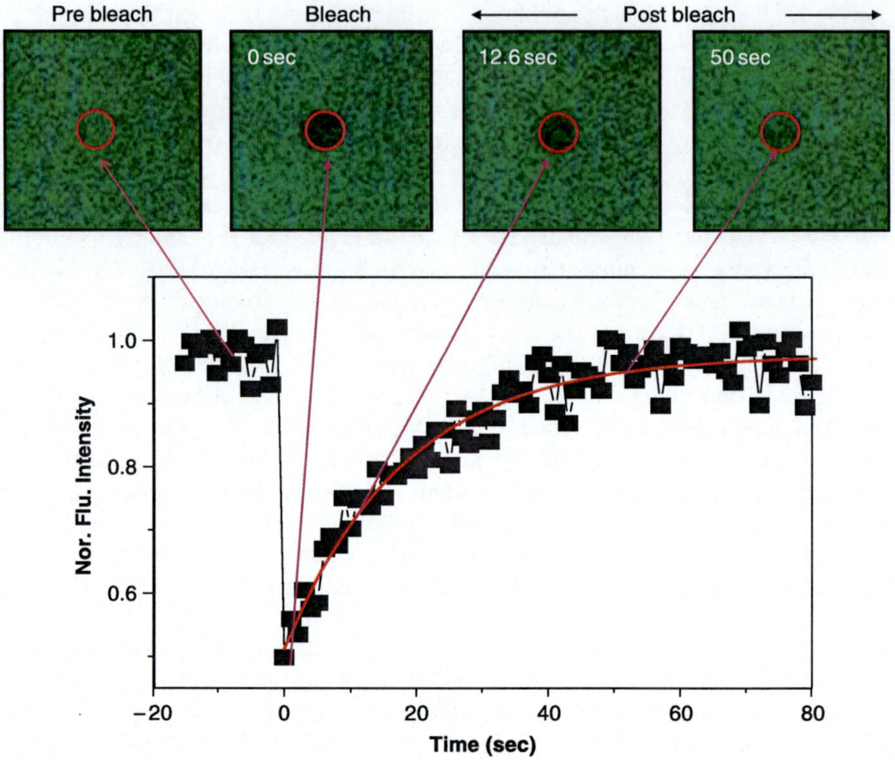

Fig. 15 *Top panel* shows the fluorescence images of DMABN in [bmim][PF_6], collected using a multiphoton confocal microscope (λ_{exc} = 730 nm, λ_{em} = 505–570 nm). The *bottom panel* represents the recovery of the normalized fluorescence within the circular ROI as a function of time. *Solid line* represents the exponential fit to the recovery data. In the *top panel*, the *circular region* represents the bleached spot. Reproduced with permission from [122]

Fig. 16 Chemical formula of 4-(*N*,*N*-dimethylamino) benzonitrile (DMABN) and 9-dicyanovinyljulolidine (DCVJ)

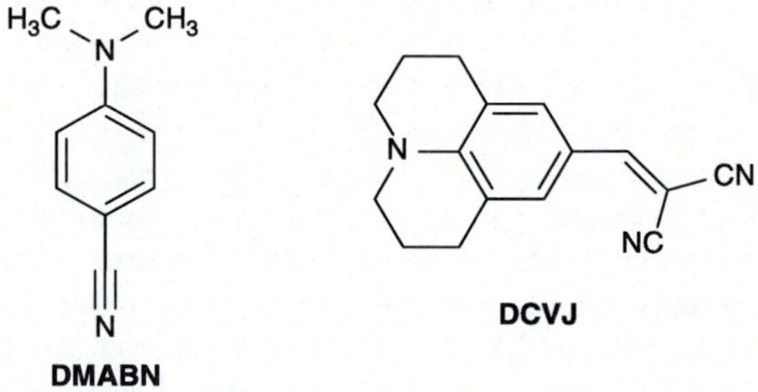

in Sect. 4.1 [69, 110] is also partly due to inhomogeneous broadening caused by the segregation of the nonpolar and polar moieties of the RTILs. Excitation wavelength-dependent solvation dynamics [53] is also perhaps the consequence of the structural heterogeneity of the RTILs. The dynamic heterogeneity evident from the dispersed kinetics of the solute rotation or nonexponential solvation dynamics in RTILs [54, 123] can also be the result of nanoscale or microscale aggregation of the polar and nonpolar segments of the constituent ions of these media.

6 Concluding Remarks

It is shown that the fluorescence response of molecular systems in RTILs provides fundamental information relating to the structure, dynamics, and several other physical characteristics of these novel media. The studies carried out so far have indicated that the microenvironments experienced by the fluorescent probe molecules in RTILs resemble those in short-chain alcohols. The dynamic fluorescence Stokes shift studies have suggested that solvent reorganization in these media is a much slower process when compared with that in conventional polar solvents due to the high viscosity of the RTILs. It is shown that slow solvation dynamics, complex nature of the RTILs and multiple interactions operational in these media can give rise to effects that are not commonly observed in conventional solvents.

Even though our understanding of these complex media has significantly improved over the past several years, a number of issues concerning these substances still remain poorly understood. For example, we do not have precise interpretation of the physical motions that contribute to the various components of the solvation dynamics despite considerable theoretical and experimental efforts. The probe dependence of the solvation dynamics is yet another issue that is not understood. While the microheterogeneous nature of these media has been indicated in several theoretical and experimental investigations, it has not yet been possible to extract the details of these domain structures from the fluorescence studies. Considering these issues and the fact that a huge number of RTILs are yet to be unexplored, it is certain that studies involving the fluorescence probe molecules in these media have a long and bright future.

Acknowledgments Some of our works presented in this chapter have been supported by the J.C. Bose Fellowship of the Department of Science and Technology, Government of India. Thanks are due to Mr. K. Santhosh for assistance with the bibliography.

References

1. Nelson WM (1998) In: Anastas PT, Williamson TC (eds) Green chemistry. Oxford University Press, Oxford
2. Tanaka K, Toda F (2000) Solvent-free organic synthesis. Chem Rev 100:1025–1074
3. Hang CT, Lianhai L, Yang Y, Wenshuo L (2002) Developing green chemistry: organometallic reactions in aqueous media. In: Abraham M, Moens L (eds) Clear solvents:

alternative media for chemical reactions and processing, ACS symposium series 819. American Chemical Society, Washington, DC, 166p
4. Kajimoto O (1999) Solvation in supercritical fluids: its effects on energy transfer and chemical reactions. Chem Rev 99:355–390
5. Seddon KR (2003) Ionic liquids: a taste of the future. Nature (Materials) 2:363–365
6. Rogers RD, Seddon KR (2003) Ionic liquids–solvents of the future? Science 302:792–793
7. Wasserscheid P, Keim W (2000) Ionic liquids-new "solutions" for transition metal catalysis. Angew Chem Int Ed 39:3772–3789
8. Wilkes JS, Levinsky JA, Wilson RA, Hussey CL (1982) Dialkylimidazolium chloroaluminate melts: a new class of room-temperature ionic liquids for electrochemistry, spectroscopy and synthesis. Inorg Chem 21:1263–1264
9. Boon JA, Levinsky JA, Pflug JL, Wilkes JS (1986) Friedel-Crafts reactions in ambient-temperature molten salts. J Org Chem 51:480–483
10. Sheldon R (2001) Catalytic reactions in ionic liquids. Chem Commun 2399–2407
11. Welton T (1999) Room-temperature ionic liquids: solvents for synthesis and catalysis. Chem Rev 99:2071–2084
12. MacFarlane DR, Meakin P, Sun J, Amini N, Forsyth M (1999) Pyrrolidinium imides: a new family of molten salts and conductive plastic crystal phases. J Phys Chem B 103:4164–4170
13. Henderson WA, Passerini S (2004) Phase behavior of ionic liquid-LiX mixures: pyrrolidinium cations and TFSI- anions. Chem Mater 16:2881–2885
14. Tao G, He L, Sun N, Kou Y (2005) New generation ionic liquids: cations derived from amino acids. Chem Commun 3562–3564
15. Fukumoto K, Yoshizawa M, Ohno H (2005) Room temperature ionic liquids from 20 natural amino acids. J Am Chem Soc 127:2398–2399
16. Ohno H, Fukumoto K (2007) Amino acid ionic liquids. Acc Chem Res 40:1122–1129
17. Holbrey JD, Reichert WM, Swatloski RP, Broker GA, Pritner WR, Seddon KR, Rogers RD (2002) Efficient, halide free synthesis of new, low cost ionic liquids: 1, 3-dialkylimidazolium salsts containing methyl- and ethyl-sulfate anions. Green Chem 4:407–413
18. MacFarlane DR, Golding J, Forsyth S, Forsyth M, Deacon GB (2001) Low viscosity ionic liquids based on organic salts of the dicyanamide anion. Chem Commun 1430–1431
19. Seddon KR (2002) Ionic liquids, industrial applications for green chemistry. American Chemical Society, Washington, DC
20. Buzzeo MC, Hardacre C, Compton RG (2004) Use of room temperature ionic liquids in gas sensor design. Anal Chem 76:4583–4588
21. Wasserscheid P, Welton T (2003) Ionic liquids in synthesis. Wiley-VCH, Weinheim
22. Brinchi L, Germani R, Savelli G (2003) Ionic liquids as reaction media for esterification of carboxylates sodium salts with alkyl halides. Tetrahedron Lett 44:2027–2029
23. Golding JJ, Macfarlane DR, Spiccia L, Forsyth GB, Skelton BW, White AH (1998) Weak intermolecular interactions in sulfonamide salts: structure of 1-ethyl-2-methyl-3-benzyl imidazolium bis[(trifluoromethyl)sulfonyl]amide. Chem Commun 1593–1594
24. Pringle JM, Golding J, Baranyai K, Forsyth CM, Deacon GB, Scott JL, MacFarlane DR (2003) The effecr of anion fluorination in ionic liquids-physical properties of a range of bis (methanesulfonyl)amide salts. New J Chem 27:1504–1510
25. Hough WL, Rogers RD (2007) Ionic liquids then and now: from solvents to materials to active pharmaceutical ingredients. Bull Chem Soc Jpn 80:2262–2269
26. Plechkova NV, Seddon KR (2008) Applications of ionic liquids in the chemical industry. Chem Soc Rev 37:123–150
27. Rogers RD, Voth GA (2007) Guest editorial – ionic liquids. Acc Chem Res 40:1077–1078
28. Carmichael AJ, Seddon KR (2000) Polarity study of some 1-alkyl-3-methylimidazolium ambient-temperature ionic liquids with the solvatochromic dye, nile red. J Phys Org Chem 13:591–595
29. Aki SNVK, Brennecke JF, Samanta A (2001) How polar are room temperature ionic liquids? Chem Commun 413–414

30. Muldoon MJ, Gordon CM, Dunkin IR (2001) Investigations of solvent-solute interactions in room temperature ionic liquids using solvatochromic dyes. J Chem Soc Perkin Trans 2: 433–435
31. Law G, Watson PR (2001) Surface tension measurements of N-alkylimidazolium ionic liquids. Langmuir 17:6138–6141
32. Huddleston JG, Visser AE, Reichert WM, Willauer HD, Broker GA, Rogers RD (2001) Characterization and comparison of hydrophilic and hydrophobic room temperature ionic liquids incorporating the imidazolium cation. Green Chem 3:156–164
33. Noda A, Hayamizu K, Watanabe M (2001) Pulsed-gradient spin echo 1H and 19F NMR ionic diffusion coefficient, viscosity, and ionic conductivity of non-chloroaluminate room-temperature ionic liquids. J Phys Chem B 105:4603–4610
34. McLean AJ, Muldoon MJ, Gordon CM, Dunkin IR (2002) Bimolecular rate constants for diffusion in ionic liquids. Chem Commun 1880–1881
35. Ngo HL, LeCompte L, Hargens L, McEwen A (2000) Thermal properties of imidazolium ionic liquids. Thermochim Acta 357:97–102
36. Dzyuba S, Bartsch RA (2002) Influence of structural variations in 1-alkyl(aralkyl)-3-methylimidazoliumhexafluorophosphates and bis(trifluoromethylsulfonyl)imides on physical properties of the ionic liquids. Chem Phys Chem 3:161–166
37. Elaiw A, Hichcock PB, Seddon KR, Srinivasan N, Tan Y-M, Welton T, Zora JA (1995) Hydrogen bonding in imidazolium salts and its implications for ambient-temperature halogenoaluminate(III) ionic liquids. J Chem Soc Dalton Trans 21:3467–3472
38. Mele A, Tran CD, Lacerda AP (2003) The structure of a room-temperature ionic liquid with and without trace amount of water: the role of C-H—O and C-H—F interactions in 1-n-butyl-3-methylimidazolium tetrafluoroborate. Angew Chem Int Ed 36:4364–4366
39. Karmakar R, Samanta A (2002) Solvation dynamics of coumarin-153 in a room temperature ionic liquid. J Phys Chem A 106:4447–4452
40. Karmakar R, Samanta A (2002) Steady-state and time-resolved fluorescence behavior of C153 and PRODAN in room-temperature ionic liquids. J Phys Chem A 106:6670–6675
41. Karmakar R, Samanta A (2003) Dynamics of solvation of the fluorescent state of some electron-donor acceptor molecules in room temperature ionic liquids, [BMIM][$(CF_3SO_2)_2N$] and [EMIM][$(CF_3SO_2)_2N$]. J Phys Chem A 107:7340–7346
42. Chakrabarty D, Hazra P, Chakraborty A, Seth D, Sarkar N (2003) Dynamics of solvent relaxation in room temperature ionic liquids. Chem Phys Lett 381:697–704
43. Ingram JA, Moog RS, Ito N, Biswas R, Maroncelli M (2003) Solute rotation and solvation dynamics in a room-temperature ionic liquid. J Phys Chem B 107:5926–5932
44. Arzhantsev S, Ito N, Heitz M, Maroncelli M (2003) Solvation dynamics of coumarin 153 in several classes of ionic liquids: cation dependence of the ultrafast component. Chem Phys Lett 381:278–286
45. Saha S, Mandal PK, Samanta A (2004) Solvation dynamics of nile red in a room temperature ionic liquid using streak camera. Phys Chem Chem Phys 6:3106–3110
46. Chowdhury PK, Halder M, Sanders L, Calhoun T, Anderson JL, Armstrong DW, Song X, Petrich JW (2004) Dynamic solvation in room-temperature ionic liquids. J Phys Chem B 108:10245–10255
47. Ito N, Arzhantsev S, Heitz M, Maroncelli M (2004) Solvation dynamics and rotation of coumarin 153 in alkylphosphonium ionic liquids. J Phys Chem B 108:5771–5777
48. Mandal PK, Samanta A (2005) Fluorescence studies in a pyrolidinium ionic liquid: polarity of the medium and solvation dynamics. J Phys Chem B 109:15172–15177
49. Paul A, Samanta A (2007) Solute rotation and solvation dynamics in an alcohol-functionalized room temperature ionic liquids. J Phys Chem B 111:4724–4731
50. Arzhantsev S, Jin H, Baker GA, Maroncelli M (2007) Measurements of the complete solvation response in ionic liquids. J Phys Chem B 111:4978–4989

51. Jin H, Baker GA, Arzhantsev S, Dong J, Maroncelli M (2007) Solvation and rotational dynamics of coumarin 153 in ionic liquids: comparisons to conventional solvents. J Phys Chem B 111:7291–7302
52. Bhargava BL, Balasubramanian S (2007) Insights into the structure and dynamics of a room temperature ionic liquid: ab initio molecular dynamics simulation studies of 1-n-butyl-3-methylimidazolium hexafluorophosphate ([bmim][PF_6]) and the [bmim][PF_6]-CO_2 mixture. J Phys Chem B 111:4477–4487
53. Adhikari A, Sahu K, dey S, Ghosh S, Mandal U, Bhattacharyya K (2007) Femtosecond solvation dynamics in a neat ionic liquid and ionic liquid microemulsion: excitation wavelength dependence. J Phys Chem B 111:12809–12816
54. Funston AM, Fadeeva TA, Wishart JF, Edward W, Castner J (2007) Fluorescence probing of temperature-dependent dynamics and friction in ionic liquid local environments. J Phys Chem B 111:4963–4977
55. Paul A, Samanta A (2007) Photoinduced electron transfer in room temperature ionic liquids: a combined laser flash photolysis and fluorescence study. J Phys Chem B 111:1957–1962
56. Paul A, Samanta A (2008) Free volume dependence of the internal rotation of a molecular rotor probe in room temperature ionic liquids. J Phys Chem B 112:16626–16632
57. Shim Y, Kim HJ (2008) Dielectric relaxation, ion conductivity, solvent rotation, and solvation dynamics in a room-temperature ionic liquid. J Phys Chem B 112:11028–11038
58. Nagasawa Y, Itoh T, Yasuda M, Ishibashi Y, Ito S, Miyasaka H (2008) Ultrafast charge transfer process of 9,9′-bianthryl in imidazolium ionic liquids. J Phys Chem B 112: 15758–15765
59. Khara DC, Paul A, Santhosh K, Samanta A (2009) Excited state dynamics of 9,9′-bianthryl in room temperature ionic liquids as revealed by picosecond time-resolved fluorescence study. J Chem Sci 121:309–315
60. Khara DC, Samanta A (2010) Rotational dynamics of positively and negatively charged solutes in ionic liquid and viscous molecular solvent studied by time-resolved fluorescence anisotropy measurements. Phys Chem Chem Phys 12:7671–7677
61. Iwata K, Kakita M, Hamaguchi H (2007) Picosecond time-resolved fluorescence study on solute–solvent interaction of 2-aminoquinoline in room-temperature ionic liquids: aromaticity of imidazolium-based ionic liquids. J Phys Chem B 111:4914–4919
62. Mali KS, Dutt GB, Mukherjee T (2008) Photoisomerization of cyanine derivatives in 1-butyl-3-methylimidazolium hexafluorophosphate and aqueous glycerol: influence of specific interactions J Chem Phys 128:124515/1–124515/9
63. Mali KS, Dutt GB, Mukherjee T (2007) Ionic liquids (special issue on ionic liquids). Acc Chem Res 40:1077–1236
64. Mali KS, Dutt GB, Mukherjee T (2007) The physical chemistry of ionic liquids (special issue on ionic liquids). J Phys Chem B 111:4639–5029
65. Mali KS, Dutt GB, Mukherjee T (2010) Physical chemistry of ionic liquids (special issue on ionic liquids). Phys Chem Chem Phys 12:1629–2032
66. Bhattacharya B, Samanta A (2008) Excited-state proton transfer dynamics of 7-hydroxyquinoline in room temperature ionic liquids. J Phys Chem B 112:10101–10106
67. Chiappe C, Pieraccini D (2005) Ionic liquids: solvent properties and organic reactivity. J Phys Org Chem 18:275–297
68. Fukuda M, Terazima M, Kimura Y (2008) Study on the intramolecular proton transfer of 4′-N, N-diethylamino-3-hydroxyflavone in imidazolium-based room temperature ionic liquids. Chem Phys Lett 463:364–368
69. Hu Z, Margulis CJ (2006) Heterogeneity in a room-temperature ionic liquid: persistent local environments and the red edge effect. Proc Natl Acad Sci USA 103:831–836
70. Kobrak MN (2008) The chemical environment of ionic liquids: links between liquid structure, dynamics, and solvation. Adv Chem Phys 139:83–135
71. Kobrak MN (2006) Characterization of the solvation dynamics of an ionic liquid via molecular dynamics simulation. J Chem Phys 125:064502–064513

72. Kobrak MN, Li H (2010) Electrostatic interactions in ionic liquids: the dangers of dipole and dielectric descriptions. Phys Chem Chem Phys 12:1922–1932
73. Lopes JNAC, Padua AAH (2006) Nanostructural organization in ionic liquids. J Phys Chem B 110:3330–3335
74. Samanta A (2010) Solvation dynamics in ionic liquids: what we have learned from the dynamic fluorescence Stokes shift studies. J Phys Chem Lett 1:1557–1562
75. Samanta A (2006) Dynamic Stokes shift and excitation wavelength dependent fluorescence of dipolar molecules in room temperature ionic liquids. J Phys Chem B 110:13704–13716
76. Mandal PK, Paul A, Samanta A (2006) Excitation wavelength dependent fluorescence behavior of the room temperature ionic liquids and dissolved dipolar solutes. J Photochem Photobiol A Chem 182:113–120
77. Mandal PK, Saha S, Karmakar R, Samanta A (2006) Solvation dynamics in room temperature ionic liqudis: dynamics stokes shift studies of fluorescence of dipolar molecules. Curr Sci 90:301–310
78. Wakai C, Oleinikova A, Ott M, Weingartner H (2005) How polar are ionic liquids? determination of static dielectric constant of an imidazolium-based ionic liquid by microwave dielectric spectroscopy. J Phys Chem B 109:17028–17030
79. Daguenet C, Dyson PJ, Krossing I, Oleinikova A, Slattery J, Wakai C, Weingartner H (2006) Dielectric response of imidazolium-based room-temperature ionic liquids. J Phys Chem B 110:12682–12688
80. Weingartner H (2008) Understanding ionic liquids at the molecular level: facts, problems, and controversies. Angew Chem Int Ed 47:654–670
81. Santhosh K, Samanta A (2010) Modulation of the excited state intramolecular electron transfer reaction and dual fluorescence of crystal violet lactone in room temperature ionic liquids. J Phys Chem B 114:9195–9200
82. Paul A, Mandal PK, Samanta A (2005) How transparent are the imidazolium ionic liquids? A case study with 1-methyl-3-butylimidazolium hexafluorophosphate, [bmim][PF_6]. Chem Phys Lett 402:375–379
83. Paul A, Mandal PK, Samanta A (2005) On the optical properties of the imidazolium ionic liquids. J Phys Chem B 109:9148–9153
84. Reichardt C (1988) Solvents and solvent effects in organic chemistry. Weinheim, VCH, Germany
85. Reichardt C (1992) Solvatochromism, thermochromism, piezochromism, halochromism and chiro-solvatochromism of pyridinium N-phenoxide betaine dyes. Chem Soc Rev 21:147–153
86. Reichardt C (1994) Solvatochromic dyes as solvent polarity indicators. Chem Rev 94:2319–2358
87. Demchenko AP, Yesylevskyy SO (2010) Probing interfaces and nanostructures with fluorescent dyes. Power and weakness of nanoscopic description. In: Demchenko AP (ed) Advanced fluorescence reporters in chemistry and biology I. Springer Series in Fluorescence 8. Springer, Heidelberg, pp 189–223
88. Kalyanasundaram K (1987) Photochemistry in microheterogeneous systems. Academic Press, Orlando
89. Reichardt C (2005) Polarity of ionic liquids determined empirically by means of solvatochromic pyridinium N-phenolate betaine dyes. Green Chem 7:339–351
90. Bagchi B, Oxtoby DW, Fleming GR (1984) Theory of the time development of the stokes shift in polar media. Chem Phys 86:257–267
91. Gvd Z, Hynes JT (1985) Time-dependent fluorescence solvent shifts dielectric friction and non equilibrium solvation in polar solvents. J Phys Chem 89:4181–4188
92. Maroncelli M, Fleming GR (1987) Picosecond solvation dynamcis of coumarin 153: the importance of molecular aspects of solvation. J Chem Phys 86:6221–6229
93. Simon JD (1988) Time-resolved studies of solvation in polar media. Acc Chem Res 21: 128–164

94. Maroncelli M, Macinnis J, Fleming GR (1989) Polar solvent dynamics and electron-transfer reactions. Science 243:1674–1681
95. Bagchi B (1989) Dynamics of solvation and charge transfer reactions in dipolar liquids. Annu Rev Phys Chem 40:115–141
96. Horng ML, Gardecki JA, Papazyan A, Maroncelli M (1995) Subpicosecond measurements of polar solvation dynamics: coumarin 153 revisited. J Phys Chem 99:17311–17337
97. Fleming GR, Cho M (1996) Chromophore-solvent dynamics. Annu Rev Phys Chem 47: 109–134
98. Lakowicz JR (1999) Principles of fluorescence spectroscopy. Kluwer Academic, Plenum, New York
99. Demchenko AP (2002) The red-edge effects: 30 years of exploration. Luminescence 17: 19–42
100. Nandi N, Bhattacharyya K, Bagchi B (2000) Dielectric relaxation and solvation dynamics of water in complex chemical and biological systems. Chem Rev 100:2013–2046
101. Bhattacharyya K (2003) Solvation dynamics and proton trnasfer in supramolecular assemblies. Acc Chem Res 36:95–101
102. Pal SK, Peon J, Zewail AH (2002) Biological water at the protein surface: dynamical solvation probed directly with femtosecond resolution. Proc Natl Acad Sci 99:1763–1768
103. Bagchi B (2005) Water dynamics in the hydration layer around proteins and micelles. Chem Rev 105:3197–3219
104. Bhattacharyya K, Bagchi B (2000) Solvation dynamics of constrined water in complex geometries. J Phys Chem A 104:10603–10613
105. Pal SK, Zhao L, Zewail A (2003) Water at DNA surfaces: ultrafast dynamics in minor groov recognition. Proc Natl Acad Sci USA 100:8113–8118
106. Ito N, Arzhantsev S, Maroncelli M (2004) The probe dependence of solvation dynamics and rotation in the ionic liquid 1-butyl-3-methylimidazolium hexafluorophosphate. Chem Phys Lett 396:83–91
107. Tokuda H, Tsuzuki S, Susan MABH, Hayamizu K, Watanabe M (2006) How ionic are room temperature ionic liquids? An indicator of the physicochemical properties. J Phys Chem B 110:19593–19600
108. MacFarlane DR, Forsyth M, Izgorodina EI, Abbott AP, Annat G, Fraser K (2009) On the concept of ionicity in ionic liquids. Phys Chem Chem Phys 11:4962–4967
109. Wang Y, Voth GA (2005) Unique spatial heterogeneity in ionic liquids. J Am Chem Soc 127: 12192–12193
110. Mandal PK, Sarkar M, Samanta A (2004) Excitation-wavelength-dependent fluorescence behavior of some dipolar molecules in room-temperature ionic liquids. J Phys Chem A 108: 9048–9053
111. Bart E, Meltsin A, Huppert D (1994) Solvation dynamics in molten salts. J Phys Chem 98:10819–10823
112. Shim Y, Duan J, Choi MY, Kim HJ (2003) Solvation in molecular ionic liquids. J Chem Phys 119:6411–6414
113. Znamenskiy V, Kobrak MN (2004) Molecular dynamics study of polarity in room-temperature ionic liquids. J Phys Chem B 108:1072–1079
114. Weber G, Shinitzky M (1970) Failure of energy transfer between identical aromatic molecules on excitation at the long wave edge of the absorption spectrum. Proc Natl Acad Sci USA 65:823–830
115. Itoh K, Azumi T (1975) Shift of the emission band upon excitation at the long wavelength absorption edge. II. Importance of the solute–solvent interaction and the solvent reorientation relaxation process. J Chem Phys 62:3431–3438
116. Valeur B, Weber G (1977) Anisotropic rotations in 1-naphthylamine existence of a red-edge transition moment normal to the ring plane. Chem Phys Lett 45:140–144
117. Lakowicz JR, Nakamoto SK (1984) Red-edge excitation of fluorescence and dynamic properties proteins and membranes. Biochemistry 23:3013–3021

118. Demchenko AP (1991) In: Lakowicz JR (ed) Topics in fluorescence spectroscopy. Plenum, New York
119. Chattopadhyay A, Mukherjee S (1993) Fluorophore environments in membrane-bound probes: a red edge excitation shift study. Biochemistry 32:3804–3811
120. Triolo A, Russina O, Bleif HJ, Cola ED (2007) Nanoscale segregation in room temperature ionic liquids. J Phys Chem B 111:4641–4644
121. Turton DA, Hunger J, Stoppa A, Hefter G, Thoman A, Walther M, Buchner R, Wynne K (2009) Dynamics of imidazolium ionic liquids from a combined dielectric relaxation and optical kerr effect study: evidence for mesoscopic aggregation. J Am Chem Soc 131: 11140–11146
122. Santhosh K, Banerjee S, Rangaraj N, Samanta A (2010) Fluorescence response of 4-(N, N′-dimethylamino)benzonitrile in room temperature ionic liquids: observation of photobleaching under mild excitation and multiphoton confocal fluorescence microscopic study of the fluorescence recovery dynamics. J Phys Chem B 114:1967–1974
123. Jin H, Li X, Maroncelli M (2007) Heterogeneous solute dynamics in room temperature ionic liquids. J Phys Chem B 111:13473–13478
124. Tomin V (2010) Physical principles behind spectroscopic response of organic fluorophores to intermolecular interactions. In: Demchenko AP (ed) Advanced fluorescence reporters in chemistry and biology I: Fundamentals and molecular design. Springer Series on Fluorescence 8. Springer, Heidelberg, pp 189–223
125. Haidekker MA, Nipper M, Mustafic A, Lichlyter D, Dakanali M, Theodorakis EA (2010) Dyes with segmental mobility: molecular rotors In: Demchenko AP (ed) Advanced fluorescene reporters in chemistry and biology I: Fundamentals and molecular design. Springer Series in Fluorescence 8. Springer, Heidelberg, pp 267–308
126. Kavarnos GJ, Turro NJ (1986) Photosensitization by reversible electron transfer: theories, experimental evidence and examples. Chem Rev 86:401–449

Fluorescence Spectroscopy in Polymer Science

Tanzeela N. Raja and Albert M. Brouwer

Abstract Polymer science is an interdisciplinary field, combining chemistry, physics, and in some cases biology. Structure, morphology, and dynamical phenomena in natural and synthetic polymers can be addressed using fluorescence spectroscopy. The most attractive aspect of fluorescent reporters is that their fluorescence parameters can give information on the nanometer length scale with an exceptional sensitivity, which allows data acquisition with submicrometer spatial resolution and millisecond time resolution. The use of fluorescent reporter molecules is, in principle, an invasive technique. Because of the large size of polymer molecules, however, small fluorescent reporter molecules of a length scale of < 2 nm can be considered a small perturbation.

Because of the enormous importance of synthetic polymers in our technology-based societies, almost every conceivable experimental technique has been applied in this field, but most of these tend to address the sample on a macroscopic scale. This chapter gives illustrative examples of the power of molecular fluorescence for investigating several microscopic aspects of polymer science.

Keywords Fluorescence spectroscopy · Macromolecules · Microscopy · Morphology · Polymers

Contents

The research of the authors in this field is part of the Dutch Polymer Institute (project # 606).

T.N. Raja and A.M. Brouwer (✉)
Van't Hoff Institute for Molecular Sciences, P.O. Box 94157, 1090 GD Amsterdam, The Netherlands
e-mail: a.m.brouwer@uva.nl

A.P. Demchenko (ed.), *Advanced Fluorescence Reporters in Chemistry and Biology III: Applications in Sensing and Imaging*, Springer Ser Fluoresc (2011) 10: 91–118,
DOI 10.1007/978-3-642-18035-4_3,

1 Introduction

Due to the diversity, heterogeneity, and complexity of polymeric materials, it is very challenging for polymer scientists to investigate processes occurring at the molecular and microscopic level in these materials. While most polymeric materials are not fluorescent, they can be investigated by embedding fluorescent probe molecules that can be used to detect properties of their immediate surroundings, and dynamical processes that change the local environment on a variety of timescales. The combination of fluorescence techniques, such as steady-state, time-resolved fluorescence, and fluorescence microscopy, provides excellent opportunities to polymer scientist to investigate structure and dynamics in polymers. Some examples are given in this chapter that deal with monitoring polymerization (kinetics and mechanism), glass transition, film formation from latex materials, characterization of morphology, etc.

We will first briefly discuss the typical fluorescence experiments and the information they can provide that is of importance for polymer science [1], and then go into a number of areas in which these different experiments have been used.

1.1 Steady-State Fluorescence

Steady-state fluorescence spectroscopy is relatively simple to perform and can provide significant information about the location and environment of a fluorescent probe. The shape of the spectrum (e.g., pyrene) [2–4] and the position of the emission maximum (for solvatochromic probes) [5–7] can provide information on local polarity and mobility. Energy transfer, as a measure of distance, between different fluorescent groups, can be detected from the emission spectrum [8–10]. Excitation with polarized light combined with polarization-sensitive detection allows to measure anisotropy, a property that can give information on the rotational mobility of a probe molecule [11, 12]. The emission intensity can be used to obtain information on dynamical processes, using properly designed probes. For example, a probe molecule can be fluorogenic, the fluorescence being turned on when it is built into a polymer [13, 14], it can be rigidochromic [15], or fluorescence can be quenched by diffusive processes [16]. It is often difficult to quantify the emission intensity from a polymer sample because the absorption is usually not known, and samples are often scattering light. Ratiometric methods overcome this problem to a large extent [17]. Dynamical processes are, generally, more accurately quantified using time-resolved spectroscopy.

1.2 *Time-Resolved Fluorescence*

The excited-state decay times of fluorescent molecules range from picoseconds to (many) nanoseconds. Dynamical processes that occur on these timescales can thus be monitored via time-resolved measurements. When dynamical quenching processes are of interest, time-resolved detection gives more reliable information than intensity measurements, which suffer from the uncertainty associated with light absorption versus scattering, which commonly occurs in polymer samples. Molecular reorientation, a measure of viscosity, is most directly measured via time-resolved anisotropy [18, 19].

In simple cases, the population of an excited state decays via first-order kinetics, that is, exponentially. In heterogeneous samples, molecules in different environments decay with different time constants, which leads to a multiexponential decay or to distributions of decay time constants. Medium reorganization on the timescale of the decay can lead to spectral evolution. In such cases it is helpful to include additional experimental variables, e.g., different excitation or emission wavelengths, to get a data set that can be subjected to global analysis [20–22].

1.3 *Fluorescence Microscopy*

In the past decades laser confocal fluorescence microscopy (LCFM) has emerged as a powerful technique to study the three-dimensional distribution of luminescent entities in biological and material sciences. In its basic form, LCFM offers a lateral diffraction limited imaging resolution of the order of a few hundred nanometers, and an axial resolution of one to several micrometers. In recent years, advanced techniques have been developed, and mainly applied to the life sciences that allow subdiffraction resolution imaging [23–25]. LCFM can gain further resolving power by the use of multiple detection channels for different colors or different polarizations, or by including the fluorescence decay time as a variable (Fluorescence Lifetime Imaging (FLIM)) [26]. For thin films, for which depth-resolved information cannot be obtained anyway, direct 2D imaging (wide field) of the fluorescence intensities allows more rapid measurements, because scanning is not necessary. Fluorescence microscopy using a near-field approach offers higher resolution than LCFM, but only at the sample surface [27].

Fluorescence microscopy is the key to single molecule spectroscopy (SMS) because of its minimized detection volume, which allows discrimination of the emission of a single molecule from the background [28]. SMS provides information on dynamics of individual molecules in heterogeneous environments. In contrast to ensemble measurements, which yield information on average properties, SMS provides information on rare events and gives access to distributions and time dependencies of properties of individual molecules. This technique can be applied to investigate various processes in polymer science, such as complicated relaxation processes of polymer chains near the glass transition temperature in amorphous solids [29–31], rotational diffusion of single molecules in polymers [32], visualization of the dynamics of polymer chains, and translational diffusion of monomers in polymerizing solution [33].

1.4 Types of Fluorescent Probes

In this chapter we will focus on fluorophores that are added as "molecular spies" to nonfluorescent samples. Polymers that are intrinsically fluorescent are outside the scope. In practice, polymer samples usually contain fluorescent impurities, which precludes the use of dyes absorbing in the UV (<350 nm), unless special precautions are taken.

For some properties of interest, any dye with a suitable brightness and spectral properties can be used. If one is interested in properties related to diffusion, the molecular size is important. Energy transfer can be studied with appropriate pairs of energy donors and acceptors [8, 34–37]. Rigidochromic dyes are typically based on a nonradiative decay process that is suppressed when the molecule's environment is rigidified, giving emission of a photon a chance in competition with the radiationless excited-state decay process [38]. Recomplexes [39] and substituted olefins [40–43] are typical examples, but also other large-amplitude excited-state processes involving electron transfer can be used [15, 31, 44].

In many cases, it is desirable to covalently attach the reporter dye to the polymer of interest. In that case, invariably some steps of organic synthesis are required. Most authors use dyes that are easily derived from commercially available ones, such as dansyl derivatives [45]. On the other hand, commercially available dyes are normally not complicated to synthesize, so modification of the starting materials in order to arrive at a copolymerizable dye is usually not difficult.

Solvatochromic fluorescent probes are dyes of which the absorption or emission color is sensitive to solvent polarity [46]. These are generally electron Donor–π–Acceptor molecules, which exhibit a relatively small dipole moment in the ground state and a larger dipole moment in excited state. Among the known solvatochromic compounds, the σ-coupled donor–acceptor systems developed by Verhoeven and coworkers are still the most sensitive [47].

2 Monitoring Polymerization

Fluorescence methods can be fruitfully used to investigate various dynamical phenomena in polymer science, such as different types of polymerization processes (including crosslinking and gel formation), irreversible film formation by the physical interpenetration and diffusion of polymer chains across interfaces, phase separation, and reversible transitions such as the glass transition. This section will focus on the monitoring of polymerization processes.

The polymerization process starts with monomers, which are often low molecular weight and low viscosity liquids, which undergo a polymerization reaction and are converted into high molecular weight polymers, with a much higher viscosity. The polymerization process may take place through different mechanisms:

1. Chain-growth reaction: addition of unsaturated monomer at the end of a growing chain.
2. Step-growth: through a stepwise reaction between functional groups of monomers.

Step-growth polymers increase in molecular weight at a very slow rate at lower conversion and reach moderately high molecular weight only at very high conversion ($> 95\%$). The mechanism and kinetics of different types of polymerization are very different but a common feature is the increase in viscosity as polymerization proceeds, which can be effectively followed by fluorescent probes that are sensitive to the mobility and rigidity of the medium.

The increase in viscosity can be probed in different ways. Emission will be anisotropic when the molecular rotation of the probe molecule is hindered on the timescale of the excited-state lifetime, typically several nanoseconds. Rigidochromic probes fluoresce weakly in a medium of low viscosity due to a radiationless decay process associated with a large-amplitude motion (molecular rotor), which is suppressed when viscosity is increased [48]. Solvatochromic fluorescent probes are also sensitive to viscosity because the stabilization of the excited-state dipole requires medium reorganization. Finally, diffusive quenching processes can be used to probe translational diffusion.

Pioneering work by Loutfy [49] established 4-(*N*, *N*-dialkylamino)benzylidene malononitriles as prototypical rigidochromic probes. The fluorescence during homogeneous-phase polymerization of styrene or acrylates turns on rather abruptly when the free volume drops below a certain value (Fig. 1). At low conversion, and at high conversion, these probes are not sensitive to the degree of polymerization.

The rapid excited-state decay through double bond rotation is related with the characteristic electronic structure [50], and is found in many Donor–π–Acceptor systems [41, 42, 51]. This kind of fluorescent probe is intrinsically also solvatochromic, so the increase in viscosity is accompanied by a shift of the emission to shorter wavelengths. The solvatochromic effect is more pronounced in more extended Donor–π–Acceptor compounds [40, 41, 52, 53].

Jager et al. monitored photopolymerization of dimethacrylates of various molecular sizes and polarities [52].

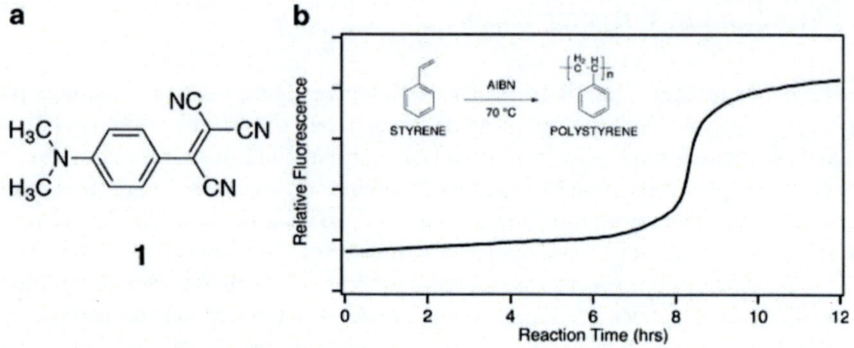

Fig. 1 (**a**) Prototypical "molecular rotor" **1**; (**b**) fluorescence intensity change as a function of styrene polymerization time [49]

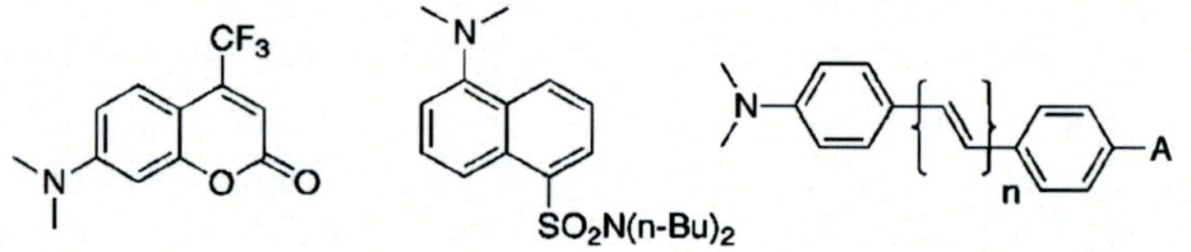

Chart 1 Chemical structures of fluorescent probes used for monitoring the degree of (photo) polymerization of diacrylates [52]

For all five solvatochromic fluorescent probes used in this study (Chart 1), the emission shifted toward shorter wavelengths upon polymerization because the dipolar excited states of the probe molecules are less stabilized in polymeric networks than in corresponding monomeric phases. Enhancement of fluorescence intensity was observed upon polymerization, which was most pronounced for 4-(dimethylamino)-4′-nitrostilbene **6**. The blue shift increases when monomers with shorter spacers were used, because this leads to a denser polymer network. This behavior was found in accordance with the fact that solvatochromic probes mainly monitor the changes in rigidity of the medium upon polymerization.

Exceptionally large fluorescence solvatochromic shifts are found in the sigma-coupled donor–acceptor systems investigated by Verhoeven and coworkers [54, 55]. The prototype 1-phenyl-4-(4-cyano-1-naphthylmethylene)piperidine (**7**) [47] has a highly dipolar excited state with a dipole moment of 25 (±2) D [56]. Depopulation of this dipolar excited state occurs via remarkably efficient radiative decay to the ground state, with quantum yields up to 85% [57]. Because of these outstanding properties compound **7** was given the nickname *FluoroProbe*.

Time-resolved spectral and polarization measurements were carried out to follow the dynamics of **7** during polymerization of methyl methacrylate (MMA). The emission maximum λ_{max} of **7** in MMA is 565 nm. Curing of MMA was carried

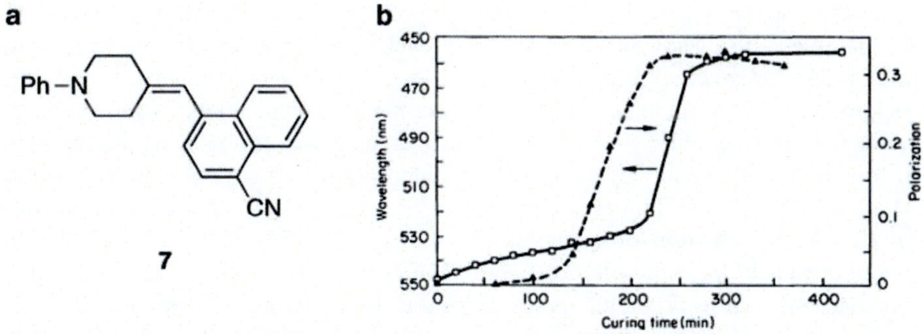

Fig. 2 **(a)** Chemical structure of 1-phenyl-4-[(4-cyano-1-naphthyl)methylene]piperidine (*Fluoroprobe*) (**7**); **(b)** fluorescence maximum (*open squares*) of **7** in PMMA and fluorescence polarization (*closed triangles*). $p = (I_{\parallel} - I_{\perp})/(I_{\parallel} - I_{\perp})$ as a function of the thermal curing time at 85°C. Reproduced with permission from [47]

out at 85°C for 15 h. Initially, a gradual hypsochromic shift occurred, but after ca. 4 h a jump of ca. 100 nm accompanied by an increase of fluorescence intensity was observed (Fig. 2). The degree of fluorescence polarization was also measured, which served as a measure for the rotational mobility of the probe molecules. The anisotropy increased rather suddenly during the curing, but well before the large shift in the emission maximum occurred: anisotropy and emission maximum probe molecular rotation and medium relaxation, respectively. The anisotropy reaches its maximum when the rotation time of the probe molecule exceeds the excited-state decay time, but the solvatochromic shift retains sensitivity to the extent of polymerization longer because it is caused by side chain rotations of the polar groups in the polymer.

For evaluation of heterogeneous systems, such as polymeric materials, where many species with different microenvironments are expected, steady-state spectroscopy has its limitations. Fluorescence lifetime as an additional observable can help to get a more accurate picture of such complicated structures. To show the combined power of steady-state and time-resolved fluorescence *Fluoroprobe* **7** was applied to study the structure of polystyrene (PS) latices and polystyrene-diglycidylmethacrylate (PS-GMA) core–shell latices [58]. The emission maximum λ_{max} of **7** in toluene is at 475 nm; in PS it is blue-shifted to 445 nm. In PS-GMA *Fluoroprobe* **7** exhibited three emission maxima at 380, 410 and 450 nm. Steady-state spectroscopy revealed that the 380 nm component is due to a decomposition product of **7** formed under the radical curing conditions. Time-resolved emission was applied to obtain a clear picture and explanation of the three emission maxima in the PS-GMA matrix. The lifetime distribution of **7** in the PS latex, quantified using the Maximum Entropy Method (MEM) [59], is clearly different from that observed for the core–shell PS-GMA latex. It was noted that in high viscosity matrices the fluorescence lifetime depends on the emission wavelength. In addition, the heterogeneity of the matrix adds to the complexity of the fluorescence lifetime data. Due to such complications it is not possible to use lifetime data in a quantitative way.

In the previous examples, the fluorescent probe molecule was simply mixed with the sample. The disadvantage of that approach is that the sensing molecule may accumulate in specific parts of the sample, e.g., in the monomer phase, and only reports on those parts rather than on the sample as a whole. This problem can be solved to a large extent by introducing fluorogenic probes, of which the fluorescence is turned on only when they are covalently incorporated into the polymer. Warman et al. [60, 61] applied fluorogenic probes to study radiation-induced polymerization of MMA. Introduction of a maleimido moiety on *Fluoroprobe* **7** leads to the nonfluorescent "maleimidofluoroprobe" (MFP) **8**. Removal of the double bond of the maleimide turns on the fluorescence of the chromophore (Scheme 1).

Warman et al. [60] monitored the course of radiation-induced polymerization of MMA in a cobalt gamma ray source by in situ fluorescence measurements. During polymerization of MMA a gradual increase in fluorescence intensity of copolymerized **8** was observed during the first 5 h. During this period the wavelength of emission remained almost constant, indicating that there was no serious increase in the viscosity of the medium (Fig. 3a). At close to 6 h elapsed time a dramatic increase in the intensity and decrease in the maximum wavelength took place, indicative of sudden rigidification of the matrix (Fig. 3).

Although the dual information provided by MFP **8** is useful, it is more difficult to extract a quantitative measure of the degree of polymerization from the data. That is more straightforward when the photophysics of the incorporated probe are unaffected by changes in viscosity or dielectric constant of the medium. This requirement is met by the pyrene derivative MPy **9**. The fluorescence of the pyrene chromophore in MPy is completely quenched but it grows on irradiation of the MMA solution as the fluorogenic conversion of MPy occurs (Scheme 1, Fig. 3b). Interestingly, both MPy and pyrene show an increase in emission intensity when the gel point was reached.

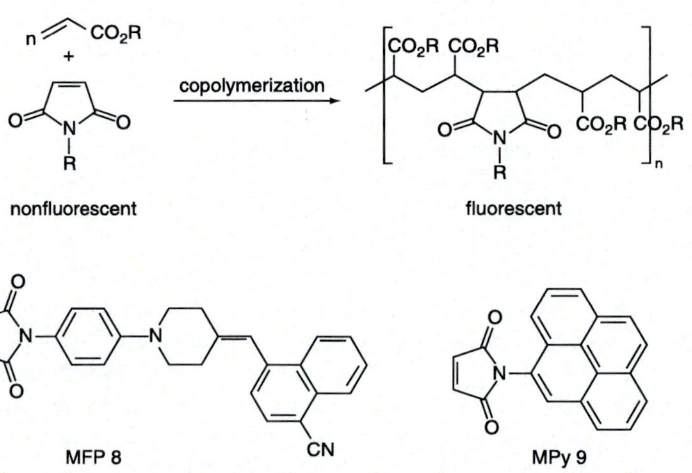

Scheme 1 Fluorogenesis of maleimide substituted probe molecules. MFP **8** and MPy **9** are nonfluorescent. Removal of the double bond in the maleimide unit turns off a quenching pathway [61]

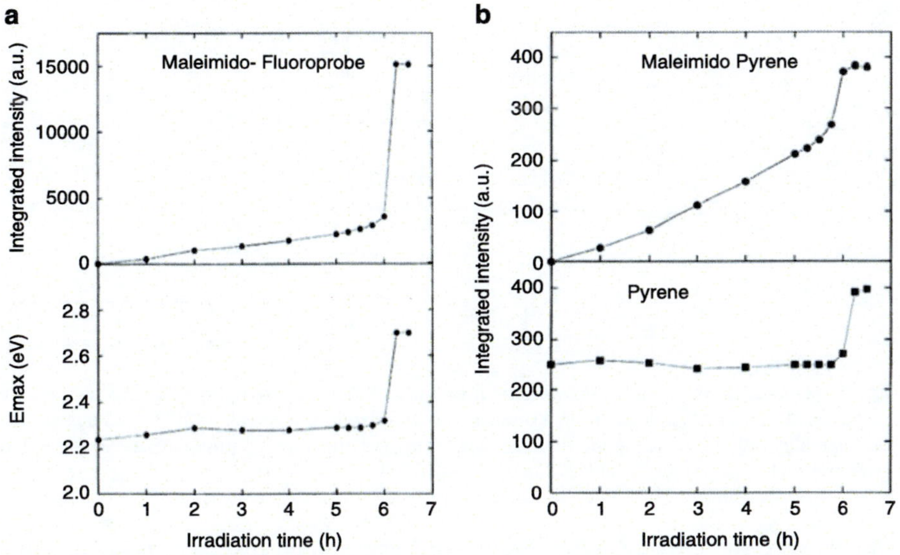

Fig. 3 (**a**) Dependence on irradiation exposure time of the wavelength maximum (*upper*) and integrated intensity (*lower*) of the fluorescence from a methyl methacrylate solution of MFP **8**; (**b**) fluorescence intensities of MPy **9** and pyrene under the same conditions. Reproduced with permission from [61]

Other pyrene derivatives also work as fluorogenic probes, e.g., *N*-(1-pyrene) methacrylamide, which was also applied for quantitative monitoring of radiation-induced polymerization of MMA by Warman and coworkers [62], or *N*-(2-anthracene)methacrylamide, which was recently used to monitor the kinetics of microemulsion polymerization [63].

These examples clearly demonstrate that application of fluorogenic probes can determine concurrently both the degree of polymerization and the changes in viscoelastic properties of the medium as polymerization proceeds.

Interestingly, Warman and coworkers recently used the fluorescence monitoring of polymerization as a tool for the detection of gamma-radiation [64].

Other phenomena that can be used to probe polymerization via the viscosity of the medium are those based on excited-state interactions of fluorophores with other molecules. Excimer formation, energy transfer, and any quenching mechanism can be used. For example, Valdes-Aguilera et al. [65] used the quenching of pyrene and its excimer by the initiator AIBN during the polymerization of MMA. The ratio of monomer to excimer emission increased during the process, because the quenching by AIBN affects mostly the excited monomer, which has a higher energy than the excimer and is much more reactive. The authors observed that the fluorescence ratio was a good measure of microviscosity around the fluorescent probe, which changes linearly with concentration up to 20–30% polymer, but beyond a certain degree of polymerization, became independent of polymer molecular weight. Thus, the methods based on diffusive quenching are particularly suitable in the early stage

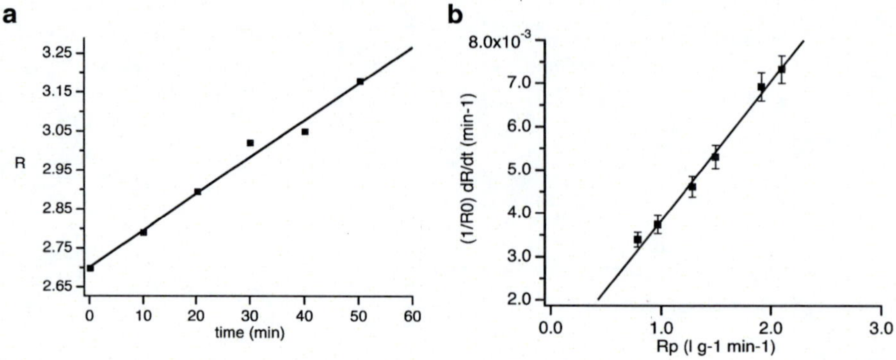

Fig. 4 (**a**) Increase in the fluorescence ratio *R* (= ratio of fluorescence intensity at 393 nm (pyrene monomer) to that at 476 nm (excimer)) during polymerization of MMA at 50°C; (**b**) correlation between the rate of increase in fluorescence ratio and the rate of polymerization measured gravimetrically. Reproduced with permission from [65]

of polymerization, where rigidochromic probes are insensitive and solvatochromic probes have a weak response (see Fig. 4a). A correlation between the rate of polymerization determined gravimetrically and that of fluorescence is presented in Fig. 4b. On the basis of this relationship, Valdes-Aguilera et al. concluded that the-rate of increase in the fluorescence ratio is a direct measure of rate of polymerization.

A related application of pyrene was explored by Pekcan and coworkers. Pekcan and Kaya [66] investigated the formation of gels when styrene in the presence of different concentrations of the cross-linker divinylbenzene (DVB) is subjected to Free-radical Crosslinking Copolymerization (FCC). Using excited-state decay time measurements of pyrene, they could monitor the consumption of styrene, which acts as an excited-state quencher, in samples with various DVB content. In this way it could be shown that early gelation takes place at high DVB content, which leads to relatively large amounts of unreacted monomer. Slow polymerization, on the other hand, results in a larger increase in the decay time of excited pyrene because fewer monomers are trapped in the gel.

3 Physical Properties

3.1 Glass Transition

Among the physical characteristics of polymers, the glass transition temperature T_g is one of the most important. Above the glass transition, in the rubbery state, the polymer has a larger free volume, the polymer chains are more mobile, and the material is softer than below T_g. Such differences in environmental properties can be probed e.g., with malachite green [44]. Like other properties, the glass transition changes when polymer films are made very thin, that is, when the thickness becomes comparable with the size of the molecules [67–69]. In the early 1990s seminal studies showed that confinement of amorphous materials at the nano-scale led to significant deviations of T_g from the bulk values. Keddie et al. used ellipsometry to characterize T_g by measuring film thickness as a function of temperature [70]. In 40 nm thick polystyrene (PS) film supported on a silica substrate a reduction in T_g was observed. A 17 nm thick PS film exhibited a 21 K reduction of T_g relative to the bulk value.

Ellison et al. [71] reported the first use of a fluorescence method to determine the effects of decreasing film thickness on T_g. They studied pyrene-doped polymer films of polystyrene (PS), poly(isobutyl methacrylate) (PiBMA), and poly(2-vinyl-pyridine) (P2VP) spin coated onto fused quartz.

Figure 5 shows the effect of temperature on the fluorescence of pyrene doped at trace levels in a PS film. The fluorescence intensity is significantly reduced with increasing temperature. This is commonly observed with molecular fluorophores because rates of nonradiative decay increase with the thermal energy in the system. Importantly, the temperature dependence of the intensity is different in the glassy and the rubbery states. T_g values were obtained from the intersection of the lines associated with the linear temperature dependences of the fluorescence intensities in the two states (Fig. 6).

Figure 6 shows the temperature dependences of the emission intensities at two wavelengths (384 and 395 nm) for PS films on fused quartz that were 24 and 487 nm thick. These results were obtained by heating the polymer films well

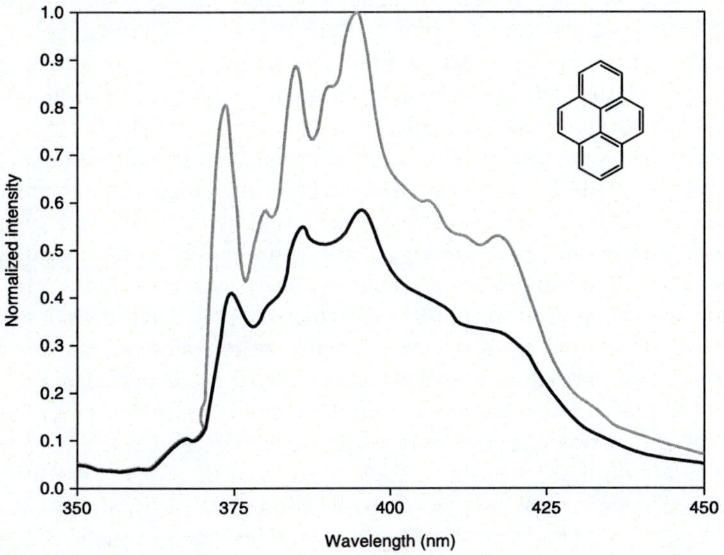

Fig. 5 Pyrene emission spectrum in a 350 nm thick polystyrene film at 398 K (*bold curve*) and 298 K (*thin curve*). From [71]

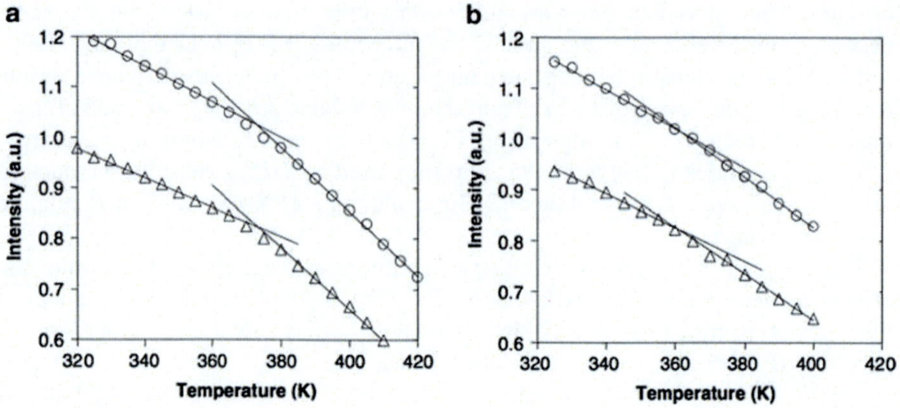

Fig. 6 Fluorescence intensity as a function of temperature monitored at emission wavelengths of 395 nm (*circles*) and 384 nm (*triangles*). Pyrene-doped PS films were (**a**) 487 nm and (**b**) 24 nm thick. Data were normalized to the intensity at 365 K and the 395 nm data were arbitrarily shifted vertically. From [71]

into the rubbery state and then measuring intensities during stepwise cooling. In the case of the 24 and 487 nm thick polystyrene samples, the apparent T_g values were 362.6 ± 2.2 K and 372.8 ± 1.4 K, respectively. The reduced T_g of the ultrathin film was attributed to the weakness of the interactions between the substrate and PS.

The temperature dependence of pyrene emission intensity for two P2VP films, one 24 nm and the other 119 nm thick, showed apparent T_gs at 398.0 ± 0.9 K and 379.3 ± 1.4 K, respectively. The bulk T_g for P2VP was 372 K (see Fig. 7). The rise of T_g with decreasing thickness of film was attributed to the strong polymer–substrate interaction between the nitrogen atoms of the pyridine units of P2VP and the hydroxyl groups at the quartz surface.

The examples show that positive as well as negative deviations of the ultrathin film-T_g from the bulk T_g can occur, which can be attributed to the interaction with the surface.

In 2008, Kim et al. [72] reported determinations of T_g in free standing single-layer polymer films using a self-referencing fluorescence intensity ratio method. The fluorescence spectral shape of copolymerized 1-pyrenylmethylmethacrylate (MApyrene) was found to change as a function of temperature, and the changing intensity ratio of two vibronic bands could be used to characterize T_g. This I_3/I_1 ratio is well known to depend on solvent polarity [2, 3]. In this case, this "solvatochromic" effect was effective only when the pyrene unit was connected to the methacrylate with a short spacer. Figure 8 shows that temperature has a strong effect on the fluorescence spectral shape and intensity of MApyrene-labeled PS. Kim et al. concluded that free standing films exhibit much stronger effects of confinement on T_g than substrate-supported films. The self-referencing fluorescence method worked in polymers other than PS, such as poly(isobutyl methacrylate)

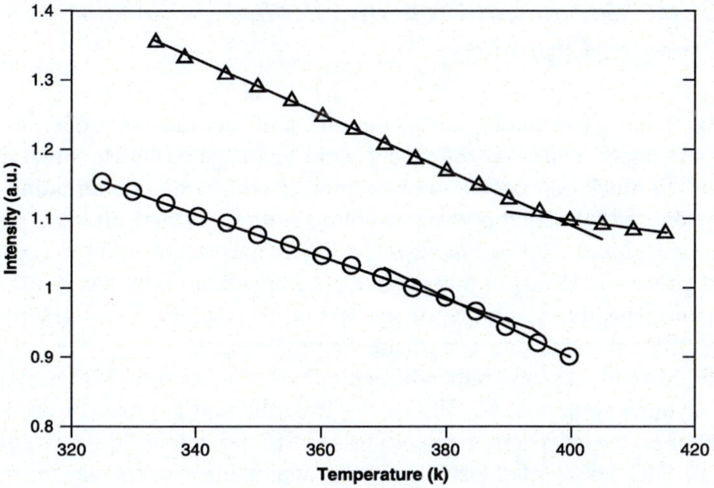

Fig. 7 Fluorescence intensity as a function of temperature monitored at an emission wavelength of 374 nm. Pyrene-doped P2VP films were 119 nm (*circles*) and 28 nm (*triangles*) thick; data normalized to 1 relative to intensity at 375 K and 400 K, respectively, and arbitrarily shifted vertically. From [71]

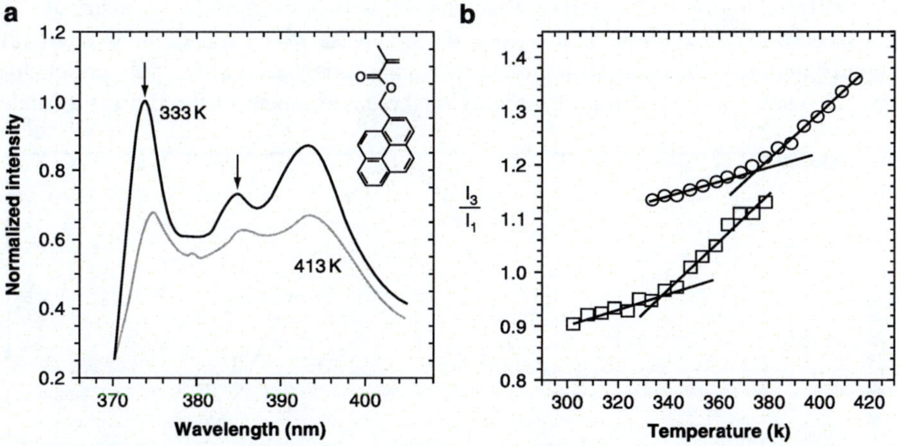

Fig. 8 (*Left*) Temperature dependence of fluorescence emission spectra of free standing films of MApyrene labeled PS, (*right*) temperature and thickness dependences of the ratio of intensities at the third and first peaks of free standing MApyrene labeled PS films; 190 nm thick film (*circles*) and 50 nm thick films (*squares*) [72]

(PiBMA) and poly(butyl acrylate) PBA. In more recent studies, the confinement effect was shown to be strongly reduced when polymer films contained additives [73] or when poly(vinylacetate) films were wet [74]. A technical advancement was the use of fluorescence decay time instead of intensity [75].

3.2 Other Mechanical Properties Probed by Solvatochromic Probes: Morphology

By using steady-state fluorescence, thermal and mechanical properties of block copolymers doped with *FluoroProbe* **7** were investigated by Hofstraat et al. [76]. They studied block copolyether–polyesters that composed of alternating blocks of 70 wt% (M_w 9,000) polybutylene terephthalate (PBT) and 30 wt% (M_w 4,000) polyethylene glycol (PEG). *Fluoroprobe* **7** was introduced into the copolymer by swelling polymeric sheets or threads in dichloromethane solutions of *Fluoroprobe*. Fluorescence spectra of **7** in homopolymers of PBT and PEG and in block copolymer PBT/PEG at room temperature are shown in Fig. 9.

The building blocks have rather different physical properties: PBT is hydrophobic, rigid, and semicrystalline, but PEG is hydrophilic and flexible. In the PEG 4000 homopolymer, the emission maximum was at 515 nm, but in PBT it was red-shifted to 540 nm. This unexpected result showed that a mobile environment exists in PBT, in which the polar ester groups can stabilize the highly dipolar excited state of **7**. Interestingly, an even more mobile environment seemed to be present in the PBT/PEG copolymer thread than in the homopolymers, as indicated by the λ_{max} of 560 nm. The emission of **7** in copolymer sheets, on the other hand, was strongly blue shifted (Fig. 10). This was taken as an indication that processing of the copolymers into sheets leads to orientation and rigidification of parts of the amorphous material.

In stretching experiments the isotropic thread showed a monotonous blue shift upon extension. This was attributed to increasing orientation of the polymer chains, which results in a reduction of mobility of the environment of the probe molecule.

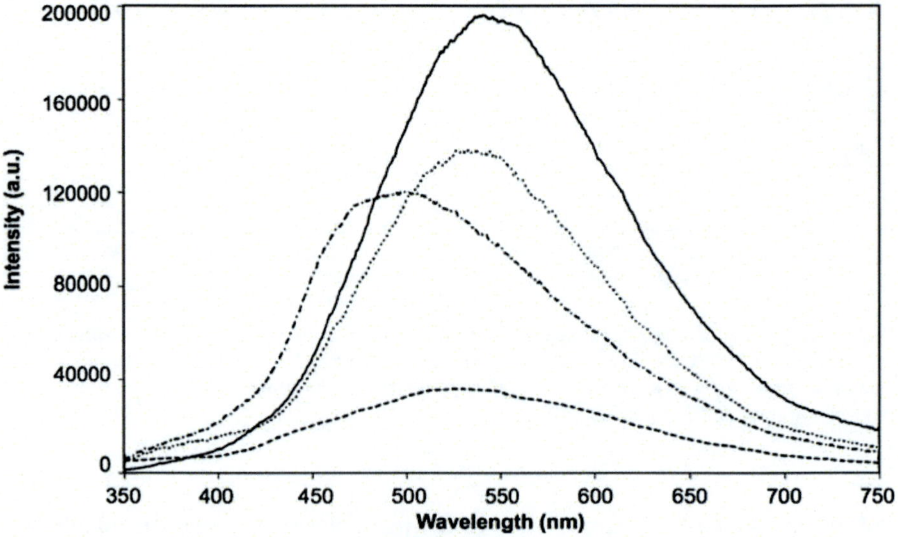

Fig. 9 Emission spectra of **7** in PBT 25000 homopolymer (*dotted line*), PEG 4000 homopolymer (*dashed/dotted line*) PBT/PEG block copolymer sheet (*dashed line*), and PBT/PEG block copolymeric thread (*solid line*). Reproduced from [76]

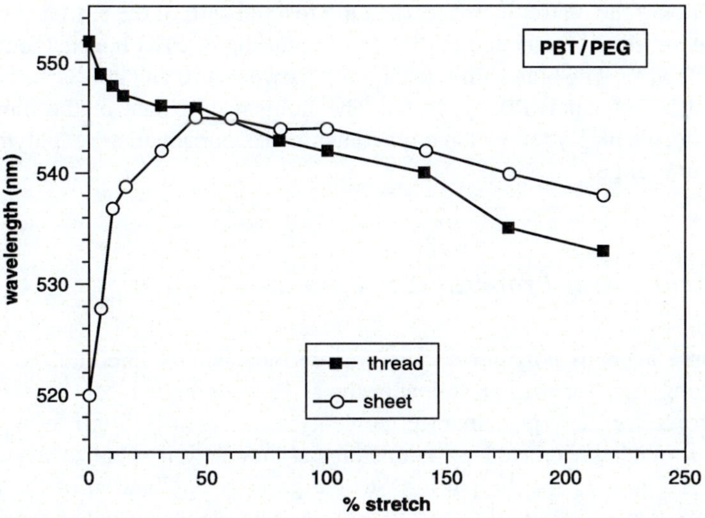

Fig. 10 Response of the λ_{max} of **7** to stretching of the PBT/PEG copolymeric thread and sheet [76]

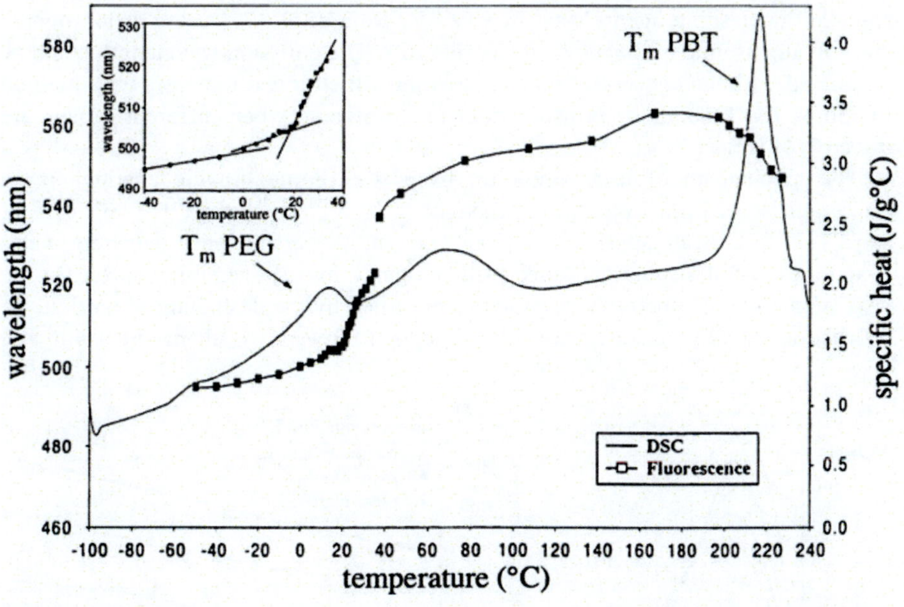

Fig. 11 Differential scanning calorimetry and temperature dependence of λ_{max} of **7** in PBT/PEG copolymer. From [76]

The fluorescence of **7** in the sheet initially showed a bathochromic shift (decreased order) upon stretching, but after 50% stretch a hypsochromic shift set in, much like in the thread.

Figure 11 shows the effects of temperature on the emission maximum, together with the specific heat (Differential Scanning Calorimetry). Going from −50 to 0°C the mobility of the matrix increases, leading to a red shift of the emission maximum of **7**. The red shift accelerated at 0°C due to melting of PEG blocks (confirmed by DSC and TMA). The blue shift encountered between 170 and 200°C was attributed to premelting of small PBT crystals. The fluorescence data of the incorporated probe molecule nicely reflect the three transitions occurring in the copolymer in this temperature range.

3.3 Mechanical Properties: From Latex to Coatings

Waterborne aqueous polymer dispersions are increasingly important in numerous applications, e.g., as paints and adhesives. Environmental concern is a major driving force for the replacement of solvent-based coatings. Film formation from e.g., latex materials created by emulsion polymerization is not a trivial matter. It requires a low T_g for the particles to merge after application of the film on a substrate and evaporation of water. In order to achieve this, organic cosolvents are added, which is an undesirable practice that needs to be minimized.

Swelling of the polymer particles in waterborne latex materials by these so-called coalescing agents was observed by Raja et al. [77] using a fluorogenic and solvatochromic probe molecule similar to **8**. The emission wavelength could be correlated with the concentration of cosolvent, which forms the basis for a method to follow the kinetics of redistribution of cosolvents when different latices are mixed [14] (Fig. 12).

The mechanism of film formation from waterborne organic coatings is an important issue. Film formation is thought to take place in three [78–80] or four steps [81, 82]. After application of an aqueous dispersion on a substrate water evaporates and the spherical latex particles come into irreversible contact. In the next stage the voids between the spheres are filled by the deshaping of particles to polyhedra. Finally, coalescence of particles followed by chain diffusion and

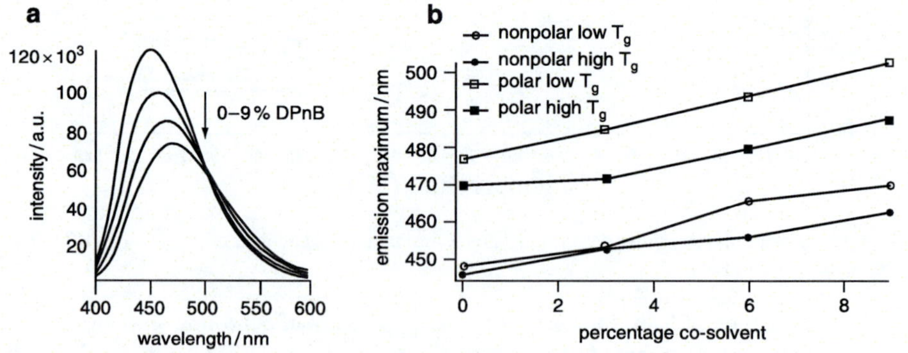

Fig. 12 (**a**) Effect of cosolvent DPnB on the emission of copolymerized probe; (**b**) dependence on the position of the emission maximum on the amount of cosolvent for different polymers [14]

entanglement leads to the formation of a coherent film. Fluorescence resonance energy transfer (FRET) is a versatile approach to investigate coalescence and interdiffusion in polymers [83]. In FRET, two kinds of fluorescent dyes are used, one is excited and acts as energy donor and the other is the acceptor. For investigation of coalescence or diffusion during latex film formation, equivalent amounts of latex particles are labeled with each of the dyes. When the particles are intact, the distance between the two dyes is large enough that energy transfer does not occur. However, when the film is formed donor and acceptor labeled latex particles come into close contact, and the dyes approach each other to a distance that is small enough for energy transfer to take place. Therefore FRET provides a clear indication of coalescence and interdiffusion of latex particles during film formation [83].

In 1990 Winnik and coworkers [84] reported the first experiments on nonradiative energy transfer to study polymer diffusion across the particle boundary during latex film formation. PMMA particles of 1 μm diameter, of which one batch was labeled with phenanthrene and the other with anthracene, were mixed, spread as a thin film and annealed above the glass transition temperature of PMMA. Energy transfer measurements effectively followed polymer diffusion across the particle boundaries during drying of the latex film. They derived a diffusion coefficient D, which characterized polymer diffusion. For PMMA particles at temperature between 403 and 450 K, diffusion coefficient values ranged from 6×10^{-18} to 9×10^{-14} $cm^2 s^{-1}$.

More recently, interdiffusion of polymers in latex film formation was studied in real time [10]. Haley et al. employed time-resolved fluorescence to measure donor fluorescence decay from 0.5 mm diameter spots at various positions in a drying film. It is well known that drying of latex films takes place from the edges of the film inward. The film drying process was temporarily arrested by sealing the film in an airtight chamber, and reducing the sample temperature to near or below the glass transition temperature T_g, which stopped polymer diffusion. Latex particles with a low T_g were composed of butyl acrylate : methyl methacrylate : methacrylic acid in 50:49:1 ratio. The donor fluorescence decay was measured at various positions across the arrested latex film. The extent of polymer diffusion was demonstrated as a function of distance from the wet–dry edge in the latex film. Energy transfer efficiency revealed a rapid evolution of film structure. Subsequently, the new technique was applied in a number of case studies of film formation at ambient temperature [85].

Turshatov et al. [86] employed time-resolved fluorescence spectroscopy to monitor intermixing of polymer particles during film formation of labeled poly(methylmethacrylate). Drying of the films was monitored by the efficiency of energy transfer between the phenanthrene donor derived from monomer **10** and nonfluorescent nitrophenylpyrrolidine acceptor (monomer **11**) (Fig. 13), as well as the scattering intensity and the lifetimes of donor units. Scattering can provide valuable information when analyzing the drying of the latex film: the disappearance of scattering marks the end of the compactification stage. The changes of the donor lifetime over time closely correlate with the changes of scattering intensity. When the interstitial water voids disappear, the light

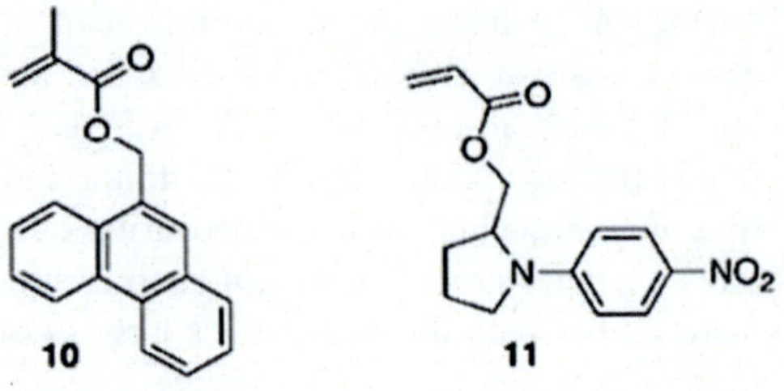

Fig. 13 Chemical structures of monomer donor and acceptor dyes used in [86]: donor phenanthrene methacrylate, Phe-MA **10**, acceptor ([1-(4-nitrophenyl)-2-pyrrolidinemethyl]acrylate, NPP-A) **11**

scattering also disappears, and the donor lifetime shows a small but sharp decrease.

On the basis of the FRET experiments, Turshatov et al. concluded that a well-intermixed layer of 20 nm develops at the interparticle junction soon after particles come into contact. This does not develop further, and thermal annealing is needed to bring about further interdiffusion. Interparticle cohesion as reflected by FRET efficiency frequently develops before the interstitial stage disappears. These findings prove that the neighboring polymer spheres develop an intermixed layer after they come into contact.

4 Microscopy: Single Molecule Studies and Imaging

Confocal microscopy has become a very popular tool for investigating many processes in biological, medical, and material science [87]. The application of a wide variety of fluorescent molecules made it possible to tag the specific position in the sample and study components and processes with submicrometer resolution. It is also possible to investigate phenomena at the single molecule level by using confocal microscopy. The main advantage of confocal microscopy is that it offers to collect serial optical sections (0.5–1.5 μm) from thick specimens up to 50 μm.

Single molecule fluorescence techniques (SMD) offer exciting new possibilities for the study of polymer dynamics [88]. Individual fluorescent molecules show the same observables as can be measured on an ensemble of molecules, such as spectrum, polarization, and decay time. SMD allows determining the distributions of these properties, but more importantly, fluctuations of the properties can be monitored on time scales from milliseconds to hours. For example, fluctuations of the emission rate of very efficient emitters could be correlated with fluctuations of the local density, in other words, the free volume, in polystyrene films well below T_g [89, 90]. Molecules embedded in polymer films show molecular reorientation jump on timescales of seconds and longer at temperatures near and above T_g [91]. These can be observed by measuring the polarization of the emitted light, but more elegantly by wide field imaging techniques which allow direct determination of the direction of the transition dipole moments [92]. Zheng et al. [93] recently applied this approach to demonstrate the rotational motion of dye molecules covalently attached to PS in very thin films. Diffusive motions can be observed with nanometer spatial resolution, which enables optical microrheology [94]. A combination of fluorescence correlation spectroscopy and wide field imaging has been used to monitor different stages of polymerization based on the mobility of copolymerizing probe molecules [33].

In the above-mentioned examples, the molecules used should be bright and photostable, and preferably not do anything other than absorb and emit light. More complicated systems, for example showing photoinduced electron transfer, can have additional value. Hofkens and coworkers measured the fluctuating intramolecular electron transfer of donor-substituted perylene imides to obtain information on fluctuations of the polymer free volume in PS near ambient temperature [95].

Siekierzycka et al. [31] showed that the fluorescence of the perylene bisimide **12** could be switched off and on reversibly when heating and cooling films across the glass transition. Interestingly, the control of photoinduced electron transfer in this case is not related to stabilization of the charge-transfer state by dipolar groups in the medium, but by the kinetic limitations imposed by the lack of free volume (Fig. 14).

Confocal microscopy as an imaging technique began to attract the interest of materials scientists in the 1990s and continues to be used on a modest scale. The polymer field so far has largely failed to pick up the spectacular developments in microscopy techniques in the past decade. There is a great potential here.

Blending of different polymers is a way to create new physical properties using existing materials. Most polymers, however, are poorly miscible, and phase separation commonly occurs. The properties will then be dependent on the spatial arrangement of the different phases, in other words, the morphology. In an early application of confocal microscopy, a depth-resolved study of films formed from a mixture of PMMA and polystyrene was carried out by Li et al. [96]. They prepared a blend of PMMA labeled with the dye *N*-methyl-*N*-(4-(7-nitrobenzo-2-oxa-1,3-

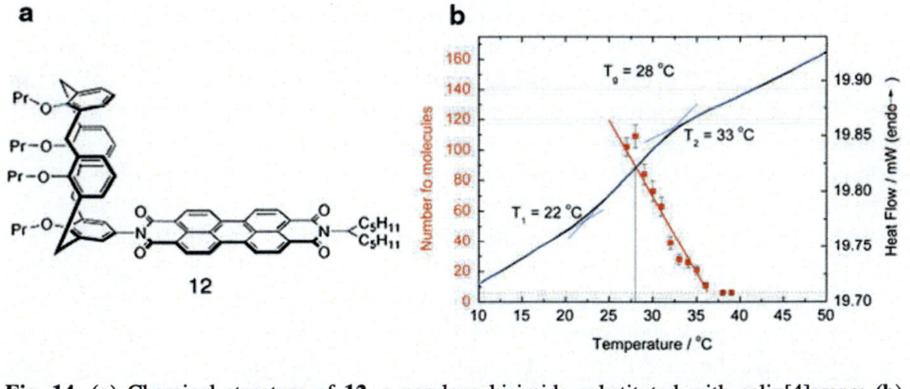

Fig. 14 (**a**) Chemical structure of **12**, a perylene bisimide substituted with calix[4]arene; (**b**) number of fluorescing molecules of **12** detected in a 26 μm × 26 μm area in a poly(vinylacetate) (pvac) matrix (*filled circle*, *left axis*) as a function of temperature and part of the DSC curve (*solid blue line*, *right axis*) with the determined T_g of the polymer. Reproduced with permission from [31]

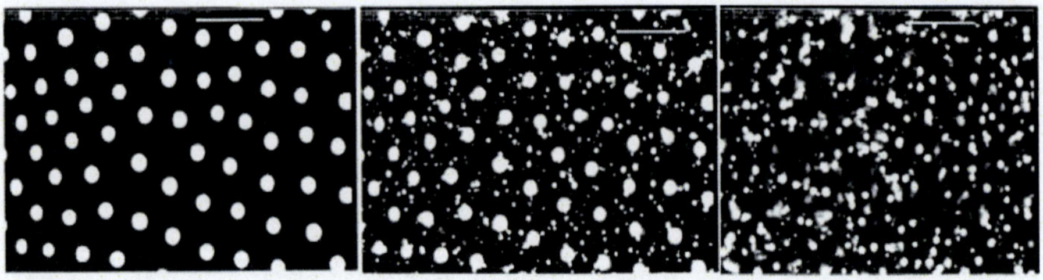

Fig. 15 Laser confocal microscopy images of (**a**) solvent cast films (from toluene) composed of 10 wt% of PMMA labeled with the dye NBD and 90% polystyrene. The *left image* is the film surface, the *middle image* is a slice of 3 μm beneath the surface and the *right image* is of a slice 6 μm beneath the surface. Reproduced from [96]

diazole))-2-aminoethylmethacrylate (NBD-MMA) with PS. Films were prepared by solvent casting and examined with a confocal fluorescence microscope (see Fig. 15). Different morphologies of the film were obtained; a surface slice of 0.5 μm thick at the air–film interface showed a uniform size of PMMA spheres of 5–6 μm with a remarkable periodicity. In a slice of the same film 3 μm below the surface PMMA particles were present in larger as well as smaller sizes. On a spatial scale of 1–2 μm a sharp transition was observed between surface morphology and the bulk morphology of the system. A further slice of 6 μm beneath the surface showed a higher content of smaller droplets with a large size dispersion. These were supposed to have arisen from phase separation through a nucleation and growth mechanism, but the regular pattern of PMMA-rich droplets on the surface could not be explained.

Winnik and coworkers continued to use confocal microscopy over the years [97–99]. Recently, they investigated the morphology and miscibility of blends of thermoplastic olefins (TPO), of practical importance for the automotive industry [100]. TPO are blends of polypropylene and impact modifiers, in this case ethylene–butene rubber (EBR) copolymers. Being very hydrophobic, TPO are difficult to paint. Therefore chlorinated polyolefins (CPO) are used as adhesion promoters. Laser scanning confocal microscopy was used to image two of the three components of the blend. The CPO component was labeled with a coumarine, excited at 351 nm, the EBR with a benzothioxanthene fluorophore (HY), excited at 488 nm. The blends were prepared by precipitation of a 1:1:1 mixture of the three polymers from solution, and annealed at a temperature (180°C) where the polymers are molten. The images (Fig. 16) show convincingly that EBR and CPO mostly localize together, forming domains within the isotactic polypropylene (iPP) material. The CPO adhesion promoter, however, is slightly miscible with iPP. In this study, other techniques were used to obtain information on the miscibility of the components, which led to a consistent picture. Of all the methods used, confocal microscopy, however, gave the most direct and visual insight into the system. The incorporation of the fluorescent labels was considered to be only a small perturbation. As a check, it was shown that the labeled polymers were fully miscible with the nonlabeled ones.

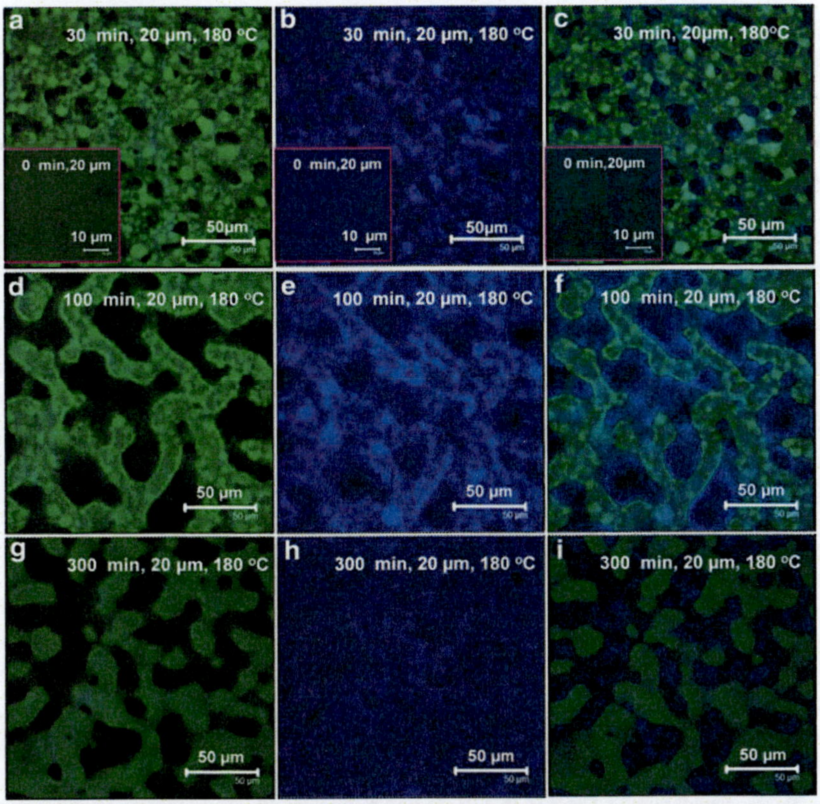

Fig. 16 Confocal images of the morphology of a blend of iPP, EBR (labeled with a benzothioxanthene HY, $\lambda_{exc} = 488$ nm), and CPO (labeled with a coumarine, $\lambda_{exc} = 351$ nm) as a function of annealing time (*left*). The *leftmost set of images* is obtained exciting at 488 nm, the middle set at 351 nm. The *third column* shows the overlapped images from the other *two columns*. Reproduced with permission from [100]

5 Conclusions

As illustrated by the examples in this chapter, fluorescence techniques have been applied successfully to polymer science over the past three decades. In this field, fluorescence is only one of the many research tools that are used. In the life sciences, on the other hand, fluorescence imaging is a major technique because of the essential role of submicrometer three-dimensional structure in the cell. In our opinion, there is plenty of opportunity to take advantage of the rapid progress in fluorescence methods in the life sciences by applying the functional dyes developed there as well as the sophisticated modern microspectroscopic techniques to polymer science. Several of the other contributions in the present series address the development of molecular markers and sensors. Many of the commercially available dyes can be readily adapted for use in polymers, or even used directly.

To modify dyes with linkers that can be copolymerized for covalent inclusion in polymers is not more difficult than to attach anchors for biological labeling. Although the questions in polymer science are different from those in life science, the use of long-wavelength excitation [101] (possibly using two photon absorption) will have advantages in the former field as well, due to the tendency of heterogeneous materials to scatter light.

Molecular fluorescent probes can give information on the dynamics of their environment on timescales from picoseconds (solvent relaxation) up to months or years (physical aging). Fluorescence-based imaging gives access to 3D-stucture and morphology information with a potential resolution down to a few tens of nanometers. These methods have great potential for polymer science. Given the present disciplinary structure of science, however, the full realization of this potential will require more collaboration of scientists from the communities of molecular fluorescence and polymer physics and chemistry.

References

1. Demchenko AP (2010) Comparative analysis of fluorescence reporter signals based on intensity, anisotropy, time-resolution and wavelength-ratiometry. In: Demchenko AP (ed) Advanced fluorescence reporters in chemistry and biology I. Springer Series on Fluorescence 8. Springer, Heidelberg, pp 3–24
2. Dong DC, Winnik MA (1982) The py scale of solvent polarities – solvent effects on the vibronic fine-structure of pyrene fluorescence and empirical correlations with ET-value and y-value. Photochem Photobiol 35:17–21
3. Kalyanasundaram K, Thomas JK (1977) Environmental effects on vibronic band intensities in pyrene monomer fluorescence and their application in studies of micellar systems. J Am Chem Soc 99:2039–2044
4. Karpovich DS, Blanchard GJ (1995) Relating the polarity-dependent fluorescence response of pyrene to vibronic coupling. Achieving a fundamental understanding of the py polarity scale. J Phys Chem 99:3951–3958
5. Demchenko AP (1982) On the nanosecond mobility in proteins – edge excitation fluorescence red shift of protein-bound 2-(para-toluidinylnaphthalene)-6-sulfonate. Biophys Chem 15:101–109
6. Paley MS, McGill RA, Howard SC, Wallace SE, Harris JM (1990) Solvatochromism. A new method for polymer characterization. Macromolecules 23:4557–4564
7. Reichardt C (1994) Solvatochromic dyes as solvent polarity indicators. Chem Rev 94: 2319–2358
8. Morawetz H (1988) Studies of synthetic polymers by nonradiative energy transfer. Science 240:172–176
9. Ni SR, Zhang P, Wang YC, Winnik MA (1994) Energy transfer studies of the boundary layer interphase in polystyrene-poly(methyl methacrylate) block copolymer films. Macromolecules 27:5742–5750
10. Haley JC, Liu Y, Winnik MA, Demmer D, Haslett T, Lau W (2007) Tracking polymer diffusion in a wet latex film with fluorescence resonance energy transfer. Rev Sci Instrum 78:084101–084105
11. Clayton AHA, Hanley QS, Arndt-Jovin DJ, Subramaniam V, Jovin TM (2002) Dynamic fluorescence anisotropy imaging microscopy in the frequency domain (rFLIM). Biophys J 83:1631–1649

12. Thomsson D, Lin HZ, Scheblykin IG (2010) Correlation analysis of fluorescence intensity and fluorescence anisotropy fluctuations in single-molecule spectroscopy of conjugated polymers. ChemPhysChem 11:897–904
13. Goes M, Lauteslager XY, Verhoeven JW, Hofstraat JW (1998) A blue excitable charge-transfer fluorescent probe and its fluorogenic derivative. Eur J Org Chem 2373–2377
14. Brouwer AM, Raja TN, Biemans K, Nabuurs T, Tennebroek R (2008) Probing of cosolvents in polymer latex materials by using solvatochromic fluorescence. Ann NY Acad Sci 1130:157–163
15. Karpiuk J (2003) Photoinduced electron transfer in malachite green lactone. Phys Chem Chem Phys 5:1078–1090
16. Maandi E, Sung CSP (2008) In situ fluorescence spectroscopic studies of polymerization of anaerobic adhesives. J Appl Polym Sci 107:3685–3693
17. Demchenko A (2010) The concept of λ-ratiometry in fluorescence sensing and imaging. J Fluoresc 20:1099–1128
18. Colby KA, Burdett JJ, Frisbee RF, Zhu L, Dillon RJ, Bardeen CJ (2010) Electronic energy migration on different time scales: concentration dependence of the time-resolved anisotropy and fluorescence quenching of lumogen red in poly(methyl methacrylate). J Phys Chem A 114:3471–3482
19. Hardison LM, Zhao XY, Jiang H, Schanze KS, Kleiman VA (2008) Energy transfer dynamics in a series of conjugated polyelectrolytes with varying chain length. J Phys Chem C 112:16140–16147
20. Beechem JM, Ameloot M, Brand L (1985) Global and target analysis of complex decay phenomena. Anal Instrum 14:379–402
21. van Stokkum IHM, Larsen DS, van Grondelle R (2004) Global and target analysis of time-resolved spectra. Biochim Biophys Acta Bioenerg 1657:82–104
22. vandeVen M, Ameloot M, Valeur B, Boens N (2005) Pitfalls and their remedies in time-resolved fluorescence spectroscopy and microscopy. J Fluoresc 15:377–413
23. Hell SW (2007) Far-field optical nanoscopy. Science 316:1153–1158
24. Hell SW (2009) Microscopy and its focal switch. Nat Methods 6:24–32
25. Ullal CK, Schmidt R, Hell SW, Egner A (2009) Block copolymer nanostructures mapped by far-field optics. Nano Lett 9:2497–2500
26. Kudryavtsev V, Felekyan S, Woźniak A, König M, Sandhagen C, Kühnemuth R, Seidel C, Oesterhelt F (2007) Monitoring dynamic systems with multiparameter fluorescence imaging. Anal Bioanal Chem 387:71
27. Sekine R, Aoki H, Ito S (2009) Conformation of single block copolymer chain in two-dimensional microphase-separated structure studied by scanning near-field optical microscopy. J Phys Chem B 113:7095–7100
28. Moerner WE, Fromm DP (2003) Methods of single-molecule fluorescence spectroscopy and microscopy. Rev Sci Instrum 74:3597–3619
29. Tomczak N, Vallée RAL, van Dijk EMHP, Garcia-Parajo M, Kuipers L, van Hulst NF, Vancso GJ (2004) Probing polymers with single fluorescent molecules. Eur Polym J 40:1001–1011
30. Gavranovic GT, Csihony S, Bowden NB, Hawker CJ, Waymouth RM, Moerner WE, Fuller GG (2006) Well-controlled living polymerization of perylene-labeled polyisoprenes and their use in single-molecule imaging. Macromolecules 39:8121–8127
31. Siekierzycka JR, Hippius C, Würthner F, Williams RM, Brouwer AM (2010) Polymer glass transitions switch electron transfer in individual molecules. J Am Chem Soc 132:1240–1242
32. Uji-i H, Melnikov SM, Deres A, Bergamini G, De Schryver F, Herrmann A, Müllen K, Enderlein J, Hofkens J (2006) Visualizing spatial and temporal heterogeneity of single molecule rotational diffusion in a glassy polymer by defocused wide-field imaging. Polymer 47:2511–2518
33. Wöll D, Uji-I H, Schnitzler T, Hotta JI, Dedecker P, Herrmann A, De Schryver FC, Müllen K, Hofkens J (2008) Radical polymerization tracked by single molecule spectroscopy. Angew Chem Int Ed 47:783–787

34. Winnik MA, Hua MH, Hougham B, Williamson B (1984) The 4-A ruler, using exciplex fuorescence to study small-molecules diffusion into nonaqueous dispersions of polymer colloids. Macromolecules 17:262–264
35. Skryshevskii YA (2010) Electronic excitation energy transfer in poly-N-epoxypropyl-3, 6-dibromocarbazole. Phys Solid State 52:1308–1314
36. Schroeder WF, Liu YQ, Tomba JP, Soleimani M, Lau W, Winnik MA (2010) Influence of ethylene glycol and propylene glycol on polymer diffusion in poly(butyl acrylate-co-methyl methacrylate) latex films. J Phys Chem B 114:3085–3094
37. Yang J, Lou XD, Spiro JG, Winnik MA (2006) Energy transfer study of the cylindrical interface formed by asymmetric isoprene-methyl methacrylate diblock copolymers bearing a dye at the junction. Macromolecules 39:2405–2412
38. Haidekker MA, Nipper M, Mustafic A, Lichlyter D, Dakanali M, Theodorakis EA (2010) Dyes with segmental mobility: molecular rotors. In: Demchenko AP (ed) Advanced fluorescence reporters in chemistry and biology 1. Springer Series on Fluorescence 8. Springer, Heidelberg, pp 267–308
39. Wu JJ, Abu-Omar MM, Tolbert SH (2001) Fluorescent probes of the molecular environment within mesostructured silica/surfactant composites under high pressure. Nano Lett 1:27–31
40. Peinado C, Salvador EF, Catalina F, Lozano AE (2001) Solvatochromic and rigidochromic fluorescent probes based on D-pi-A diaryl ethylene and butadiene derivatives for UV-curing monitoring. Polymer 42:2815–2825
41. Peinado C, Bosch P, Martin V, Corrales T (2006) Photoinitiated polymerization in bicontinuous microemulsions: fluorescence monitoring. J Polym Sci A Polym Chem 44:5291–5303
42. Abdel-Mottaleb MSA, El-kady MY, Loutfy RO, Winnik FM (1990) Flexible Styrylcoumarin fluorescence probe for polymerization. J Photochem Photobiol A 53:387–396
43. Willets KA, Ostroverkhova O, He M, Twieg RJ, Moerner WE (2003) Novel fluorophores for single-molecule imaging. J Am Chem Soc 125:1174–1175
44. Ye JY, Hattori T, Nakatsuka H, Maruyama Y, Ishikawa M (1997) Microscopic dynamics of the glass transition investigated by time-resolved fluorescence measurements of doped chromophores. Phys Rev B Condens Matter 56:5286–5296
45. Shea KJ, Saski DY, Stoddard GJ (1989) Fluorescent probes for evaluating chain solvation in network polymers. An analysis of the solvatochromic shift of dansyl probe in macroporous styrene-divinylbenzene and styrene-diisopropenylbenzene copolymers. Macromolecules 22:1722–1730
46. Callis PR (2010) Electrochromism and solvatochromism in fluorescence response of organic dyes: a nanoscopic view. In: Demchenko AP (ed) Advanced fluorescence reporters in chemistry and biology I. Springer Series on Fluorescence 8. Springer, Heidelberg, pp 309–330
47. van Ramesdonk HJ, Vos M, Verhoeven JW, Möhlmann GR, Tissink NA, Meesen AW (1987) Intramolecular charge-transfer fluorescence as a mobility probe in poly(methylmethacrylate). Polymer 28:951–956
48. Bosch P, Catalina F, Corrales T, Peinado C (2005) Fluorescent probes for sensing processes in polymers. Chem Eur J 11:4314–4325
49. Loutfy RO (1986) Fluorescence probes for polymer free volume. Pure Appl Chem 58: 1239–1248
50. Swalina C, Maroncelli M (2010) Nonradiative deactivation in benzylidene malononitriles. J Phys Chem C 114:5602–5610
51. Lichlyter DJ, Haidekker MA (2009) Immobilization techniques for molecular rotors-towards a solid-state viscosity sensor platform. Sens Actuators B Chem 139:648–656
52. Jager WF, Volkers AA, Neckers DC (1995) Solvatochromic fluorescent probes for monitoring the photopolymerization of dimethyacrylates. Macromolecules 28:8153–8158
53. Jager WF, Kudasheva D, Neckers DC (1996) Organic donor-pi-acceptor salts: a new type of probe for monitoring photopolymerization processes. Macromolecules 29:7351–7355
54. Mes GF, de Jong B, van Ramesdonk HJ, Verhoeven JW, Warman JM, de Haas MP, Horsman-van den Dool LEW (1984) Excited-state dipole-moment and solvatochromism of highly fluorescent rod-shaped bichromophoric molecules. J Am Chem Soc 106:6524–6528

55. Verhoeven JW (2005) Sigma-coupled charge-transfer probes of the fluoroprobe and fluorotrope Type. In: Geddes CD, Lakowicz JR (eds) Part A: Topics in fluorescence spectroscopy. Advanced concepts in fluorescence sensing Part A: Small molecule sensing, vol 9. Springer. Heidelberg
56. Jenneskens LW, Verhey HJ, van Ramesdonk HJ, Witteveen AJ, Verhoeven JW (1991) Intramolecular charge-transfer fluorescence of 1-phenyl-4-[(4-cyano-1-naphthyl)methylene]piperidine as a morphology probe in alpha, omega-diacetyl poly(ethylene glycol) matrices. Macromolecules 24:4038–4040
57. Scherer T, Hielkema W, Krijnen B, Hermant RM, Eijckelhoff C, Kerkhof F, Ng AKF, Verleg R, van der Tol EB, Brouwer AM, Verhoeven JW (1993) Synthesis and exploratory photophysical investigation of donor-brige-acceptor systems derived from N-substituted 4-piperidones. Recl Trav Chim Pays-Bas 112:535–548
58. Hofstraat JW, Verhey HJ, Verhoeven JW, Kumke MU, Li G, Hemmingsen SL, McGown LB (1997) Fluorescence lifetime studies of labelled polystyrene latices. Polymer 38:2899–2906
59. Livesey AK, Brochon JC (1987) Analyzing the distribution of decay constants in pulse-fluorimetry using the Maximum Entropy Method. Biophys J 52:693–706
60. Warman JM, Abellon RD, Verhey HJ, Verhoeven JW, Hofstraat JW (1997) Maleimido-fluoroprobe: a dual-purpose fluorogenic probe of polymerization dynamics. J Phys Chem B 101:4913–4916
61. Warman JM, Abellon RD, Luthjens LH, Suykerbuyk JWA, Verhey HJ, Verhoeven JW (1999) In situ monitoring of the radiation-induced polymerisation of methylmethacrylate using fluorogenic molecular probes. Nucl Instrum Methods Phys Res B 151:361–366
62. Frahn MS, Abellon RD, Jager WF, Luthjens LH, Warman JM (2001) Synthesis and characterization of a new fluorogenic probe molecule N-(1-pyrene)methacrylamide for monitoring radiation-induced polymerization. Nucl Instrum Methods Phys Res B 185:241–247
63. Feng HK, Dan Y, Zhao Y (2010) A sensitive fluorescence method for monitoring the kinetics of microemulsion polymerization. Can J Chem 88:185–191
64. Warman JM, de Haas MP, Luthjens LH (2009) High-energy radiation monitoring based on radio-fluorogenic co-polymerization. I: small volume in situ probe. Phys Med Biol 54: 3185–3200
65. Valdes-Aguilera O, Pathak CP, Neckers DC (1990) Pyrene as a fluorescent-probe for monitoring polymerization rates. Macromolecules 23:689–692
66. Pekcan O, Kaya D (2001) Fast transient fluorescence (FTRF) technique for monitoring free-radical crosslinking copolymerization (FCC) of styrene (S) with various divinylbenzene (DVB) contents. Polymer 42:7865–7871
67. de Gennes PG (2000) Glass transitions in thin polymer films. Eur Phys J E 2:201–203
68. Priestley RD, Ellison CJ, Broadbelt LJ, Torkelson JM (2005) Structural relaxation of polymer glasses at surfaces, interfaces and in between. Science 309:456–459
69. Rittigstein P, Priestley RD, Broadbelt LJ, Torkelson JM (2007) Model polymer nanocomposites provide an understanding of confinement effects in real nanocomposites. Nat Mater 6:278–282
70. Keddie JL, Jones RAL, Cory RA (1994) Size-dependent depression of the glass-transition temperature in polymer-films. Europhys Lett 27:59–64
71. Ellison CJ, Kim SD, Hall DB, Torkelson JM (2002) Confinement and processing effects on glass transition temperature and physical aging in ultrathin polymer films: novel fluorescence measurements. Eur Phys J E 8:155–166
72. Kim S, Roth CB, Torkelson JM (2008) Effect of nanoscale confinement on the glass transition temperature of free-standing polymer films: novel, self-referencing fluorescence method. J Polym Sci B Polym Phys 46:2754–2764
73. Ellison CJ, Ruszkowski RL, Fredin NJ, Torkelson JM (2004) Dramatic reduction of the effect of nanoconfinement on the glass transition of polymer films via addition of small-molecule diluent. Phys Rev Lett 92:095702

74. Kim S, Mundra MK, Roth CB, Torkelson JM (2010) Suppression of the T_g-nanoconfinement effect in thin poly(vinyl acetate) films by sorbed water. Macromolecules 43:5158–5161
75. Mundra MK, Ellison CJ, Rittigstein P, Torkelson JM (2007) Fluorescence studies of confinement in polymer films and nanocomposites: glass transition temperature, plasticizer effects, and sensitivity to stress relaxation and local polarity. Eur Phys J Spl Top 141:143–151
76. Hofstraat JW, Veurink J, Gebben B, Verheij HJ, Verhoeven JW (1998) Charge-transfer fluorescent probes applied to the characterization of thermal and mechanical properties of polymers. J Fluoresc 8:335–342
77. Raja TN, Brouwer AM, Biemans K, Nabuurs T, Tennebroek R (2010) Detection of coalescing agents in water borne latex emulsions using an environment sensitive fluorescent probe. Photochem Photobiol Sci 9:975–984
78. Kara S, Pekcan O, Sarac A, Arda E (2006) Film formation stages for poly(vinyl acetate) latex particles: a photon transmission study. Colloid Polym Sci 284:1097–1105
79. Pekcan O, Canpolat M, Gocmen A (1993) Variation in optical density during latex film formation: interdiffusion of fluorescence labeled polymers. Polymer 34:3319–3321
80. Wang YC, Winnik MA (1993) Polymer diffusion across interfaces in latex films. J Phys Chem 97:2507–2515
81. Keddie JL, Meredith P, Jones RAL, Donald AM (1995) Kinetics of film formation in acrylic lattices studied with multiple angle of incidence ellipsometry and environmental SEM. Macromolecules 28:2673–2682
82. Routh AF, Russel WB (2001) Deformation mechanisms during latex film formation: experimental evidence. Ind Eng Chem Res 40:4302–4308
83. Felorzabihi N, Froimowicz P, Haley JC, Bardajee GR, Li BX, Bovero E, van Veggel FCJA, Winnik MA (2009) Determination of the Förster distance in polymer films by fluorescence decay for donor dyes with a nonexponential decay profile. J Phys Chem B 113:2262–2272
84. Pekcan O, Winnik MA, Croucher MD (1990) Fluorescence studies of polymer colloids. 25. Fluorescence studies of coalescence and film formation in poly(methyl methacrylate) nonaqueous dispersion particles. Macromolecules 23:2673–2678
85. Haley JC, Liu YQ, Winnik MA, Lau W (2008) The onset of polymer diffusion in a drying acrylate latex: how water initially retards coalescence but ultimately enhances diffusion. J Coat Technol Res 5:157–168
86. Turshatov A, Adams J, Johannsmann D (2008) Interparticle contact in drying polymer dispersions probed by time resolved fluorescence. Macromolecules 41:5365–5372
87. Diaspro A (ed) (2002) Confocal and two-photon microscopy: foundations, applications and advances. Wiley, New York
88. Wöll D, Braeken E, Deres A, De Schryver FC, Uji-i H, Hofkens J (2009) Polymers and single molecule fluorescence spectroscopy, what can we learn? Chem Soc Rev 38:313–328
89. Vallee RAL, Tomczak N, Vancso GJ, Kuipers L, van Hulst NF (2005) Fluorescence lifetime fluctuations of single molecules probe local density fluctuations in disordered media: a bulk approach. J Chem Phys 122:114702–114709
90. Vallee RAL, Paul W, Binder K (2007) Single molecule probing of the glass transition phenomenon: simulations of several types of probes. J Chem Phys 127:154903
91. Adhikari AN, Capurso NA, Bingemann D (2007) Heterogeneous dynamics and dynamic heterogeneities at the glass transition probed with single molecule spectroscopy. J Chem Phys 127:1140581–1140589
92. Dedecker P, Muls B, Deres A, Uji-i H, Hotta J, Sliwa M, Soumillion JP, Müllen K, Enderlein J, Hofkens J (2009) Defocused wide-field imaging unravels structural and temporal heterogeneity in complex systems. Adv Mater 21:1079–1090
93. Zheng ZL, Kuang FY, Zhao J (2010) Direct observation of rotational motion of fluorophores chemically attached to polystyrene in its thin films. Macromolecules 43:3165–3168
94. Kandar AK, Bhattacharya R, Basu JK (2010) Interfacial microrheology as a tool to study viscoelastic transitions in nanoconfined soft matter. Phys Rev E 81:0415041–04150100

95. Gronheid R, Stefan A, Cotlet M, Hofkens J, Qu JQ, Müllen K, Van der Auweraer M, Verhoeven JW, De Schryver FC (2003) Reversible intramolecular electron transfer at the single-molecule level. Angew Chem Int Ed 42:4209–4214
96. Lin L, Sosnowski S, Chaffey CE, Balke ST, Winnik MA (1994) Surface morphology of a polymer blend examined by laser confocal fluorescence microscopy. Langmuir 10:2495–2497
97. Moffitt M, Rharbi Y, Chen W, Tong JD, Winnik MA, Thurman DW, Oberhauser JP, Kornfield JA, Ryntz RA (2002) Stratified morphology of a polypropylene/elastomer blend following channel flow. J Polym Sci B Polym Phys 40:2842–2859
98. Tong JD, Moffitt M, Huang XY, Winnik MA, Ryntz RA (2001) Use of a dye-labeled ethylene-butene copolymer as a tracer in laser scanning confocal fluorescence microscopy studies of thermoplastic olefins. J Polym Sci A Polym Chem 39:239–252
99. Vorobyova O, Winnik MA (2001) Confocal microscopy studies of shear-induced coalescence in blends of poly(butyl methacrylate) and poly(2-ethylhexyl methacrylate). J Polym Sci B Polym Phys 39:2317–2332
100. Deng KQ, Felorzabihi N, Winnik MA, Jiang ZH, Yin ZH, Yaneff PV, Ryntz RA (2009) Investigation of morphology and miscibility of isotactic polypropylene, ethylene-butene copolymer and chlorinated polyolefin blends via LSCFM, SEM, WAXD, and DMA. Polym Adv Technol 20:235–245
101. Patsenker LD, Tatarets AL, Terpetschnig EA (2010) Long-wavelength probes and labels based on cyanines and squaraines. In: Demchenko AP (ed) Advanced fluorescence reporters in chemistry and biology I. Springer Series on Fluorescence 8. Springer, Heidelberg, pp 65–104

Fluorescence Probing in Structurally Anisotropic Materials

From Liquid Crystals to Macromolecules, Micelles and Lipid Bilayers

Semen O. Yesylevskyy and Alexander P. Demchenko

Abstract Acquisition and analysis of information on structure and dynamics of nanometer-sized objects are a strong challenge for fluorescence probing technique. As it was discussed in the first chapter of this volume, the nanoscopic dimension itself, its structural heterogeneity and the peculiarities of their surface–interface properties put serious restrictions on the application of methods that were successful when applied to neat solvents. Even more challenging is the application of fluorescence probing to the nanosize materials with strong structural anisotropy. In addition to subnanoscale gradients of polarity and mobility, these systems demonstrate anisotropic molecular fields acting on fluorescence reporter dyes. Ordered arrangements of charged and dipolar molecules can generate strong electric fields. Similarity of these effects prompted us to provide their unified discussion. Fluorescence probing in ultrathin films and monolayers, surfactant micelles, phospholipid bilayers and, finally, biological macromolecules – proteins and nucleic acids – will be discussed here. We critically analyze nanoscopic description of these systems based on physical theory and empirical correlations and try to draw the link to high-resolution structural data and to the results of quantum mechanical and molecular dynamics computations.

Keywords Anisotropic materials · Fluorescence probes · Liquid crystals · Micelles · Nucleic acids · Proteins · Ultrathin films

---

S.O. Yesylevskyy (✉)
Institute of Physics, National Academy of Sciences of Ukraine, Prospect Nauki, 46, Kyiv 03039, Ukraine
e-mail: yesint3@yahoo.com

A.P. Demchenko
A.V. Palladin Institute of Biochemistry, National Academy of Sciences of Ukraine, Leontovicha st. 9, Kyiv 01601, Ukraine
e-mail: alexdem@ukr.net

A.P. Demchenko (ed.), *Advanced Fluorescence Reporters in Chemistry and Biology III: Applications in Sensing and Imaging*, Springer Ser Fluoresc (2011) 10: 119–158,
DOI 10.1007/978-3-642-18035-4_4,

Contents

1 Introduction

With the booming development of nanotechnology, there is a strong demand to characterize the structural and dynamic properties of nanosize materials of different origin and composition. In the first chapter of this volume [1], we provided critical analysis of different trends in the application of fluorescence method to these systems. The simplest and presently most popular approach is the molecular probing based on empirical correlations between the interactions and dynamics on molecular level and the parameters characterizing macroscopic continuous media. The examples are the "polarity" and "viscosity" that are derived from calibration based on macroscopic measurements of these parameters in neat solvents. We discussed the range of applicability of this approach and its limitations that are especially noticeable when these properties exhibit the gradients that are steeper than molecular dimensions of the fluorescence reporter.

Meantime, molecular-scale analysis based on physical theory that is also used in the studies of these systems requires application of models that are still remote from complete description of spectroscopic behavior. Their benefit is the direct inclusion into analysis of the properties of the probing dye and in deriving its interactions with surrounding molecules and of their motions in the form of molecular relaxations. The medium is still described on the level of quasi-continuous approximation, which ignores molecular structure and, definitely, the structural and dynamic anisotropy originated by it. This approach known in physics as "mesoscopic" can be called as "nanoscopic" when applied to nanoscale phenomena. *Molecular dynamics* (MD) simulations were useful in analysis of these systems in order to shed light on the applicability of this approach and to reveal its benefits and limitations. From this discussion, we have seen that in materials with very high surface-to-volume ratio, many properties are determined by structure and dynamics at the surface and cannot be derived or predicted from the known properties of correspondent bulk materials.

Notably, there are materials that are very important for basic science and practical applications, in which, in addition to subnanoscale gradients of polarity and mobility, there is a strong *structural anisotropy*. It influences their many properties including location and orientation of probing dyes. In anisotropic media, fluorescence emission becomes more complex than in the conventional solvents. Since the fluorescent probes always sense the properties of their local environment, an intrinsic anisotropy of this environment influences the fluorescence response and leads to a number of nontrivial effects. These effects include static charge–dipole and dipole–dipole interactions of the probing dye with surrounding groups of atoms. In organized media with highly ordered location and orientation of charged and dipolar groups, these influences can be approximated by electrostatic potentials located at interfaces. Rotational and translational dynamics in these systems contain, in addition to diffusional, the electric field-induced component. Anisotropy in molecular interactions and dynamics that includes strong *electric field gradients* can be sensed by probing dyes displaying the property of *electrochromism*. This is in contrast to isotropic media, in which spatial averaging of local interactions can justify their description in terms of "polarity" and "viscosity".

All inhomogeneous media with structural anisotropy can be classified into two broad classes. The media of the first class are *anisotropic but continuous* at the nanoscopic length scales comparable to the size of the fluorescent probes. The most prominent examples of such media are liquid crystals, which are homogeneous but feature strong molecular ordering. The media of the second class are not only *anisotropic* but also *inhomogeneous* already at the nanoscopic level. This class includes Langmuir–Blodgett and polymer "layer-by-layer" films, surfactant micelles, lipid bilayers, proteins, DNA molecules, etc.

In the case of homogeneous anisotropic media, the macroscopic description may be sufficient in many cases. However, such scalar macroscopic properties as the dielectric constant or refractive index become tensors. When such media are also inhomogeneous, such description fails. The macroscopic characteristics are not applicable to them even in the tensor form because the local properties depend on the environment and change significantly at the nanometer length scale. These systems are the most challenging for physical modeling. The atomistic description provided by molecular dynamics (MD) simulations in combination with different quantum mechanics (QM) and molecular mechanics (MM) methods bring understanding to spectroscopic response obtained in such systems. These techniques could complement the fluorescent data, which lack such fine-grained spatial resolution.

Therefore in the present chapter we decided to combine the results of fluorescence probing obtained for the ultrathin films and monolayers, surfactant micelles, phospholipid bilayers and, finally, biological macromolecules – proteins and nucleic acids. Our tools will be the comparative analysis of different approaches for gaining information on nanoscale structure and dynamics in these systems. Where appropriate, we will try to find connection with the high-resolution structural data and of the results of quantum mechanics and molecular dynamics simulations.

2 Liquid Crystals

Liquid crystals are highly anisotropic materials, in which elongated or disk-like organic molecules have no positional order, but they self-align to attain the long-range directional order with their roughly parallel axes. There are several types of liquid crystals, which differ by the types of molecules and the characteristics of their ordering. The molecules could be ordered in one (nematics) or two (smectics) directions or be organized into chiral spirals (cholesterics). All liquid crystals possess large molecular anisotropy and intermolecular ordering described by the preferred orientation vector (so-called director). The centers of masses of the molecules are randomly distributed as in a liquid but still maintain long-range directional order as in crystal. Still, molecules in these structures retain the freedom to flow.

Nematic phases are attractive model systems for the investigation of anisotropic intermolecular interaction. The host provides a relatively rigid, more packed and anisotropic surrounding for the dye molecules influencing their spectral properties [2–4]. The mean value of dielectric constant in the nematic phase is usually taken as:

$$\bar{\varepsilon} = \frac{1}{3}\left(\varepsilon_{\|} + 2\varepsilon_{\perp}\right),$$

where $\varepsilon_{\|}$ and $\varepsilon_{\perp}$ are dielectric constants parallel and perpendicular to the molecular axis, respectively [5].

It was shown that interfacial ordering and induction of structural anisotropy in nematic liquid crystals may influence substantially the fluorescence spectra, as it was shown for incorporated ICT dyes [5]. The absorption and emission wavelengths of the dyes in liquid crystals are longer than that expected from the value of the mean dielectric constant. This suggests that the dye molecules experience a more polar environment along the $\varepsilon_{\|}$ direction rather than that expressed by mean dielectric constant. The stronger solvation energy in one direction and unusual character of relaxations suppresses the dye electron transfer reactions [6].

The guest–host interactions in the liquid crystals, such as the interactions of the fluorescent dyes with surrounding molecules, are being studied for decades [7]. The dyes, which exhibit the pleochroic behavior (dependence of the absorption spectrum on molecular orientation with respect to the polarization of light) are especially interesting in this respect. The controlled color switching in the nematic liquid crystal can be modulated by the electric field orienting the guest dye

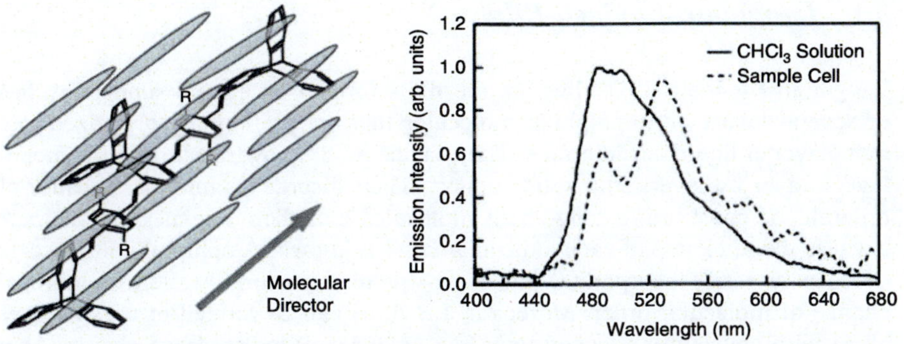

Fig. 1 Suggested molecular packing model in liquid crystal solution of the triptycene polymeric dye (*left*) and dramatic influence of ordering on the emission spectrum of the dye (*right*). The reference spectrum in chloroform solution is smooth (*solid line*), while the spectrum in the liquid crystal (*dashed line*) shows a "dip" relating to the photonic band of the liquid crystal. Figure from [8]

molecules in desired direction. The comparison of anisotropic and isotropic environments revealed systematic spectral shifts, which were correlated with the solvent polarity and anisotropy [5]. Incorporating the dyes into the liquid crystals (doping by the dyes) could also be used to increase the ordering in the liquid crystalline matrix. Particularly, such techniques allow creating the media for high-quality lasers [8] (Fig. 1).

The extensive theoretical study [9] showed that the interactions of a quadrupole created by the solute and the dipoles of the solvent can be treated in a rigorous way, which allows describing the behavior of real liquid crystals [10]. Although such theories have little practical importance now, they can lead to the development of novel analysis and simulation techniques in the future.

3 Ultrathin Films

Ultrathin films such as the Langmuir–Blodgett (LB) films, self-assembled monolayers (SAMs), and layer-by-layer films have a strong potential for various applications, including construction of molecular devices, generation of nonlinear optical elements, and designing of nanosensors. Many of these functions require incorporation of organic dyes with their regular arrangement on molecular scale. Incorporation endows them with new functions, such as the generation of second-order nonlinear optical response or the production of fluorescence signal necessary for sensor performance. Particularly SAMs of dansyl, immobilized on glass could be used as highly selective fluorescence sensors for nitrobenzene [11]. The dyes incorporated into ultrathin films can also be used as fluorescent probes to characterize film properties [12, 13].

3.1 Langmuir–Blodgett Films

Langmuir–Blodgett (LB) films are the films formed on a solid support by one or several layers of amphiphilic molecules that are preliminary organized into monolayer at liquid–air interface. They can be built up by transferring the monolayers, layer by layer, onto solid support. Their disorder is minimal because of ordering of constituting long-chain molecules and their thickness is accurate because the thickness of each molecular layer is known. Amphiphilic fluorescent dyes can be easily incorporated into these structures vectorially and with minimal change of molecular order. Moreover, LB films can be formed of organic dyes alone with covalently attached long (e.g., octadecyl) hydrocarbon chains. As it was recently shown, the proper "tail–head–tail" attachment avoids self-quenching and allows achieving enhanced fluorescence emission [14]. The studies of anisotropy show high degree of orientation and immobilization of guest dye molecules. Meantime, even in these highly ordered systems the heterogeneity of emission can sometimes be observed, so that characterization of their interactions and dynamics is complicated. The dual fluorescence emission was obtained from organized Nile Red aggregates formed inside the LB films and in binary solvent mixtures [15]. The splitting of absorption spectrum into two maxima was explained by a superposition of two types of monomeric molecules with planar and perpendicular orientation of diethylamino group relative to the chromophore plain [16].

There are two common types of aggregates, which could be observed in LB films [17]. The so-called *J*-aggregates exhibit intense, very narrow, and red-shifted absorption bands that are known as *J*-bands [18]. The *H*-aggregates exhibit blue-shifted absorption maxima called the *H*-bands. *J*-aggregates are common for LB films of many dyes, particularly they were studied extensively for merocyanine [19]. *H*-aggregates were observed for bridged polar stilbene (2-(4′-((perfluorooctyl)-sulfonyl) phenyl)-6-(*N*,*N*-dimethylamino) benzofuran). The change of excitation and emission spectra in such films with the change of solvent polarity reflects the charge transfer in the excited state and is explained with the exciton model [20]. The long-chain merocyanine dyes form both types of aggregates depending on the composition of LB monolayer [21].

The PBBO probe (well-known derivative of oxazole chromophore) could also be incorporated into LB films. The complete demixing of the binary components of the monolayers was observed, which leads to the formation of clusters and aggregates of PBBO molecules in the films. *J*-type aggregates of PBBO molecules in LB films were detected as well as their excimeric emission [22].

The spectral shifts in LB films are usually explained by the formation of excitonic states. However, alternative mechanisms are also possible. Particularly, a thorough spectroscopic study of well-known zwitterionic chromophore C16H33Q-3CNQ complemented by theoretical modeling suggested that large blue-shift of the absorption band in the LB film with respect to solution is attributed to the increased polarity of the environment, rather than to excitonic effects [23].

The molecular environment of polymer LB films of poly(*N*-dodecyl acrylamide) and poly(*tert*-pentyl acrylamide), which are copolymerized with 1-pyrenylmethyl acrylate (PyMA) luminescent probe, was studied by several experimental techniques including fluorescence spectroscopy. It was shown that the orientation of pyrene ring in the monolayers changes from perpendicular to parallel with the subtle change of the polymer chain length. These findings show that the molecular orientation of luminescent molecules can be controlled easily by small changes of the length of side chains in polymer LB assembly [24]. The molecular motion of pyrene chromophore is restricted in the polymer LB films in all cases [25].

The packing of dye molecules in the films could change with time, which leads to the changes of their spectral characteristics. Particularly, the shift of the excimer-like emission band from 602 to 636 nm was observed in the aging DBPI-stearic acid mixed LB films [26].

Mixtures of derivatives of coumarin-2 and coumarin-343 can self-assemble on silicon wafers to form monolayers in which light harvesting and energy transfer (the phenomena explained in [27]) take place. It was shown that photophysical properties of these aggregates, such as donor quenching and fluorescence amplification, could be tuned by varying the molar ratio of the donor and acceptor chromophores on the surface or by increasing the number of coumarin-2 chromophores on the donor adsorbate through the use of a specific linker [28].

LB films allow creating the local electric fields by designed strength and orientation of constituting dipoles [29]. This opens many possibilities for studying electrochromic properties of fluorescent dyes in these systems. The Stark effect spectroscopy of dyes oriented in thin solid films is broadly used to characterize these dyes, in particular, to determine their excited-state dipole moments. Here the goal should be different: to describe electrostatic anisotropy of the films in order to understand their absorption and other properties and also to model the effects of field gradients existing in more complex and flexible materials, such as micelles and biomembranes.

3.2 *Layer-by-Layer Polyelectrolyte Multilayers*

Polyelectrolytes are the charged polymers with acid or basic ionizable groups. They can form multilayers made by alternate layers of oppositely charged polyelectrolyte molecules. The technique known as layer-by-layer assembly allows obtaining stable multilayers stabilized by electrostatic interactions. The major interest in these films is their application in areas where their composition and thickness on the nanometer scale has to be tuned. The studies of their internal dynamics and water content must explain their functional properties, such as permeability and elasticity [30]. As well as LB films, they can provide unidirectional orientation of fluorescence dyes.

Fluorescence probing of polarity in these multilayers is presently limited by the application of pyrene with reference to Py empirical scale [12]. This is reasonable, since pyrene is a neutral molecule and must be equally distributed between the layers of different charges. The techniques of immobilization of pyrene on solid surfaces have been developed [31], which provides additional possibilities for studying the coverage of flat surfaces and nanoparticles with polyelectrolyte layers.

The structure and dynamics of the polyelectrolyte multilayers is not well understood. Despite this fact there are only few attempts to model them [32]. Due to very large size of these systems, conventional all-atom simulations are not practical. The coarse-grained models with continuous solvation could be applied, but their success in reproducing such experimental observations as the stability and linear film growth in aqueous solutions is still limited [33]. Further advance in the simulation techniques may provide the means, which could be used complementary to spectroscopic studies.

The self-assembling of monolayers and multilayers based on electrostatic interactions allow producing a noncentrosymmetric arrangement of nonlinear optical chromophores to yield thin films with high nonlinear optical susceptibility. Such materials find many applications in the development of optical devices and sensors. Formation of these layers around silver nanoparticles results in strong enhancement of these nonlinear optical effects due to localized surface plasmon resonance (collective oscillations of the conduction electrons at optical frequencies) improving dramatically these materials [34]. Strong demand for optimization of their properties will stimulate their characterization with fluorescence methods.

4 Micelles

Micelles are the simplest self-organized assemblies that originate because the constituting amphiphilic surfactants tend to expose their hydrophilic part to water and to keep the hydrophobic portion away from water. *Direct micelles* are formed in bulk water with the formation of inner nanovolume with low-polar surfactant chains. *Reverse micelles* are the micelles formed in hydrophobic solvent (e.g., hexane) that stretch out their hydrophobic parts with polar head groups pointing inward. Inner volume of direct micelles can incorporate water-insoluble hydrophobic compounds making them soluble, and reverse micelles can solubilize molecules of water or other polar compounds by incorporating them into their inner volume. Numerous applications in lubrication and separation technologies, in laundry and other practical areas, stimulate the study of these systems on a molecular level.

Micelles are very flexible dynamic structures. They are formed above a certain critical micelle concentration (CMC) of monomeric surfactant molecules (usually in micromolar range) and their size and shape are concentration-dependent. Solubilization of various types of solutes can change their properties. Many fluorescent dyes incorporate easily into micelles and they are frequently used to detect CMC by the change of polarity or quenching properties of their environment. Structural anisotropy in micelles is observed on the level of interacting surfactant molecules that align with polar heads facing one direction and hydrophobic tails – the other. The polar heads can be neutral, cationic, anionic, or zwitterionic. Interacting noncovalently with molecules constituting the micelle, the small dye molecules can distribute to different locations and redistribute during fluorescence lifetime, demonstrating spatial and temporal heterogeneity of their response, but their preferential location and orientation often exists. Designing larger molecules for which specific location can be provided is important for this research because of the presence of strong gradients of their many properties on molecular scale.

4.1 Direct Micelles

All arsenals of fluorescence techniques that involve both empirical correlations and physical modeling [1] have been applied to the studies of micelles. Very popular is the analysis based on empirical correlations of spectral response with polarity [35]. There are many dynamic information collected about the rates of molecular relaxations in the probe dye environments [36–38].

4.1.1 Polarity and Hydration

Many practical applications require determination of these parameters. Meantime, presently applied methods experience the following problems:

Imprecise fluorophore location. When measured in heterogeneous systems with strong molecular-scale gradients of polarity (which are the micelles), the obtained fluorescence parameters are almost meaningless if there is no information on probe location [39].

The latter information is often obtained from fluorescence response of probe itself and it lacks sufficient precision. It can be based, for example, on the effects of additionally introduced collisional quenchers [40, 41]. However, their quenching efficiencies may differ from the correspondent bulk values with their still imprecise locations within the structures [42]. In addition, probes themselves can be dispersed between different locations [43] and even distributed into aqueous phase [44, 45]. For instance, pyrene [46] and Prodan [40], which are often claimed as hydrophobic probes, are easily quenched by iodine anion; this suggests their mostly interfacial location. Thus, low own polarity and low solubility in water does not guarantee that the probe will be incorporated into central hydrophobic part of micelle. Covalent attachment of long hydrophobic chain also does not guarantee the chromophore location in the hydrophobic core. Thus, 4-aminophthalimide moiety despite its covalent linking with the nonpolar tail of the long alkane surfactant chain localizes itself in the interfacial region of the micelle by folding up of the chain [47].

One can make the probe very similar to that of surfactant molecule, with polar head and apolar tail. Examples are the functional indole [48] and *trans*-stilbene [49] derivatives. The fluorophore may be a part of the polar head, as NBD-fatty acids [50] but can also be incorporated into aliphatic hydrocarbon tail, as in anthroyl derivatives of fatty acids [51]. In addition, the probes can be made completely hydrophobic by attaching one or two hydrophobic tails [49, 52]. This approach is not fully developed, and the main problem here is that the hydrophobicity increased by substituents is not always sufficient to ensure the definite chromophore location in flexible hydrophobic areas of micelle.

Low information content of probe response. Totally apolar probes (such as anthracene or 2,3-dimetylnaphtalene [53]) are polarity-insensitive. Meantime, even in the case of the most responsive and informative "environment-sensitive" probes, like Prodan, the wavelength shift of fluorescence spectrum is the only parameter that is usually recorded. It can be sensitive to all types of noncovalent interactions including hydrogen bonding [54], which was actually confirmed in micelle studies [55]. In polarity sensing, it depends not only on dipole–dipole interactions with the environment but also on the rate of reorganization of these interactions (dielectric relaxations) on the scale of fluorescence lifetime [56]. Local electric fields produced by proximal charges generate, additionally, the so-called electrochromic shifts [57]. Interfacial electrostatic fields produced in micelles by charges located at the interface must influence the binding of charged and dipolar compounds [58]. These fields influence the rates of excited-state reactions in micelles [59–61], but the lack of tools to discriminate the effects of polarity, H-bonding, and electric fields does not allow providing detailed analysis of these effects. Without possibility of locating probes in deep hydrophobic areas some authors claim about the presence of significant amount of hydration water in these areas [39].

With the focus on increasing the information content of fluorescence response, the researchers try to develop fluorophores that possess excited-state reactions that allow achieving two, instead of one, first excited states exhibiting different interactions with the environment. As a result, two well-separated fluorescence emission bands with different sensitivity to particular intermolecular interactions have to be obtained. For instance, one band should be sensitive mostly to polarity and the other one to hydrogen bonding. And since the two states are connected by an excited-state reaction, relative intensities of two bands may contain information on this reaction and thus on perturbation of this reaction by the environment. Intramolecular charge transfer (ICT) [62–64] and excited-state intramolecular proton transfer (ESIPT) [56, 65] were suggested as the reactions that could provide two-color response and suggested promising prospect in their application.

As a generic approach for different applications as micelle and biomembrane probes, a series of 3-hydroxychromone (3HC) and particularly 3-hydroxyflavone (3HF) derivatives were suggested. They exhibit ESIPT reaction with initially excited normal (N*) band and ESIPT reaction product tautomer (T*) band demonstrating strong solvent-dependent variations of their relative intensities. With 3HF as a probe, Sarkar and Sengupta [56] observed dramatic transformations of the two-band spectra of 3HF on the formation of micelles of Triton X-100, sodium dodecyl sulfate (SDS), and cetyltrimethyl ammonium bromide (CTAB). However, it is known that the sensitivity to polarity of parent 3HF is low, and the majority of this effect may be attributed to disruption of hydrogen bonds with water on probe incorporation into micelles. Introduction of electron-donor dialkylamino group into

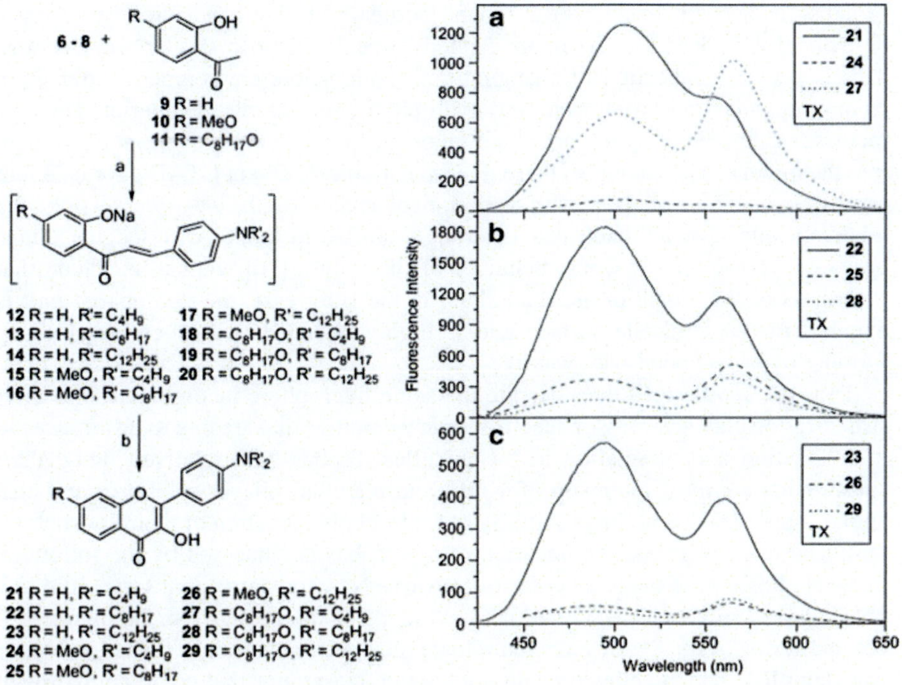

Fig. 2 The scheme of synthesis with intermediates (**6–20**) and final products (**21–29**) of synthetic hydrophobic 3HF probes for micelle studies (*left*) and their fluorescence spectra in neutral Triton X-100 micelles formed in water (*right*) [68]

phenyl ring of 3HF increased dramatically the dipole moment of N* (but not of T*) state and made the correspondent fluorescence band highly sensitive to polarity [66] and to electrostatic fields [67] in their environment.

A series of uncharged 3HF derivatives with aliphatic hydrophobic tails were synthesized and applied in the study of micelles [68]. Together with high hydrophobicity of 3HF itself this could guarantee the probe location in the low-polar core of micelles (Fig. 2).

We undertook a series of experiments on incorporation of synthesized dyes into micelles formed by cationic surfactant CTAB and neutral Triton X-100. Surprisingly, despite similarity of structures of these dyes we got very divergent and therefore very interesting results on their behavior in these micelles.

In Triton X-100 micelles, all the dyes incorporate well into micelles and show increase in intensity from almost zero level observed in water (where most of them have a very low solubility) to a definite level which reaches maximum at concentrations corresponding to surfactant CMC. This level remains constant on further increase in surfactant concentration (data not shown). Meantime, the integral intensity and its distribution between N* and T* bands varies dramatically (Fig. 2a–c) and the probes with the longest alkyl chains (of 12 carbon atoms) exhibit low intensity probably due to their aggregated state. There is an important regularity upon increase of substituent chain length: the N* band compared to T* band intensity changes dramatically. It decreases its relative intensity and moves to shorter wavelengths. This is a strong indication of changing the chromophore environment – transition from more polar and hydrated sites to hydrophobic sites in micelle interior. Thus the increase of the size of hydrophobic tail favors deeper location of fluorophore moiety into the micelle. The studies of these dyes in cationic detergent CTAB gave different results. In the line of dyes with variation of chain length, we observe only the relatively intensive N* band that is strongly shifted to longer wavelengths. Thus, based on known spectroscopic behavior of these dyes [70], we can conclude that regardless of the length of attached chain in the latter case, the fluorophore part is exposed to polar micelle surface and is highly hydrated (with 4-carbonyl group forming hydrogen bond with water).

Thus the synthesized dyes despite the same fluorophore incorporated into their structures display a variety of spectroscopic properties that are due to differences in their location and orientation in the micelles. Probes with relatively long alkyl substituents are most interesting for different micellar, bilayer structures and also apolar interfaces due to the possibility of hydrophobic location of their fluorophore unit. The low polarity of the location of the probes is witnessed by the following facts: (1) Position of the most polarity-sensitive N* band corresponds to that in low polar aprotic solvents. (2) There is no evidence of the presence of the hydrogen-bonded forms of the dyes (that should have the spectrum strongly shifted to longer wavelengths). The micelle cores do not contain the groups that could be hydrogen bond donors, so the only expected partners for the formation of these bonds could be hydration water. Thus the probes are completely inaccessible to contact with water molecules. (3) On the additions of iodine ions that are known as strong water-soluble collisional fluorescence quenchers, the quenching effect is not observed. This fact is especially demonstrative for CTAB micelles since I^- ions can substitute Br^- ions in the polar interface layer thus increasing substantially its local concentration. Thus in our knowledge this is the first example of successful incorporation of polarity-sensitive dye into anhydrous hydrophobic core of aqueous micelle providing a tool for studies on solubilization of different compounds in this interior.

4.1.2 Dynamics in Micellar Environments

As it is discussed in this book [1], the two fluorescence methods that are commonly applied in the studies of molecular dynamics are the time-resolved spectroscopy of molecular relaxations and the time-resolved anisotropy. When applied to micelles, both of them commonly use the formalism developed in the studies of neat solvents. The lack of information on probe location and anisotropy in structures and in interactions with their molecular environment make imprecise this extremely valuable information. For revealing the mechanisms of probe response, the influence of additional parameters, such as variation of solvent composition or the temperature, is commonly applied. Thus, it was found that the solvation dynamics in SDS micelles formed in D_2O solution is about 1.2 times slower than that in these micelles formed in H_2O, which could be due to slower process of the hydrogen bond dynamics in the hydration layer around the micelle [37].

The temperature effect on solvation dynamics has been investigated in two neutral micelles, namely, Triton X-100 (TX100) and Brij-35 (BJ-35), using dynamic fluorescence Stokes-shift method, to explore the role of micellar size and hydration on the solvation process. In the course of relaxation the fluorescence spectra move as a function of time, and this motion can be expressed with the aid of function of *C*(*t*). It allows connecting the motion of spectrum to a full scale of this event and making comparisons of relaxation rates obtained with different dyes.

$$C(t) = \frac{\nu(t) - \nu(\infty)}{\nu(0) - \nu(\infty)}. \tag{1}$$

Here $\nu(t)$ is the position of spectrum in cm^{-1} as a function of time, and $\nu(0)$ and $\nu(\infty)$ are these values extrapolated to time zero and infinity correspondingly.

In TX-100, the temperature effect on $C(t)$ function is not only very strong but shows an unusual inversion around 298 K. On the contrary, for the BJ-35 micelle, the temperature effect is not that significant. This could be due to the temperature-dependent changes in micellar size and hydration, which are reported to be very large for TX-100 but very marginal for BJ-35 [36].

The rotational and translational diffusion coefficients do not follow the trend predicted by Stokes–Einstein equation, indicating that the dynamics in micelles is non-Brownian [71]. Particularly, the study of the phenosafranin (PSF) dye embedded into the micelles formed by anionic surfactants showed an enhancement in the fluorescence anisotropy and rotational relaxation time of the probe in the micellar environment in comparison to bulk liquid. This suggests that the fluorophore is located in the areas of restricted mobility inside the micelles [72]. An extensive time-resolved study of cationic, anionic, and neutral micelles containing different fluorescent dyes was reported. It was shown that the fluorescence anisotropy decay is caused by rotational and translational diffusion of the dye in the micelle coupled with the rotational motion of the micelle as a whole [73].

4.1.3 Micelles of Ionic Liquids

Ionic liquids, most commonly comprised of organic cations and inorganic anions, become increasingly popular in recent years [74]. These liquids could promote micelle formation or form micelles themselves in water. Solvation dynamics of the coumarin 153 dye was recently studied in the 1-Cetyl-3-vinylimidazolium bromide and 1-cetyl-3-vinylimidazolium NTf2 ionic liquid–water systems. It was shown that the coumarin is located in the peripheral ionic double-layer of the micelle (so-called Stern layer). The solvation dynamics of the probe suggests the presence of highly confined water molecules in the interface of the micelles [75].

Atomistic MD simulations are proved to be extremely useful in determining the properties of the bulk and micellar phases of ionic liquids [76, 77]. The formation of the micelles constitutes a significant challenge for MD simulations due to their large size and large formation times. However, the usage of coarse-grained MD simulations allowed to overcome these difficulties and to model spontaneous formation of the micelles in 1-*n*-decyl-3-methylimidazolium bromide/water mixtures [78].

There are surprisingly few MD studies of solvation dynamics of the fluorescent probes in ionic liquids. The solvation study of coumarin 153 in 1-butyl-3-methylimidazolium hexafluorophosphate could be mentioned [79]. It was found that solvation dynamics is dominated by complex collective ionic motion.

4.2 *Reverse Micelles*

Reverse (inverted) micelles are formed in hydrophobic solvent (e.g., hexane) by assembling the polar head groups of surfactant molecules that become protected from their unfavorable solvation. Such structures form polar interior that can solubilize polar compounds including water molecules. Micelles formed in the complete absence of water or other polar molecules are called "empty". In their presence the two interfaces are formed, the surfactant headgroup region facing a water pool and the surfactant tail region facing the apolar solvent. AOT (bis(2-ethylhexyl)sodium sulfosuccinate) is an anionic amphiphile with an hydrophobic double chain (Fig. 3), and because of small head to tail area it can readily form reverse micelles.

4.2.1 Probing the Reverse Micelle Interface

The heterogeneity of fluorescence reporter location and orientation in the micelle brings a lot of uncertainty in the analysis of their response. Thus, a popular polarity-sensitive probe Prodan is distributed in different micelle regions and yields fluorescence response from multiple locations simultaneously [80]. Four principal microenvironments for Prodan were detected, including an inner water pool, a bound water region, the AOT interface, and the surrounding hydrocarbon solvent phase. The excited-state equilibrium in the population of locally excited and charge transfer states of the probe, depending on its location (Fig. 4), produces differences in spectral position and lifetime. The two states exhibit different solvation times (2 and 0.4 ns respectively) [69].

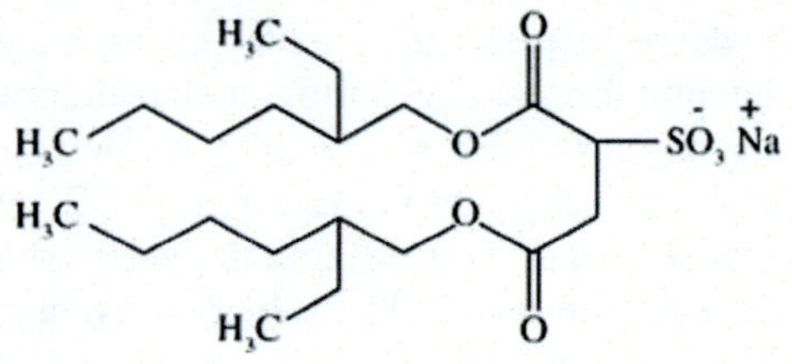

Fig. 3 AOT (bis(2-ethylhexyl)sodium sulfosuccinate) – the prominent compound, which forms reverse micelles easily

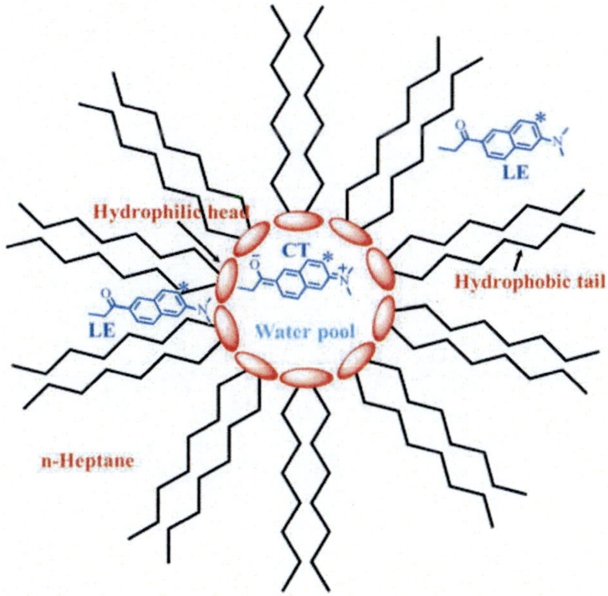

Fig. 4 The probe Prodan in reverse micelle formed by 1,4-bis-(2-ethylhexyl) sulfosuccinate (AOT) in *n*-heptane/water environment. "LE" and "CT" are the locally excited and charge-transfer emissive states of Prodan, respectively. Picture from [69]

The ability of smart 3HC dyes to provide multiparametric information on their intermolecular interactions was explored in the studies of AOT reverse micelles [81]. From Fig. 5 that compares the fluorescence spectra in micelles with that obtained in neat hydrophobic solvents one can derive that the position of polarity sensitive N* band corresponds to that of hexane but its relative intensity to the second T* band is much higher. Since the T* band is formed as a result of excited-state intramolecular proton transfer (ESIPT), one can infer that the dye is located in hydrophobic environment but its ESIPT reaction is suppressed. Since the positions of the absorption spectra of the dyes in 0.1 M AOT/hexane are the same as those in neat hexane, no significant change in the ground-state interactions is expected. Meantime, their 4-carbonyl and 3-hydroxyl groups can participate in intermolecular specific interactions with nonsolvated ionic headgroups at the micellar interface increasing the barrier for the ESIPT reaction. Strongly increased (compared to that in hexane) fluorescence quantum yields witnesses for high rigidity of probe environment.

Addition of water to AOT/hexane reverse micelles results in dramatic transformation of the fluorescence spectra (Fig. 6). The most spectacular is the decrease in relative intensity of the N* band accompanied by its less significant red shift. Such decrease is exactly the opposite of the common solvent polarity effect that could be expected with incorporation of highly polar water molecules. Since the two bands are significantly separated (up to 106 nm), this provides an almost total switch of emission color from blue–green to orange–red. The 3HC probes FE and FA are neutral and relatively low-polar molecules (log *P* 2.87 and 2.54, respectively). From their spectra we can witness the complete absence of hydrogen bonding of their carbonyl groups with water, which is the only proton donor component in this

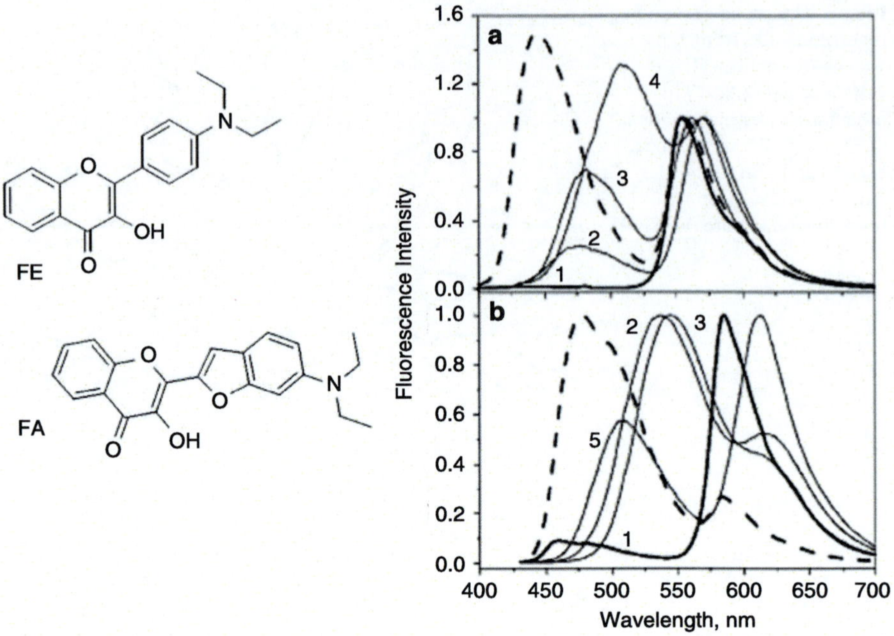

Fig. 5 Fluorescence spectra of probe FE (**a**) and FA (**b**) in 0.1 M AOT/hexane reverse micelles (*dashes*) and in neat organic solvents (*solid curves*). Solvents were hexane (1), ethyl acetate (2), chloroform (3), acetonitrile (4), and toluene (5). Note that according to the position of the short-wavelength N* band, the probe environment is strongly hydrophobic but its relative intensity is dramatically higher [81]

system. Gradual addition of water produces different effect: the spectra become similar to that in hexane. This result suggests that the probes are squeezed out from micelle to hexane solvent.

4.2.2 Water in Reverse Micelles and Solvation Dynamics

Reverse micelles incorporating water molecules are the subject of frequent studies to determine the viscosity, rigidity, and proximity within their interior mostly on the basis of empirical correlations with the data obtained in neat solvents [82, 83]. There are not so many attempts to apply time-resolved spectroscopy of molecular relaxations to these systems. Meantime, it was observed that solvation dynamics in the interior of reverse micelles is often dominated by a slow 0.1–0.3 ns time scale component arising due to geometric restrictions to rotations of polar solute and dye molecules [84].

Slow relaxation dynamics at the interface of reverse micelles was also observed with the detection of Red Edge effect of several fluorescent probes, particularly, of NBD derivative of cholesterol [82].

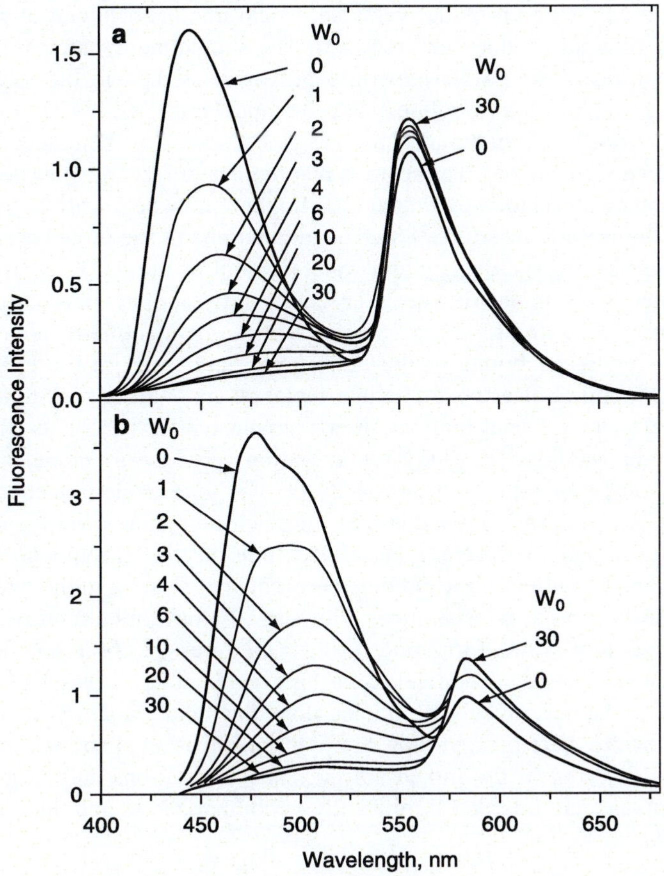

Fig. 6 Fluorescence spectra of probe FE (**a**) and FA (**b**) in 0.1 M AOT/hexane reverse micelles at increasing the water/AOT molar ratios, W_0. Figure from [81]

Based on spectroscopic behavior of coumarin dyes, it was inferred that the water molecules close to the surfactant head groups are more restricted compared to those in the core of the water pool that is formed at water/AOT molar ratio higher than 7 [36]. The alignment of water molecules near the interface and its dynamics may be different from that in water bulk because of geometric hindrance and different hydrogen-bond patterns [85]. In large reverse micelles, the dynamics of water could be separated into two ensembles: slow interfacial water and bulk-like core water. As the reverse micelle size decreases, the influence of the interface increases and the bulk-like phase disappears when the micelle size reaches 4 nm [86]. These results obtained by vibrational spectroscopy were supported by time-resolved fluorescence studies of a new dye exhibiting ICT reaction, in which nanosecond component of molecular relaxations was observed [87]. Solvation dynamics was shown to depend strongly on water pool size.

Computer simulation provides valuable insight into the behavior of water molecules in the micelle-confined environment. The short-time dynamics (up to 2 ps), which is not easily accessible experimentally was studied in the model system containing a simple diatomic solute inside the model micelles. It was shown that the short-time relaxation is dominated by water and occurs at the solute sites where hydrogen bonds are broken. The relative magnitudes of fast and slow contributions to the solvent response for a particular chromophore may be sensitive to its location relative to the micelle interface [88]. The simulations of the model system, which matches the size, charge density and water content of the AOT reverse micelles, revealed that water molecules near the cavity interface are immobilized by the counter ions. Three structural regions of water can be identified: water trapped in the ionic layer, water bound to the ionic layer, and water in the bulk-like core region. The mobility of water in the interfacial layers is greatly restricted for both translational and rotational motions, in agreement with experiments [89]. Similar model system was used to study the solvation response of model anionic and cationic chromophores in the water cavity of AOT-like model micelles. Dramatic differences in the solvation responses for these chromophores were revealed. This could be attributed to different initial locations of the chromophores and the pronounced slow motion of the cationic chromophore relative to the interface [90].

Advances in simulation techniques allowing all-atom simulations of the reverse micelles containing water, formamide or their mixtures are observed only in recent studies. Their structure was found to be less regular than it was thought before. Orientation of the trapped solvent dipoles along the radial directions increases from the center toward AOT head groups. Preferential solvation of sodium ionic species by water is detected in the mixed solvent. The translations and rotations of the confined solvents are retarded significantly in comparison to bulk liquids (up to 50 times) [91].

5 Phospholipid Bilayers

High structural anisotropy combined with steep (on molecular level) gradients of hydration, polarity, and electrostatic potentials makes molecular probing in biomembranes very challenging. These problems are clearly seen already in phospholipid bilayer formed of a single or of limited number of components. In biological membranes and phospholipid bilayers as their models, the gradients of polarity extend on the same or even shorter length scale as the fluorophore sizes [92]. Correspondingly, the microenvironment of a solute varies from a highly hydrophobic environment inside the alkyl chains of the bilayer to a very polar environment at the interface with aqueous milieu. Phospholipid segmental mobility in the membrane exhibits strong gradients increasing toward the chain ends as well as diffusional mobility of incorporated small molecules. Strong structural anisotropy results in anisotropic orientation of incorporated dyes. Depth-dependent and anisotropic binding of hydration water contributes additionally to this complexity. Moreover, ordered charges and dipoles in bilayer structure together with adsorbed ions generate strong local electric fields. Meantime, there are many authors that do not notice such difficulties and try to derive "polarity" and "fluidity" values from fluorescence dyes, the location and preferential orientation of which is not determined with certainty or even not specified. This could lead to misinterpretation of many results in view of the fact that the probes may exhibit heterogeneity of location sites, distribution between H-bonded and non-H-bonded forms and, even more, relocation during the excited-state lifetime.

These membrane properties allow applying an approximation based on nanoscopic resolution in one dimension, along the membrane normal. An implicitly or explicitly applied averaging operates along the membrane plane at different depths. Already on such level of detail the common for description macroscopic variables such as polarity, viscosity or hydration loose their meaning unless it becomes specified, to which particular site they refer and what kind of modeling brings their particular values. MD simulations and molecular probing are presently seen as the only methods that can allow providing in a consistent manner the analysis of structure, dynamics, and interactions leading to this quasi-continuous description [93].

5.1 Hydration, Polarity, and Fluidity as a Function of Probe Location

With the improvement and development of new research techniques, the data obtained in old days need to be revised. This refers to the response of "polarity-sensitive" dyes in spectral, time-resolved, anisotropy and wavelength-selective domains. There are very popular probes sensitive to structural transformations in the membrane, such as 6-propionyl-2-(dimethylamino)naphthalene (Prodan) and 2-dimethylamino-6-lauroylnaphthalene (Laurdan). What do they sense? The change of molecular order and of the rate of dielectric relaxations of their location site or their relocation to a different site? [94] With this type of dyes, the problem is complicated in view of switching their emission between LE and ICT states [95]. Of these two emissions, only emission from ICT state exhibits dielectric–relaxational shift of emission from 415 to 435 nm [96]. The interplay between these two emissions is strongly polarity dependent, and since the gradient of polarity is steeper than the fluorophore dimension, ambiguity in interpretation is inevitable. When the dye location is heterogeneous, then the observation of Red Edge effects as the measure of the rate of dielectric relaxations in the membrane is no longer reliable [97], since the dependence of fluorescence spectra on excitation wavelength (λ_{ex}) may be also the result of such heterogeneity. The λ_{ex}-dependent rates of emission decays can be observed in these cases [98] (Fig. 7).

Distribution between different locations was described for several small dye molecules. Fitting to bi-modal function, it was interpreted as the distribution of free dye and its H-bonded complex with water molecule [100].

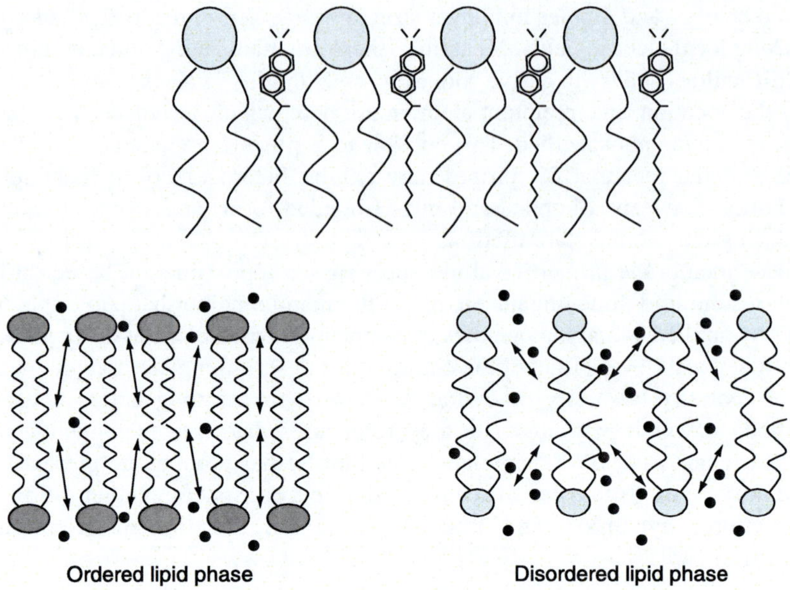

Fig. 7 Cartoon illustrating the location of Laurdan probe in the lipid membrane (*top*). The rearrangement of water molecules (*black dots*) localized around the Laurdan dipoles (*arrows*) may be responsible for the red shift observed in different lipid phases. Figure from [99]

This complex is more polar and occupies more shallow position in the membrane. Relocation of such dyes to different sites may occur in the case of change in order and mobility in the membrane, but interpretation of such changes is difficult. Dielectric relaxations in membranes are commonly related to the induced motions of water molecules as they are smallest dipoles possessing highest mobility [101]. Their dynamics is strongly retarded at interfaces [102], but the actual distance scale is not known. If this occurs on the scale of sizes of water molecules, then this scale is much shorter than the spatial range of reactive field sensed by a fluorophore.

Intramembrane location of a fluorophore needs a special probe design. In our experience, the most precise location and orientation of fluorophore at the desired site can be achieved only with uncharged and relatively nonpolar dye molecules that allow chemical modifications at their two opposite sides. The dyes of 3HC family satisfy to these requirements [103]. Addition of hydrophobic tails determines the membrane affinity and, if necessary, vertical orientation in the bilayer, whereas the attached charged and polar groups fix the fluorophore location relative to bilayer surface. With their aid, together with increased sensitivity to polarity and hydration at the site of their location there appears the possibility for more selective probing of surface (Ψ_s), dipole (Ψ_d), and transmembrane (Ψ_t) components of total electrostatic potential of the membrane [93].

5.2 *Dielectric Relaxations in Membranes*

Truly isotropic arrangement of disordered dipoles, the relaxations of which is thought to determine the spectral shifts, does not exist in membranes. Molecular ordering and generation of static molecular-scale electric fields determines the pattern of dielectric relaxations that is usually derived from the time-dependent shifts of the spectra. Their signatures change dramatically across the bilayer influencing both the solvation energies and relaxation rates. Instructive in this respect is the result obtained already two decades ago [104, 105]. The popular polarity-sensitive probes 1.8-ANS (l-anilinonaphthalene-8-sulfonate) and 2.6-TNS (2-toluidinonaphthalene-6-sulfonate) were studied upon incorporation into model phospholipid vesicles by a combination of fluorescence techniques. They involved the site-selective quenching in time-resolved spectroscopy, time-resolved anisotropy, and observation of Red Edge effects. The aromatic parts of these dyes are hydrophobic. But owing to the presence of a negatively charged sulfate group, they are located in the region of the heads of phospholipids at the polar surface layer of the membrane bilayer. We clearly observed the time-dependent motion of their spectra indicating some dynamic process. Meantime, the absence of Red Edge effect showed us that it was not a dielectric relaxation (which should be faster than the emission) but some other process that extends to a longer time scale.

What was this process? The emission decays exhibit remarkable differences upon application of hydrophobic quencher that quenches fluorescence "from inside" the membrane and of ionic quenchers that operate "from outside". This fact showed us that the dyes change their location in the course of emission: being more polar in the excited state they move along the gradient of polarity out from the membrane. Quite a different result was obtained for structurally similar dyes phenylnaphthylamines, l-AN and 2-AN, that possess no net charge and locate deeper in the membrane, on the level of glycerol skeletons of phospholipids. The observation of Red Edge effects indicated much slower (by orders of magnitude) dielectric relaxation.

The wavelength-selective decay of anisotropy demonstrated a new interesting phenomenon: the dielectric relaxation proceeds by induced rotation of the dye molecules [104]. In membrane they are surrounded by less mobile large and anisotropic phospholipids forming the bilayer. Thus, we learned that the charged probes located at the interface and neutral probes located deeper in the bilayer sense different dynamics occurring on different time scales. Therefore, precise location of the dyes is extremely important.

5.3 *Nanoscopic Heterogeneity in the Plane of Bilayer*

Nanoscopic heterogeneity in the plane of bilayer presents a new problem requiring proper approach. This problem is clearly seen in the description of "rafts" – the organized rigid domain structures formed with participation of steroids and sphingolipids [106]. Some dyes are incorporating into rafts and some are not. Covalent binding of the dye to raft-forming lipid does not guarantee its inclusion into raft [107]. The evidence is accumulating that microdomains in the membranes of living cells are essentially different from rafts in model membranes due to their fluid-mosaic organization with inclusion of proteins and minor membrane components [108]. Their size is much smaller and their internal organization can be different due to incorporation of proteins. When they are smaller than the wavelength of light, they cannot be resolved by common light spectroscopy, but the image formed by fluorescent probes can indicate predominant lipid phases.

When applied to model membranes, fluorescent dyes like Prodan and Laurdan demonstrate lower polarity and hydration in the raft phase [109], but it is not unlikely that they occupy different location. Design of more sophisticated 3HC probe F2N8 with the attachment of hydrophobic tail and, on the opposite side, of two charges complementing the lipid charges, allows establishing its more definite location in the membrane. Using this probe it was confirmed that hydration is lower in rafts than in the gel phase, which can be due to exclusion of water by cholesterol in the glycerol region of the membrane [110].

Meantime even in this case, the probe can change its location and orientation in the bilayer. Observation of the probe fluorescence in giant phospholipid vesicles under polarized excitation showed that in the fluid phase a uniform fluorescence intensity was observed all over the vesicle. This showed a variety of its orientations. In contrast, in the raft phase, maximum fluorescence intensity is observed in the regions where the light polarization is perpendicular to the bilayer plane. This suggests that the high order of this phase imposes a vertical orientation to the fluorophore [111].

These results show that lipid order in the membrane influences a number of structural and dynamic parameters that can be derived from the response of fluorescence probes. Extracting one of these parameters and calibrating its value against a set of data obtained in homogeneous media may bring significant errors in interpretation of the data. Molecular dynamics simulations of hydrated bilayer structures incorporating the probes are expected to lead to their adequate description [93]. However, currently there are no dedicated molecular dynamics studies, which address the issues of probe location in the rafts or compared to other biomembrane domains.

6 Proteins

Ability for many protein macromolecules of three-dimensional structures with atomic resolution lays a fruitful background for correlations of spectroscopic response with the structure. Important structural and dynamic information can be obtained from *intrinsic fluorophores*, particularly from Trp residues emitting in the near-UV [112]. Fluorescence response can be enhanced when proteins are modified at specific sites with the inclusion of *synthetic fluorophores*. The flexibility of protein structures and their dependence on external conditions and intermolecular interactions stimulated the development of fluorescence techniques to characterize these changes. This allows obtaining not only the information important for biology, but also to test and further develop fluorescence techniques.

6.1 Probing Protein Polarity

The protein polarity is the classical problem in biophysics, which is being studied for decades. There are several approaches to describing this property. *Explicit atomistic treatment*, which is commonly employed in MD simulations, is the most precise and accurate, but it is too detailed for the majority of practical applications. Another extreme is the application of *empirical polarity scale* where the amino acids are roughly subdivided into polar and nonpolar classes. Typical for research of 40–50 years ago, this simple approach is still frequently used in interpretation of spectroscopic data.

The spectra of the protein Trp residues depend strongly on their solvent exposure. The changes in the folding states of the protein may expose previously internally located amino acids to the solvent, which often leads to dramatic shifts of fluorescence spectra to longer wavelengths. Indeed, in folded proteins the fluorescence band maximum is typically found in the region of 325–335 nm and for unfolded proteins with Trp exposed to water the fluorescence spectrum is at longer wavelengths of 350–353 nm [112]. The quantum yield of the Trp fluorescence is also surprisingly different in different proteins and does not correlate with these spectral shifts. It is determined by different factors, particularly by the proximity of electron-transfer quenchers.

The statical ordering of charged and dipolar groups around fluorescence emitters uses a more precise alternative to "polarity" language that can explain the diversity in positions of Trp bands and their quantum yields in proteins. Recently, this problem was addressed by applying the QM/MM simulation technique [113, 114]. It was shown that there is a 30-fold difference in Trp fluorescence quantum yields and lifetimes depending on the rate of electron transfer between excited indole ring of Trp and the nearest backbone amides. Such strong dependence on the protein environment arises from local electric potential difference between the Trp ring and acceptor amide and depends also on location of nearby charges and local hydrogen bonding.

In experimental studies, the nanoscopic description, which keeps high spatial resolution around the sites of interest but averages fast motions around it, is a preferable way of describing the polarity effects in proteins. Meantime, the application of theory of solvatochromism to Trp absorption and emission is difficult because of the involvement of two electronic transitions that overlap in absorption spectra and of its participation in specific interactions including H-bonding. The solvatochromic organic dyes as fluorescent probes can provide better description of environment properties being incorporated into the protein structure or adsorbed at the binding sites.

The ratiometric dual-emission 3-hydroxychromone (3HC) probes are proved to be useful in determining the polarities of the binding sites in proteins. The probe FA binds with high affinity (~10^7 M^{-1}) at a well-determined site of serum albumin molecule and responds to binding by dramatic enhancement of its emission. In bovine serum albumin (BSA), this probe demonstrates the presence of both N* and T* bands in fluorescence emission, with the predominance of the N* band [115]. The position of this band witnesses for the probe binding in the environment of very low polarity. Meantime, when the ratio of relative intensities of N* and T* bands is plotted to a polarity function obtained for a series of neat low-polar solvents, a strong deviation is observed showing a shift in ESIPT equilibrium toward the product of this reaction. Interpretation in polarity terms alone cannot explain this effect, and therefore some local interactions must be involved. It is also very interesting that the function that characterizes electronic polarizability in probe environment is relatively high, corresponding to effective refractive index $n = 1.54$, which further demonstrates the screening from water and suggests closeness to the binding site of aromatic side groups of tryptophan, tyrosine, and phenylalanine.

6.2 *Solvation Response as a Measure of Protein Dynamics*

Natural fluorescence of the Trp residues is the most common and widely used tool for studying solvation dynamics and the dynamic properties of the local protein environment. Aqueous environment is natural for many protein molecules, and a surprising result of earlier studies was the fact that the rates of dielectric relaxations within their structures are by 2–3 and more orders of magnitude slower than in surrounding solvent. This allowed observing static and dynamic Red Edge effects even in single-Trp proteins, where the structural heterogeneity was not present [116, 117].

Thus, on nanosecond time scale these small bodies of the size of 10–100 nm behave as nanoscopic high-viscosity liquids and even solids. Due to structural constrains, *dielectric relaxations* in proteins are very complicated and hierarchical that leads to commonly observed nonexponential decays of fluorescence emission. A simple dielectric continuum description of the proteins is clearly insufficient, even though such a description has been widely used to correlate experimental data [118–120].

Protein molecules respond to electrostatic perturbations through complex reorganizations of their structure including rotations and displacements of the charged and polar groups, as well as of solvent molecules. These solvation responses occur both in the protein interior and on its exposed surface. The spectrum of response times in proteins is broad, it extends from femtoseconds to nanoseconds. Despite increasing amount of studies, the exact mechanisms of solvation response are still not well understood.

The dielectric response of proteins is usually measured by monitoring the time-dependent Stokes shift of the chromophore located in the site of interest inside the protein. The interpretation of these experiments depends on the solvation correlation function, which describes the time-dependence of the Stokes shift and hence the dielectric response of the medium to a change in charge distribution (1). The methodological basis for constructing solvation correlation functions correctly was developed recently [121]. Meantime, the presence of static Red Edge effects in many studied proteins [112] demonstrates that high statical order cannot be disrupted or reorganized on a time scale of emission.

The picture of slow dynamics, which is dominated by highly collective motions of the protein, solvent and the chromophore itself is now well established and confirmed in a number of MD and experimental studies [122]. Numerous studies using natural Trp fluorescence (see below) also support this view.

The spectrum of relaxation times observed in the time-resolved spectroscopic experiments with Trp fluorescence is broad and covers several orders of magnitude from picoseconds to tens of nanoseconds. This is usually attributed to specific water dynamics at the protein–water interface, which influences the chromophore [123]. There is clear evidence that the first layers of water possess different molecular arrangement and dynamics than the bulk water. Solvation dynamics of such water molecules displays an almost bulk-water-like ultrafast component and a surprising slow component at the 100–1,000 ps time scale. Fluorescence probing at the interface shows that the ultrafast component arises from an extended hydrogen bond network while the ultraslow component originates from binding of water molecules to biological macromolecules. Moreover, such slowing down of dynamics of water dipoles is not regular along the protein surface. The sites important for biocatalysis possess static electric fields that have to be supported by slow-water dynamics.

Whereas the fast relaxation times reflect the dynamics of water molecules around the Trp residues, the origin of slower relaxations is sometimes questioned. Particularly, the decay times, ranging from 10 ps to several nanoseconds, which are usually attributed to extremely slow hydration water motions, may correspond to internal protein dynamics and the solvent polarization effects instead [124]. This view is also supported by complementary experimental and MD studies of apomyoglobin [125]. Experimental studies of the tryptophan fluorescence and MD simulations reveal relaxation times of 5 and 87 ps. MD simulations allowed separating the relaxations of water shell and the protein by freezing selectively the motions of the protein or solvent atoms. They revealed the crucial role of protein motions in such slow dynamic response. Observation of slow water dynamics in MD simulations requires protein flexibility, regardless of whether the slow Stokes-shift component results from the water or protein contribution. The initial phases of dynamics (few picoseconds) represent fast local reorientations and translations of hydrating water molecules, while subsequent slow relaxation involves strongly coupled collective water–protein motions.

Further insight into the surface water dynamics around the myoglobin molecule was obtained using a Trp scan by site-specific mutations. Placing the Trp residue into 29 different positions in a protein globule allowed observing two distinct phases of water dynamics in the hydration layer. The first phase (1–8 ps) represents

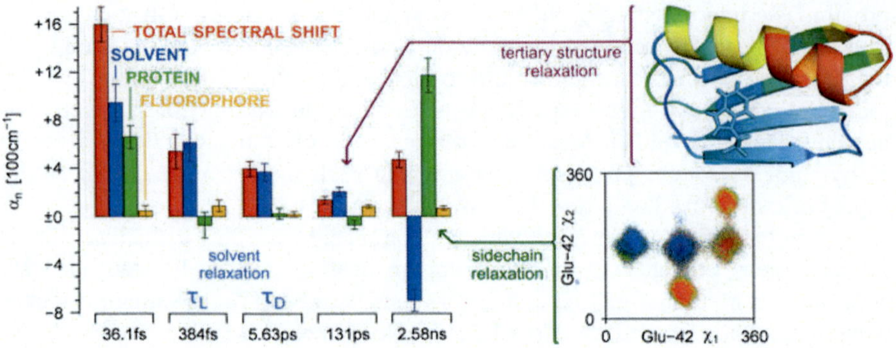

Fig. 8 Contributions of the protein, solvent and the fluorophore to the total spectral shift in GB1 protein. Five different relaxation components are shown. Figure from [127]

initial local relaxation, while the second slow phase (20–200 ps) corresponds to collective restructuring of hydration shell [126].

Very recent methodological advances in MD simulations allowed even better separation of the solvent, protein, and fluorophore contributions to the spectral shifts. The solvent response of Trp-43 residue in GB1 protein was studied using these novel techniques. Precise quantitative contributions of the solvent and protein to each of relaxation components were computed. The longest relaxation times (131 ps) is attributed to the changes in the tertiary structure of the globule (Fig. 8) [127].

Synthetic fluorescent amino acids can be incorporated into multiple sites in the protein globule in order to create a kind of solvation response map of the globule using this approach. Particularly, the solvation kinetics at multiple sites throughout the sequence of the B1 domain of streptococcal protein G (GB1) was monitored using the synthetic fluorescent amino acid Aladan [128].

Aladan was incorporated into seven different GB1 sites ranging from that buried within the protein core to fully solvent-exposed on the protein surface, and also located within different secondary structure elements including β-sheets, helices, and loops (Fig. 9). The time-dependent Stokes shifts were measured over the femtosecond to nanosecond time scales by fluorescence up-conversion and time-correlated single photon counting techniques. It was shown that all sites exhibited an initial, ultrafast Stokes shift on the subpicosecond time scale. With the increase of probe distance from the surface, the dynamics of the solvation response become slower. The long-time dynamics is believed to be caused to a large extent by restricted movements of the surrounding protein residues.

It is very difficult to determine the contributions of the protein and the solvent to solvation dynamics in such experiments. However, this task is trivial in MD simulations. The complementary MD study of solvation dynamics in 11 different sites in GB1 protein was performed recently [129]. It was shown that the polar solvation dynamics are position-dependent and highly heterogeneous. The solvent contributions are found to vary from negligible after a few picoseconds to dominant

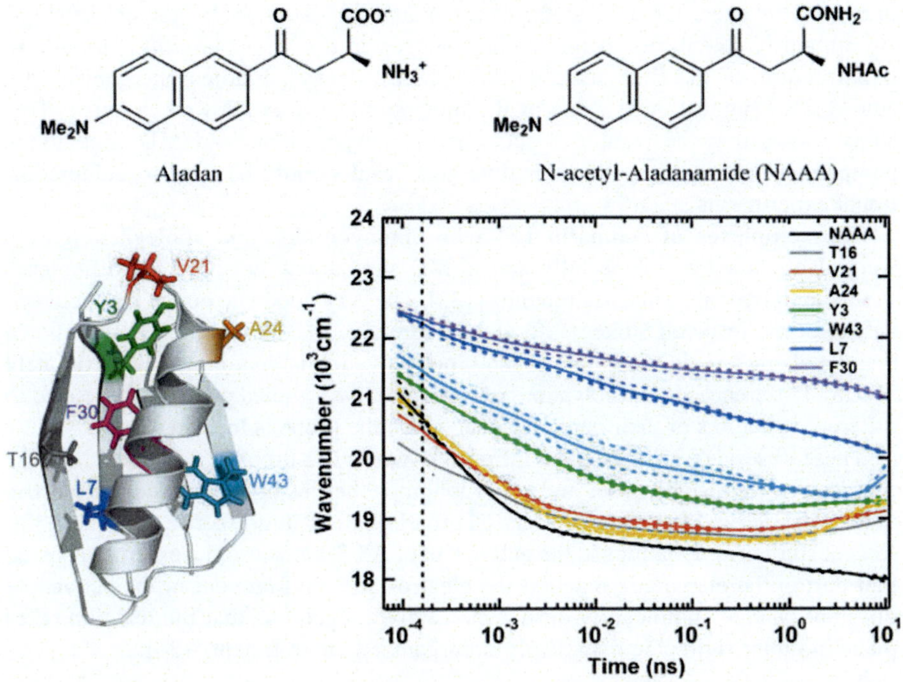

Fig. 9 Synthetic fluorescent amino acid Aladan and its soluble derivative *N*-acetyl-Aladanamide, which is used as a reference (*top*); sites of incorporation of Aladan into the GB1 protein (*bottom left*) and dynamic Stokes shift of Aladan at these sites (*bottom right*). Figure from [128]

on a scale of hundreds of picoseconds. The latter components are caused by coupled motions of hydration water and protein conformational dynamics.

6.3 Dynamics at the Ligand Binding Sites

The dynamics of the ligand binding sites in proteins is extremely important because of its crucial role in the protein functioning. However, probing the dielectric response in the binding sites and interpretation of the results may be challenging due to highly heterogeneous and often unusual environment in such sites. Thus, the well-defined system for comparative testing of the applied techniques is required in such studies. One of these systems is the complex of myoglobin (in either apo- or holo-form) with the coumarin 153 dye, which was characterized extensively by the spectroscopic and NMR techniques [130].

As an example of these studies, the measurements of dielectric response of coumarin 153 dye attached to the wild-type and mutant apomyoglobins in combination with MD simulations were performed [131]. An agreement between experimental and simulated solvation relaxation functions was excellent for the wild-type protein. However, the results for the mutants appeared to be rather different. Additional NMR studies of the mutants suggested that the coumarin is mobile in the structure on the time scale of fluorescence decays. Fluorescence anisotropy studies also suggest that in one of the mutants the dye is allowed to move. This study shows that the results of spectroscopic experiments and MD simulations complement each other, and their diverging results could be used to guide additional experiments.

The complexes of coumarin 153 with apomyoglobin and apoleghemoglobin were also studied experimentally and in MD simulations to reveal the differences in solvation dynamics in these proteins [132]. The very good agreement was obtained between time-resolved Stokes shifts and the results of MD simulations. In was shown that the characteristic time scale of the probe solvation is very rapid (approximately 300 fs). Differences in the solvation relaxation for two studied proteins are related to different structures of their heme pockets where the probe is located.

These observations of fast and ultrafast dynamics in proteins do not exclude the presence of high static order and static inhomogeneities of intramolecular electric fields revealed by Red Edge effects of protein bound fluorescence ligands [133]. Recent studies of dynamics in the active site of Δ5-3-ketosteroid isomerase showed that during the enzymatic reaction the electrostatic environment in the active site does not change significantly during the catalytic cycle, so that the reaction takes place in rather rigid electrostatically preorganized environment [134].

6.4 Probing Protein–Lipid Systems

Membrane proteins represent one of the most heterogeneous and anisotropic environments for the fluorescent probing. The probe incorporated into the membrane protein as well as intrinsic Trp can contact with surrounding lipid bilayer at different depths. As a result their environment may range from the aqueous interface outside the membrane to highly hydrophobic core of the bilayer. In addition, these fluorophores may exhibit very strong electrochromism due to the presence of biomembrane electrostatic potentials [93]. Thus the site-directed labeling of the membrane protein allows probing the protein–lipid–water interface at different depths. The solvatochromic Badan fluorescent label, which could be placed into different sites rather easily, is commonly used in such studies (Fig. 10).

This technique was recently used in the studies of M13 major coat protein embedded into lipid bilayers [135]. Specific spectral decomposition method allowed determining the contributions to the Stokes shifts from water, dynamics of the label, and the local protein environment (Fig. 10). It was shown that the Badan labels form the hydrogen bonds with water molecules even in the hydrophobic core of the membrane. This suggests deep penetration of water into the membrane at the protein–lipid interface or strong noncovalent binding of water molecules to the probe, which may drive them to bilayer center. It was shown that the hydration level and local polarity of the membrane depend on the headgroup

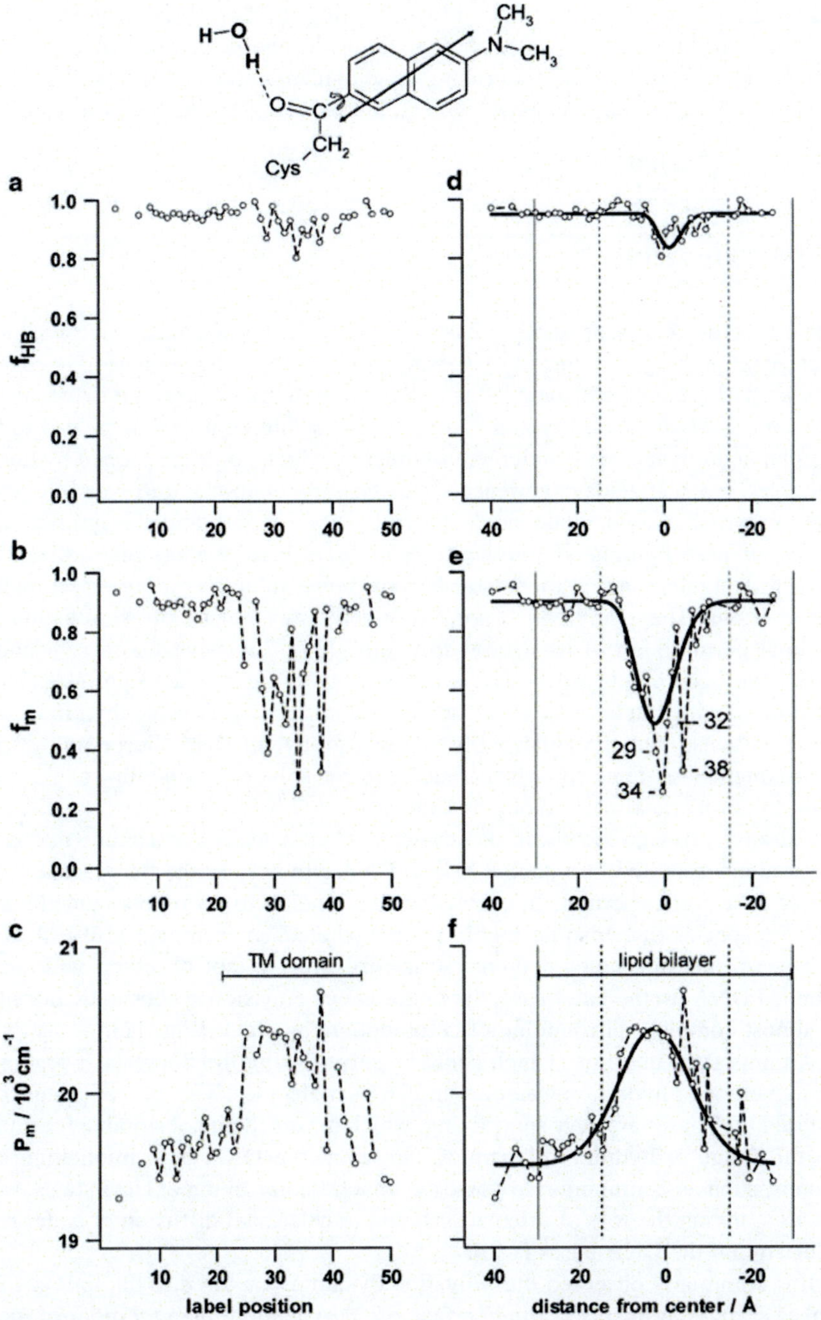

Fig. 10 Badan fluorescent label covalently linked to the cysteine sulfur and hydrogen-bonded to a water molecule as it appears in the labeled protein (*top*). Various Badan fluorescence parameters are shown for 40 different mutants versus label positions (**a–c**) and versus distance from bilayer center (**d–f**). The *solid vertical lines* represent the full bilayer thickness (59.5 Å) while the dotted lines represent the acyl chain/glycerol backbone interfaces. Figure from [135]

charge and the lipid chain length [135]. Labeling the M13 protein in several key positions varying from the water phase to the hydrophobic center of the membrane provides a "nanoscale ruler" for probing protein–membrane interactions. This allows precise profiling of the dynamics and distribution of water molecules in membrane systems [136].

7 Nucleic Acids

Nucleic acids represent another case of highly heterogeneous environment for fluorescent probing. In contrast to the majority of proteins they are heavily charged, thus both solvation and ionic effects play an extremely important role in their functioning. High flexibility and a variety of possible conformations lead to very complex and diverse set of microenvironments, which could be probed by fluorescent dyes. Nucleic acids experience a variety of reversible and irreversible perturbations, which may include nucleobase damage, depurination/depyrimidination events, or base flipping. Fluorescent nucleic acid base analogs that are sensitive to their local microenvironment have become powerful tools for investigating these perturbations. They can be incorporated directly into the site of examination with minimal perturbation of its natural structure [137]. Emissive fluorescent nucleosides have been employed to assess the polarity of nucleic acid grooves. Thus a furan-containing nucleoside, which serves as a minimally invasive thymine analog, was used to estimate the polarity of the major groove of DNA. Surprisingly apolar environment was detected, which reflects lower solvent availability of the major groove than it was usually thought [138].

Extremely low polarity was also detected when 3-hydroxychrimone fluorescent dye was incorporated into double-helical DNA structure as the fluorescent derivative of polyamine spermin. In contrast, on its binding to a single-stranded nucleic acid the spectra demonstrate highly polar hydrated environment [139]. A strong change of emission color demonstrating this effect is not observed with single-stranded DNA. In the latter case, the fluorescence emission spectrum demonstrates an almost complete surrounding of chromophore by water (Fig. 11).

An important question, which could be solved using the fluorescent probing, is the dynamics of hydration water around DNA molecules. The excited states of the natural DNA bases are very short living, which makes them not suitable for probing the full range of hydration dynamics. However, the DNA base mimicking compounds, such as 2-aminopurine possesses much longer living excited states, which allows studying the role of orientational and translational diffusion of water molecules around the DNA bases [140].

It is commonly observed that solvation dynamics captured in the time-resolved Stokes-shift experiments at the interface of DNA is more than an order of magnitude slower than in bulk solution. The interpretation of this phenomenon is still controversial. Previously it was believed that dramatic slowing down of the solvating water diffusion causes this effect. However, there is growing evidence that the DNA

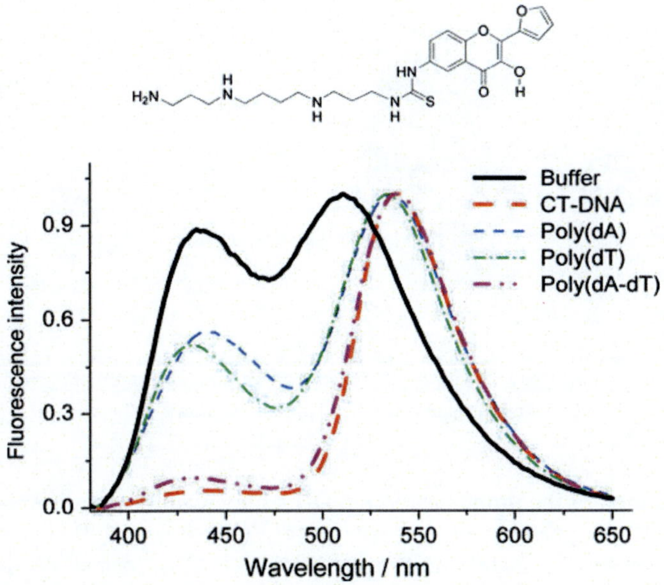

Fig. 11 Spectra of polyamine spermin derivative (*shown at top*) in the buffer solution and in the complex with different DNA strands [139]

itself is responsible for the longest observed relaxation times. The solvation dynamics experiments of the fluorescent probe H33258 bound to DNA followed by MD simulations were performed to clarify this issue [141]. The calculated time scales for the solvation response of H33258 in solution (0.17 and 1.4 ps) and bound to DNA (1.5 and 20 ps) are consistent with experiment (0.2 and 1.2 ps, 1.4 and 19 ps, respectively). Deconvolution of the calculated response in MD simulations revealed that solvating water molecules are relatively mobile, only slowing by a factor of 2–3 in comparison to bulk solution, and it is the DNA motion that was responsible for the long-time component [141, 142].

Another set of extensive time-resolved Stokes-shift experiments complemented by molecular dynamics simulations were performed recently to study the whole range of dynamics in DNA. One of the bases of DNA segment was substituted by the coumarin-102 dye. An extensive methodology was developed to compare experimental results with the data of all-atom MD simulations. An excellent agreement between both techniques was achieved. It was shown that the dynamics spreads over many decades in time with no clear separation into discrete time scales. The simulations confirm the existence of very slow relaxation in DNA extending out to 5 ns and longer. Fluctuations of the hydration shell dominate the dynamics even at the long time scales. The later could be explained by the confinement of water molecules in the grooves and by their electrostatic interaction with the phosphates. In contrast to the water, the DNA and counterion contributions have well-defined relaxation times of 30 and 200 ps respectively [143].

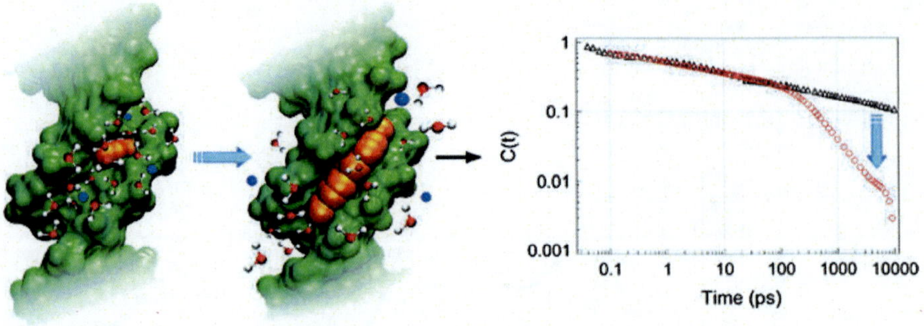

Fig. 12 The schematic representations of the groove-bound (*right*) or the base-stacked (*left*) probes embedded into DNA double helix. Dramatic difference in the slow dynamics in the time span of 100 ps to 10 ns measured by these probes is shown (blue vertical arrow at right). Figure from [144]

The variation of the probe location in DNA may lead to very different results of the Stokes-shift dynamics experiments [144]. The Stokes-shift dynamics of the fluorescent probe 4′,6-diamidino-2-phenylindole (DAPI), located inside the minor groove of DNA, was compared to the results from base-stacked probe that replaces a DNA base pair. It was shown that the dynamics extends over five decades in time from 100 fs to 10 ns in both cases. The dynamics measured with either the groove-bound or the base-stacked probes are similar in the time span of 100 fs to 100 ps but differ substantially from 100 ps to 10 ns. The authors claim that the groove-bound water molecules inside DNA are mainly responsible for the slow dynamics seen in native DNA [144] (Fig. 12).

It is clearly seen that there are two alternative views on the origin of slow dynamic response in DNA. Some authors claim that the slowest dynamics is dominated by the confined water molecules with severely retarded motions, while the other attributes this dynamics to the motions of DNA itself. Further advances in the methodology and the usage of extensive MD simulations may help to resolve this controversy.

8 Conclusions

The request for proper description and understanding, on molecular level, of systems with nanoscale heterogeneity and structural anisotropy is presently strong and it will further increase facing the needs of life sciences and modern technologies. In this study, we tried to analyze the properties of these systems as they are seen with fluorescence probes. In all of them, structural anisotropy results in an appearance of local electric fields that are not space-averaged and may lead to dramatic spectroscopic effects. Dynamics in these systems are modulated by these fields and differ dramatically from isotropic and even anisotropic diffusional motions.

In order to provide proper description of molecular-scale interactions and dynamics, two major trends are presently seen. One is an empirical approach that is based on comparison of output signal with some database composed of the results obtained on well-characterized systems. As such reference databases, the series of absorption and fluorescence spectra obtained in fluid solvents of different polarity and viscosity are frequently used. The scientists must be aware of essential conceptual limitations of this approach. Structural and electrostatic anisotropy together with restricted anisotropic dynamics containing strong ordered dielectric component in the studied media makes such comparison unreliable. Moreover, the output result will inevitably depend on the properties of the probing dye and will not be quantitatively reproduced with the probe change. The only possibility to overcome these limitations is to use as the reference the same system in well-specified experimental conditions and to study the structural and dynamic variations when these conditions change. Such experiments are typical with biological macromolecules and biomembranes.

The other trend is based on physical modeling of intermolecular interactions. In its classical form based on Onsager–Debye model and considering only "universal" interactions [1], it can serve only as a first approximation in our description. Spatial averaging of molecular interactions required in its application does not allow in-depth analysis of the discussed systems. Since the vectorial local electric fields are probably the most important characteristics of the microenvironment, new extensions in this approach are needed. The power in this approach is not only in direct account of the properties of probing dyes but also in the ability to provide description of dynamic properties in terms of the rates of dielectric relaxations.

All the above discussion demonstrates clearly that in inhomogeneous systems with high anisotropy the spectroscopic response depends dramatically on the properties of probing dye. Important is its molecular and electronic structure and the change of this structure on electronic excitation plus location and orientation in the studied system and local interactions that it exhibits. Essential are its ground-state distribution and excited-state dynamics. When fluorescence method is applied, this multicoordinate response is reduced to a very limited number of parameters. Molecular dynamics simulations can be seen as the means to integrate this response being able to observe the analyzed structure with atomic resolution at equilibrium together with its fluctuations around this equilibrium. The results of this integration are expected to be seen in nearest future.

References

1. Demchenko AP, Yesylevskyy SO (2010) Interfacial behavior of fluorescent dyes. Power and weakness of nanoscopic description. In: Demchenko AP (ed) Advanced fluorescence reporters in chemistry and biology III: Applications. Springer Series on Fluorescence 10. Springer, Heidelberg, pp 3–62
2. Ghanadzadeh A, Sariri R, Bahrpima K (2001) Aggregate formation of R6G in anisotropic solvent. Spectrochim Acta 57:155–163

3. Ghanadzadeh A, Zanjanchi M, Tirbandpay R (2002) The role of host environment on the aggregative properties of some ionic dye materials. J Mol Struct 616:167–174
4. Iwanaga H, Naito K, Effenberger F (2000) Oligothiophen dyes for guest- host liquid crystal displays. Liq Cryst 27:115–123
5. Tajalli H, Gilani AG, Zakerhamidi MS, Tajalli P (2008) The photophysical properties of Nile red and Nile blue in ordered anisotropic media. Dyes Pigment 78:15–24
6. Kapko V, Matyushov DV (2006) Dynamical arrest of electron transfer in liquid crystalline solvents. J Phys Chem B 110:13184–13194
7. Heilmeier GH, Castellano JA, Zanoni LA (1969) Guest-host interactions in nematic liquid crystals. Mol Cryst Liq Cryst 8:293–304
8. Araoka F, Shin K-C, Takanishi Y, Ishikawa K, Takezoe H (2003) How doping a cholesteric liquid crystal with polymeric dye improves an order parameter and makes possible low threshold lasing. J Appl Phys 94:279–283
9. Celebre G, Ionescu A (2010) Orientational mechanisms in liquid crystalline systems. 1. A reaction field analytical description of the interaction between the electric quadrupole moment of a probe solute and the electric field gradient of a nematic solvent. J Phys Chem B 114:228–234
10. Celebre G, Ionescu A (2010) Orientational mechanisms in liquid crystalline systems. 2. The contribution to solute ordering from the reaction field interaction between the solute electric quadrupole moment and the solvent electric field gradient. J Phys Chem B 114:235–241
11. Ding L, Kang J, Lü F, Gao L, Yin X, Fang Y (2006) Fluorescence behaviors of 5-dimethylamino-1-naphthalene-sulfonyl-functionalized self-assembled monolayer on glass wafer surface and its sensing properties for nitrobenzene. Thin Solid Films 515:3112–3119
12. Tedeschi C, Möhwald H, Kirstein S (2001) Polarity of layer-by-layer deposited polyelectrolyte films as determined by pyrene fluorescence. JACS 123:954–960
13. Caminati G, Gabrielli G, Ricceri R, Turro C, Turro NJ (1996) Vectorial PET in LB bilayers and in vesicles: nanosecond fluorescence and laser flash-photolysis investigation. Thin Solid Films 284–285:718–722
14. Rajesh K, Rajendra K, Radhakrishnan TP (2010) Fluorescence enhancement in Langmuir-Blodgett films: role of amphiphile structure, orientation, and assembly. J Phys Chem B 114:849–856
15. Dutta AK, Kamada K, Ohta K (1996) Langmuir-Blodgett films of nile red: a steady-state and time-resolved fluorescence study. Chem Phys Lett 258:369–375
16. Alekseeva VI, Marinina LE, Savvina LP, Ibrayev NK, Zikirina AM (2005) Spectral and luminescent properties of Nile Red dye in Langmuir-Blodgett films. Mol Cryst Liq Cryst 427:159–166
17. Eisfeld A, Brigg JS (2006) The J- and H-bands of organic dye aggregates. Chem Phys 324:376–384
18. Losytskyy MY, Yashchuk VM (2010) Fluorescent J-aggregates and their biological applications. In: Demchenko AP (ed) Advanced fluorescence reporters in chemistry and biology. Springer Series on Fluorescence 9. Springer, Heidelberg, pp 135–157
19. Kuroda S (2004) J-aggregation and its characterization in Langmuir-Blodgett films of merocyanine dyes. Adv Colloid Interf Sci 111:181–209
20. Le Breton H, Bennetau B, Dunogues J, Letard J-F, Lapouyade R, Vignau L, Morand J-P (1995) Structural characterization of Langmuir-Blodgett films of a bridged polar Stilbene: 2-(4′-((Perfluorooctyl)-sulfonyl)phenyl)-6-(N, N-dimethylamino)benzofuran. Langmuir 11: 1353–1360
21. Ray K, Nakahara H (2001) Spectroscopic and structural studies of cyanine dyes in functionalized Langmuir–Blodgett films: electron and energy transfer processes. Jpn J Appl Phys 40:5095–5104
22. Hussain SA, Deb S, Bhattacharjee D (2005) Spectroscopic characterizations of non-amphiphilic 2-(4-biphenylyl)-6-phenyl benzoxazole molecules at the air–water interface and in Langmuir–Blodgett films. J Luminesc 114:197–206

23. Terenziani F, Painelli A, Girlando A (2004) From solution to Langmuir–Blodgett films: spectroscopic study of a Zwitterionic dye. J Phys Chem B 108:10743–10750
24. Matsui J, Mitsuishi M, Miyashita T (1999) Characterization of the molecular environment of polymer Langmuir–Blodgett films using a pyrene fluorescent probe. Macromolecules 32:381–386
25. Matsui J, Mitsuishi M, Miyashita T (2001) Molecular orientation and motion of pyrene molecules at the interface of polymer LB films. Stud Surf Sci Catal 132:573–576
26. Dutta AK, Kamada K, Ohta K (1996) Langmuir–Blodgett films of nonamphiphilic N, N′-Bis (2, 5-di-tert-butylphenyl)-3, 4, 9, 10-perylenedicarboximide: a spectroscopic study. Langmuir 12:4158–4164
27. Demchenko AP (2010) Collective effects influencing fluorescence emission. In: Demchenko AP (ed) Advanced fluorescence reporters in chemistry and biology II. Springer Series on Fluorescence 9. Springer, Heidelberg, pp 107–132
28. Chrisstoffels LAJ, Adronov A, Fréchet JMJ (2000) Surface-confined light harvesting, energy transfer, and amplification of fluorescence emission in chromophore-labeled self-assembled monolayers. Angew Chem 39:2163–2167
29. Blinov LM, Palto SP, Udal'yev AA, Yudin SG (1989) Measurements of local fields of molecular dipoles in Langmuir-Blodgett films. Thin Solid Films 179:351–357
30. Klitzing RV (2006) Internal structure of polyelectrolyte multilayer assemblies. Phys Chem Chem Phys 8:5012–5033
31. Mazur M, Blanchard GJ (2005) Probing intermolecular communication with surface-attached pyrene. J Phys Chem B 109:4076–4083
32. Cerdà JJ, Qiao B, Holm C (2009) Modeling strategies for polyelectrolyte multilayers. Eur Phys J 177:129–148
33. Cerdà JJ, Qiao B, Holm C (2009) Understanding polyelectrolyte multilayers: an open challenge for simulations. Soft Matter 5:4412–4425
34. Chen K, Durak C, Heflin JR, Robinson HD (2007) Plasmon-enhanced second-harmonic generation from ionic self-assembled multilayer films. Nano Lett 7:254–258
35. Banerjee D, Das PK, Mondal S, Ghosh S, Bagchi S (1996) Interaction of ketocyanine dyes with cationic, anionic and neutral micelles. J Photochem Photobiol A Chem 98:183–186
36. Satpati AK, Kumbhakar M, Nath S, Pal H (2009) Influence of confined water on the photophysics of dissolved solutes in reverse micelles. ChemPhysChem 10:2966–2978
37. Shirota H, Castner EW (2000) Solvation in highly nonideal solutions: a study of aqueous 1-propanol using the coumarin 153 probe. J Chem Phys 112:2367
38. Tamoto Y, Segawa H, Shirota H (2005) Solvation dynamics in aqueous anionic and cationic micelle solutions: sodium alkyl sulfate and alkyltrimethylammonium bromide. Langmuir 21:3757–3764
39. Novaki LP, El Seoud OA (2000) Microscopic polarities of interfacial regions of aqueous cationic micelles: effects of structures of the solvatochromic probe and the surfactant. Langmuir 16:35–41
40. Cramb DT, Beck SC (2000) Fluorescence quenching mechanisms in micelles: the effect of high quencher concentration. J Photochem Photobiol A Chem 134:87–95
41. Moro F, Goni FM, Urbaneja MA (1993) Fluorescence quenching at interfaces and the permeation of acrylamide and iodide across phospholipid-bilayers. FEBS Lett 330: 129–132
42. Moore SA, Moore SE, Glenn KM, Bhattacharya SC, Palepu RM (2005) Fluorescence quenching of Safranine T by inorganic anions in Tween micelles. Can J Chem Revue Canadienne De Chimie 83:2067–2072
43. Shannigrahi M, Bagchi S (2004) Dual probe solubilisation in two distinct regions of pure and mixed micelles: a pico-second time resolved fluorescence study. Chem Phys Lett 396:367–371
44. Maciejewski A, Kubicki J, Dobek K (2005) Shape and position of 4-aminophthalimide (4-AP) time-resolved emission spectra (TRES) versus sodium dodecyl sulfate SDS concentration in

micellar solutions: the partitioning of 4-AP in the micellar phase and in water surrounding the micelles. J Phys Chem B 109:9422–9431
45. Quintana SS, Moyano F, Falcone RD, Silber JJ, Correa NM (2009) Characterization of multifunctional reverse micelles' interfaces using hemicyanines as molecular probes. II: Effect of the surfactant. J Phys Chem B 113:6718–6724
46. Gratzel M (2007) Photovoltaic and photoelectrochemical conversion of solar energy. Philos Trans A Math Phys Eng Sci 365:993–1005
47. Saroja G, Samanta A (1996) Photophysical studies on a fluorescence probe labelled fatty acid: chain folding in a micellar environment. J Chem Soc Faraday Trans 92:2697–2701
48. Schore NE, Turro NJ (1974) Novel and versatile fluorescence probe for structure of micelles – 11-(3-hexyl-1-indolyl)undecyltrimethylammonium bromide. J Am Chem Soc 96:306–308
49. Schmehl RH, Whitten DG (1980) Intramicellar electron-transfer quenching of excited-states – determination of the binding constant and exchange-rates for dimethylviologen. J Am Chem Soc 102:1938–1941
50. Chattopadhyay A, Mukherjee S, Raghuraman H (2002) Reverse micellar organization and dynamics: a wavelength-selective fluorescence approach. J Phys Chem B 106:13002–13009
51. Blattler T, Huwiler C, Ochsner M, Stadler B, Solak H, Voros J, Grandin HM (2006) Nanopatterns with biological functions. J Nanosci Nanotechnol 6:2237–2264
52. Schore NE, Turro NJ (1975) Novel fluorescent-probe for micellar systems – 1, 3-dialkylindoles. J Am Chem Soc 97:2488–2496
53. Jobe DJ, Verrall RE, Skalski BD (1993) Effects of short-chained alcohols on the intramicellar fluorescence quenching of 2, 3-dimethylnaphthalene by the counterion in mixed micelles of alcohol and decyltrimethylammonium bromide. Langmuir 9:2814–2819
54. Suppan P, Ghoneim N (1997) Solvatochromism. Royal Society of Chemistry, Cambridge, UK
55. Maciejewski A, Kubicki J, Dobek K (2006) Different sources of 4-aminophthalimide solvation dynamics retardation inside micellar systems. J Colloid Interf Sci 295:255–263
56. Chakrabartya D, Chakrabortya A, Setha D, Hazraa P, Sarkar N (2004) Dynamics of solvation and rotational relaxation of Coumarin 153 in 1-butyl-3-methylimidazolium hexafluorophosphate [bmim][PF6]-water mixtures. Chem Phys Lett 397:469–474
57. Liptay W (1969) Electrochromism and solvatochromism. Angew Chem Int Ed 8:177–188
58. DelaCruz JL, Blanchard GJ (2003) Understanding the balance between ionic and dispersion interactions in aqueous micellar media. J Phys Chem B 107:7102–7108
59. Panja S, Chowdlury P, Chakravorti S (2003) Exploring the location and orientation of 4-(N, N-dimethylamino) cinnamaldehyde in anionic, cationic and non-ionic micelles. Chem Phys Lett 368:654–662
60. Chowdhury P, Chakravorti S (2004) Effects of micelles on excited state intramolecular proton transfer activities of 2-hydroxy 1-naphthaldehyde. Chem Phys Lett 395:103–108
61. Prado-Gotor R, Jimenez R, Perez-Tejeda P, Lopez-Lopez M, Sanchez F (2001) Electron transfer reactions in micellar systems: separation of the true (unimolecular) electron transfer rate constant in its components. Chem Phys 263:139–148
62. Fayed TA, Etaiw Sel D (2006) Fluorescence characteristics and photostability of benzoxazole derived donor-acceptor dyes in constrained media. Spectrochim Acta A Mol Biomol Spectrosc 65:366–371
63. Krishnamoorthy G, Dogra SK (1999) Twisted intramolecular charge transfer emission of 2-(4′-N, N-dimethylaminophenyl)benzimidazole in micelles. J Colloid Interf Sci 213:53–61
64. Singh UC, Kollman PA (1986) A combined ab initio quantum mechanical and molecular mechanical method for carrying out simulations on complex molecular systems: applications to the $CH_3Cl + Cl-$ exchange reaction and gas phase protonation of polyethers. Journal of Computational Chemistry 7(6):718–730
65. Dennison SM, Guharay J, Sengupta PK (1999) Intramolecular excited-state proton transfer and charge transfer fluorescence of a 3-hydroxyflavone derivative in micellar media. Spectrochim Acta A Mol Biomol Spectrosc 55:903–909

66. Klymchenko AS, Demchenko AP (2003) Multiparametric probing of intermolecular interactions with fluorescent dye exhibiting excited state intramolecular proton transfer. Phys Chem Chem Phys 5:461–468
67. Klymchenko AS, Demchenko AP (2002) Electrochromic modulation of excited-state intramolecular proton transfer: the new principle in design of fluorescence sensors. J Am Chem Soc 124:12372–12379
68. Ozturk T, Klymchenko AS, Capan A, Oncul S, Cikrikci S, Taskiran S, Tasan B, Kaynak B, Ozbeyd S, Demchenko AP (2007) New 3-hydroxyflavone derivatives for probing hydrophobic sites in microheterogeneous systems. Tetrahedron 63:10290–10299
69. Adhikary R, Barnes CA, Petrich JW (2009) Solvation dynamics of the fluorescent probe PRODAN in heterogeneous environments: contributions from the locally excited and charge-transferred states. J Phys Chem B 113:11999–12004
70. Caarls W, Celej MS, Demchenko AP, Jovin TM (2009) Characterization of coupled ground state and excited state equilibria by fluorescence spectral deconvolution. J Fluoresc 20:181–190
71. Bales BL, Harris FL, Peric M (2009) Electron paramagnetic resonance line shifts and line shape changes due to spin exchange of nitroxide free radicals in liquids. 7. Singly charged surfactant nitroxide. J Phys Chem A 113:9295–9303
72. Das P, Chakrabarty A, Mallick A, Chattopadhyay N (2007) Photophysics of a cationic biological photosensitizer in anionic micellar environments: combined effect of polarity and rigidity. J Phys Chem B 111:11169–11176
73. Maiti NC, Krishna MMG, Britto PJ, Periasamy N (1997) Fluorescence dynamics of dye probes in micelles. J Phys Chem B 101:11051–11060
74. Samanta A (2010) Fluorescence probing of the physicochemical characteristics of the room temperature ionic liquids. In: Demchenko AP (ed) Advanced fluorescence reporters in chemistry and biology III: Applications. Springer Series on Fluorescence 10. Springer, Heidelberg, pp 65–89
75. Mukherjee P, Crank JA, Halder M, Armstrong DW, Petrich JW (2006) Assessing the roles of the constituents of ionic liquids in dynamic solvation: comparison of an ionic liquid in micellar and bulk form. J Phys Chem A 110:10725–10730
76. Maginn EJ (2009) Molecular simulation of ionic liquids: current status and future opportunities. J Phys Condens Matter 21:373101
77. Bhargava BL, Klein ML (2009) Aqueous solutions of imidazolium ionic liquids: molecular dynamics studies. Soft Matter 5:3475–3480
78. Bhargava BL, Klein ML (2009) Formation of micelles in aqueous solutions of a room temperature ionic liquid: a study using coarse grained molecular dynamics. Mol Phys 107: 393–401
79. Kobrak MN (2006) Characterization of the solvation dynamics of an ionic liquid via molecular dynamics simulation. J Chem Phys 125:64502
80. Karukstis KK, Frazier AA, Martula DS, Whiles JA (1996) Characterization of the microenvironments in AOT reverse micelles using multidimensional spectral analysis. J Phys Chem 100:11133–11138
81. Klymchenko AS, Demchenko AP (2002) Probing AOT reverse micelles with two-color fluorescence dyes based on 3-hydroxychromone. Langmuir 18:5637–5639
82. Kelkar DA, Chattopadhyay A (2004) Depth-dependent solvent relaxation in reverse micelles: a fluorescence approach. J Phys Chem B 108:12151–12158
83. Correa NM, Levinger NE (2006) What can you learn from a molecular probe? new insights on the behavior of C343 in homogeneous solutions and AOT reverse micelles. J Phys Chem B 110:13050–13061
84. Kondo M, Heisler IA, Meech SR (2010) Reactive dynamics in micelles: auramine O in solution and adsorbed on regular micelles. J Phys Chem B 114:12859–12865

85. Waluyo I, Nordlund D, Bergmann U, Pettersson LGM, Nilsson A (2009) Increased fraction of weakened hydrogen bonds of water in aerosol OT reverse micelles. J Chem Phys 131: 03110
86. Moilanen DE, Fenn EE, Wong D, Fayer MD (2009) Water dynamics in large and small reverse micelles: from two ensembles to collective behavior. J Chem Phys 131:014704
87. Biswas R, Rohman N, Pradhan T, Buchner R (2008) Intramolecular charge transfer reaction, polarity, and dielectric relaxation in AOT/water/heptane reverse micelles: pool size dependence. J Phys Chem B 112:9379–9388
88. Faeder J, Ladanyi BM (2001) Solvation dynamics in aqueous reverse micelles: a computer simulation study. J Phys Chem B 105:11148–11158
89. Faeder J, Ladanyi BM (2000) Molecular dynamics simulations of the interior of aqueous reverse micelles. J Phys Chem B 104:1033–1046
90. Faeder J, Ladanyi BM (2005) Solvation dynamics in reverse micelles: the role of headgroup-solute interactions. J Phys Chem B 109:6732–6740
91. Pomata MH, Laria D, Skaf MS, Elola MD (2008) Molecular dynamics simulations of AOT-water/formamide reverse micelles: structural and dynamical properties. J Chem Phys 129:244503
92. Epand RM, Kraayenhof R (1999) Fluorescent probes used to monitor membrane interfacial polarity. Chem Phys Lipids 101:57–64
93. Demchenko AP, Yesylevskyy SO (2009) Nanoscopic description of biomembrane electrostatics: results of molecular dynamics simulations and fluorescence probing. Chem Phys Lipids 160:63–84
94. Kusube M, Matsuki H, Kaneshina S (2005) Effect of pressure on the Prodan fluorescence in bilayer membranes of phospholipids with varying acyl chain lengths. Colloids Surf B Biointerf 42:79–88
95. Kozyra KA, Heldt JR, Gondek G, Koviek P, Heldt J (2004) Influence of DPPC liposome concentration on the fluorescence properties of PRODAN and LAURDAN. Z. Naturforsch 59a:809–818
96. Rowe BA, Neal SL (2006) Photokinetic analysis of PRODAN and LAURDAN in large unilamellar vesicles from multivariate frequency-domain fluorescence. J Phys Chem B 110:15021–15028
97. Demchenko AP (2008) Site-selective Red-Edge effects, Chap. 4. Methods Enzymol 450:59–78
98. Sena P, Satohb T, Bhattacharyyaa K, Tominaga K (2005) Excitation wavelength dependence of solvation dynamics of coumarin 480 in a lipid vesicle. Chem Phys Lett 411:339–344
99. Sanchez SA, Tricerri MA, Gunther G, Gratton E (2007) Laurdan generalized polarization: from cuvette to microscope. In: Modern research and educational topics in microscopy. Formatex, Spain
100. Demchenko AP, Mely Y, Duportail G, Klymchenko AS (2009) Biophysical properties of lipid membranes by environment-sensitive fluorescent probes. Biophys J 96:3461–3470
101. Jurkiewicz P, Sykora J, Olzynska A, Humpolickova J, Hof M (2005) Solvent relaxation in phospholipid bilayers: principles and recent applications. J Fluoresc 15:883–894
102. Nandi N, Bhattacharyya K, Bagchi B (2000) Dielectric relaxation and solvation dynamics of water in complex chemical and biological systems. Chem Rev 100:2013–2046
103. Klymchenko AS, Duportail G, Ozturk T, Pivovarenko VG, Mely Y, Demchenko AP (2002) Novel two-band ratiometric fluorescence probes with different location and orientation in phospholipid membranes. Chem Biol 9:1199–1208
104. Gakamsky DM, Nemkovich NA, Rubinov AN (1992) Wavelength-dependent rotation of dye molecules in a polar solution. J Fluoresc 2:81–92
105. Demchenko AP, Shcherbatska NV (1985) Nanosecond dynamics of charged fluorescent probes at the polar interface of a membrane phospholipid bilayer. Biophys Chem 22:131–143
106. Feigenson GW (2007) Phase boundaries and biological membranes. Annu Rev Biophys Biomol Struct 36:63–77

107. Shaw JE, Epand RF, Epand RM, Li ZG, Bittman R, Yip CM (2006) Correlated fluorescence-atomic force microscopy of membrane domains: structure of fluorescence probes determines lipid localization. Biophys J 90:2170–2178
108. Fan J, Sammalkorpi M, Haataja M (2010) Formation and regulation of lipid microdomains in cell membranes: theory, modeling, and speculation. FEBS Lett 584:1678–1684
109. Parasassi T, Di Stefano M, Loiero M, Ravagnan G, Gratton E (1994) Cholesterol modifies water concentration and dynamics in phospholipid bilayers: a fluorescence study using Laurdan probe. Biophys J 66:763–768
110. M'Baye G, Mely Y, Duportail G, Klymchenko AS (2008) Liquid ordered and gel phases of lipid bilayers: fluorescent probes reveal close fluidity but different hydration. Biophys J 95:1217–1225
111. Klymchenko AS, Oncul S, Didier P, Schaub E, Bagatolli L, Duportail G, Mély Y (2009) Visualization of lipid domains in giant unilamellar vesicles using an environment-sensitive membrane probe based on 3-hydroxyflavone. Biochim Biophys Acta 1788:495–499
112. Demchenko AP (1986) Ultraviolet spectroscopy of proteins. Springer, Berlin
113. Callis PR, Liu T (2004) Quantitative prediction of fluorescence quantum yields for tryptophan in proteins. J Phys Chem B 108:4248–4259
114. Callis PR, Vivian JT (2003) Understanding the variable fluorescence quantum yield of tryptophan in proteins using QM-MM simulations. Quenching by charge transfer to the peptide backbone. Chem Phys Lett 369:409–414
115. Ercelen S, Klymchenkoa AS, Demchenko AP (2003) Novel two-color fluorescence probe with extreme specificity to bovine serum albumin. FEBS Lett 538:25–28
116. Demchenko AP (1988) Red-edge-excitation fluorescence spectroscopy of single-tryptophan proteins. Eur Biophys J 16:121–129
117. Demchenko AP, Gryczynski I, Gryczynski Z, Wiczk W, Malak H, Fishman M (1997) Intramolecular dynamics in the environment of the single tryptophan residue in staphylococcal nuclease. Biophys Chem 48:39–48
118. Sharp KA, Honig B (1990) Electrostatic interactions in macromolecules: theory and applications. Annu Rev Biophys Biophys Chem 19:301–332
119. Schutz CN, Warshel A (2001) What are the dielectric "constants" of proteins and how to validate electrostatic models? Proteins Struct Funct Bioinformatics 44:400–417
120. Simonson T (2001) Macromolecular electrostatics: continuum models and their growing pains. Curr Opin Struct Biol 11:243–252
121. Bose S, Adhikary R, Mukherjee P, Song X, Petrich JW (2009) Considerations for the construction of the solvation correlation function and implications for the interpretation of dielectric relaxation in proteins. J Phys Chem B 113:11061–11068
122. Nilsson L, Halle B (2005) Molecular origin of time-dependent fluorescence shifts in proteins. PNAS 102:13867–13872
123. Bhattacharyya K (2008) Nature of biological water: a femtosecond study. Chem Commun (Camb):2848–2857
124. Halle B, Nilsson L (2009) Does the dynamic Stokes shift report on slow protein hydration dynamics? J Phys Chem B 113:8210–8213
125. Li T, Hassanali AA, Kao YT, Zhong D, Singer SJ (2007) Hydration dynamics and time scales of coupled water-protein fluctuations. J Am Chem Soc 129:3376–3382
126. Zhang L, Wang L, Kao Y-T, Qiu W, Yang Y, Okobiah O, Zhong D (2007) Mapping hydration dynamics around a protein surface. PNAS 104:18461–18466
127. Toptygin D, Woolf TB, Brand L (2010) Picosecond protein dynamics: the origin of the time-dependent spectral shift in the fluorescence of the single Trp in the protein GB1. J Phys Chem B 114:11323–11337
128. Abbyad P, Shi X, Childs W, McAnaney TB, Cohen BE, Boxer SG (2007) Measurement of solvation responses at multiple sites in a globular protein. J Phys Chem B 111:8269–8276
129. Golosov AA, Karplus M (2007) Probing polar solvation dynamics in proteins: a molecular dynamics simulation analysis. J Phys Chem B 111:1482–1490

130. Mukherjee P, Halder M, Hargrove MS, Petrich JW (2006) Characterization of the interactions of fluorescent probes with proteins: coumarin 153 and 1, 8-ANS in complex with holo- and apomyoglobin. Photochem Photobiol 82:1586–1590
131. Bose S, Adhikary R, Barnes CA, Fulton DB, Hargrove MS, Song X, Petrich JW (2010) Comparison of the dielectric response obtained from fluorescence upconversion measurements and molecular dynamics simulations for coumarin 153-apomyoglobin complexes and structural analysis of the complexes by NMR and fluorescence methods. DOI: 10.1021/jp1008225
132. Halder M, Mukherjee P, Bose S, Hargrove MS, Song X, Petrich JW (2007) Solvation dynamics in protein environments: comparison of fluorescence upconversion measurements of coumarin 153 in monomeric hemeproteins with molecular dynamics simulations. J Chem Phys 127:055101
133. Demchenko AP (2002) Red-edge effects: thirty years of exploration (review). Luminescence 17:19–42
134. Childs W, Boxer SG (2010) Solvation response along the reaction coordinate in the active site of ketosteroid isomerase. J Am Chem Soc 132:6474–6480
135. Koehorst RB, Spruijt RB, Hemminga MA (2008) Site-directed fluorescence labeling of a membrane protein with BADAN: probing protein topology and local environment. Biophys J 94:3945–3955
136. Koehorst RB, Laptenok S, van Oort B, van Hoek A, Spruijt RB, van Stokkum IH, van Amerongen H, Hemminga MA (2010) Profiling of dynamics in protein-lipid-water systems: a time-resolved fluorescence study of a model membrane protein with the label BADAN at specific membrane depths. Eur Biophys J 39:647–656
137. Wilhelmsson LM (2010) Fluorescent nucleic acid base analogues. Q Rev Biophys 43:159–183
138. Sinkeldam RW, Greco NJ, Tor Y (2008) Polarity of major grooves explored by using an isosteric emissive nucleoside. ChemBioChem 9:706–709
139. Shvadchak VV, Klymchenko AS, de Rocquigny H, Mély Y (2008) Sensing peptide-oligonucleotide interactions by a two-color fluorescence label: application to the HIV-1 nucleocapsid protein. Nucleic Acid Res 37:e25
140. Pal SK, Peon J, Zewail AH (2002) Ultrafast decay and hydration dynamics of DNA bases and mimics. Chem Phys Lett 363:57–63
141. Furse KE, Corcelli SA (2008) The dynamics of water at DNA interfaces: computational studies of Hoechst 33258 bound to DNA. J Am Chem Soc 130:13103–13109
142. Furse KE, Corcelli SA (2010) Molecular dynamics simulations of DNA solvation dynamics. J Phys Chem Lett 1:1813–1820
143. Sen S, Andreatta D, Ponomarev SY, Beveridge DL, Berg MA (2009) Dynamics of water and ions near DNA: comparison of simulation to time-resolved stokes-shift experiments. J Am Chem Soc 131:1724–1735
144. Pal N, Verma SD, Sen S (2010) Probe position dependence of DNA dynamics: comparison of the time-resolved stokes shift of groove-bound to base-stacked probes. JACS 132:9277–9279

Part III
Fluorescence Reporters in Biosensing

Optimized Dyes for Protein and Nucleic Acid Detection

Sergiy M. Yarmoluk, Vladyslava B. Kovalska, and Kateryna D. Volkova

Abstract Fluorescent homogeneous detection is widely used in modern biomedical techniques for analysis and quantification of nucleic acids and proteins. This method is based on the ability of low-fluorescent dye to bind noncovalently with target biomolecule with significant increase of dye's emission intensity. A wide range of probes for homogeneous detection developed during last decades are reviewed here. Series of cyanine dyes were developed for using in visualization of nucleic acids in living cells and detection of amplification products in real-time PCR. Besides, the cyanines, and triphenylmethane dyes that are able to detect certain nucleic acid structures (double stranded, triplex, and quadruplex DNA) and styrylcyanine dyes for two-photon excited fluorescent detection and imaging of DNA are described. Dyes applied for nonspecific proteins detection belong to different classes, among them are complexes of Ru^{2+}, merocyanines, and trimethine cyanines. Moreover, cyanine dyes sensitive to amyloid β-pleated protein formations and albumin-specific squaraine dyes are discussed here.

Keywords Fluorescent detection · Homogeneous probes · Nucleic acids · Proteins

Contents

S.M. Yarmoluk (✉), V.B. Kovalska, and K.D. Volkova
Institute of Molecular Biology and Genetics, National Academy of Sciences of Ukraine, 150 Zabolotnogo St, 03143 Kyiv, Ukraine
e-mail: sergiy@yarmoluk.org.ua

A.P. Demchenko (ed.), *Advanced Fluorescence Reporters in Chemistry and Biology III: Applications in Sensing and Imaging*, Springer Ser Fluoresc (2011) 10: 161–200,
DOI 10.1007/978-3-642-18035-4_5,

1 Introduction

Approach of fluorescent homogeneous detection is widely used in modern biomedical techniques for analysis and quantification of nucleic acids and proteins and could be applied in real-time PCR, gel electrophoresis, capillary electrophoresis, flow cytometry, microscopy, blotting, etc. This method is based on the ability of fluorescent dye to bind noncovalently with target biomolecule with significant changes of dye's spectral properties [1].

The most extensively used homogeneous detection systems are based on the increase of the emission intensity of the dye.

Here, we report the recently developed efficient fluorescent dyes for homogenous detection of nucleic acids and proteins in various applications. A large part of these dyes belongs to the various families of cyanines. A wide range of unsymmetrical mono- and polymethine dyes are efficiently used for postelectrophoretic visualization of nucleic acids, visualization of nucleic acids in living cells, and detection of amplification products in real-time PCR. Symmetric trimethine cyanines containing substituents in polymethine chain also have shown their applicability as nucleic acids sensitive fluorescent probes.

Besides, in this chapter the dyes able to detect certain nucleic acids structures are described. Among them are the dyes with crescent shape of molecules – unsymmetric and symmetric cyanine dyes with long polymethine chain. Such dyes are preferentially groove binders and thus are sensitive to double-stranded AT-rich DNA sequences. The representatives of porphyrins, triphenylmethanes, and cyanines have shown their ability to interact with high selectivity with noncanonical DNA structures – quadruplexes and triplexes. Here also described styrylcyanine dyes which are developed as effective two-photon excited (TPE) probes for detection and imaging of DNA.

Special attention in this chapter is paid to the probes proposed for the mostly used technique in proteomics – nonspecific detection of proteins (usually in the presence of denatured agent sodium dodecyl sulfate). The wide range of highly efficient fluorescent dyes, which belong to different classes, such as metalocomplexes, merocyanines, and polymethine cyanines, developed for this application, is described. Also, cyanine dyes that are able to recognize the particular tertiary structure, amyloid β-pleated proteins formation, are discussed here. The squaraine dyes that demonstrate noticeable specificity to albumin proteins are depicted in the chapter as well.

2 Dyes for Nucleic Acids Detection

2.1 Binding of Ligands to Nucleic Acids

Interaction of small molecules with double-stranded (ds) DNA can occur via three modes (Fig. 1).

First mode of small molecule/DNA complex formation involves external electrostatic binding between them. DNA duplex contains a negatively charged sugar phosphate so that a positively charged ligand binds externally to the double helix. This type of interactions is generally not dependent on DNA sequence.

Another mode describes interactions between ligand and DNA via intercalation. Molecules that are flat, generally aromatic or heteroaromatic ones, bind to DNA by inserting and stacking themselves between the base pairs of the duplex. Intercalation is a noncovalent interaction in which the molecule is held perpendicular to the helix axis. This causes the elongation of DNA duplex.

Groove binding is the third mode. The major and minor grooves are different when it comes to electrostatic potentials, hydrogen bonding, steric effects, and degree of hydration. Minor groove-binding molecules are often crescent shaped and reported to be cationic ligands that interact with DNA by forming hydrogen bonds to the base pairs, which in turn stabilize the complex [2].

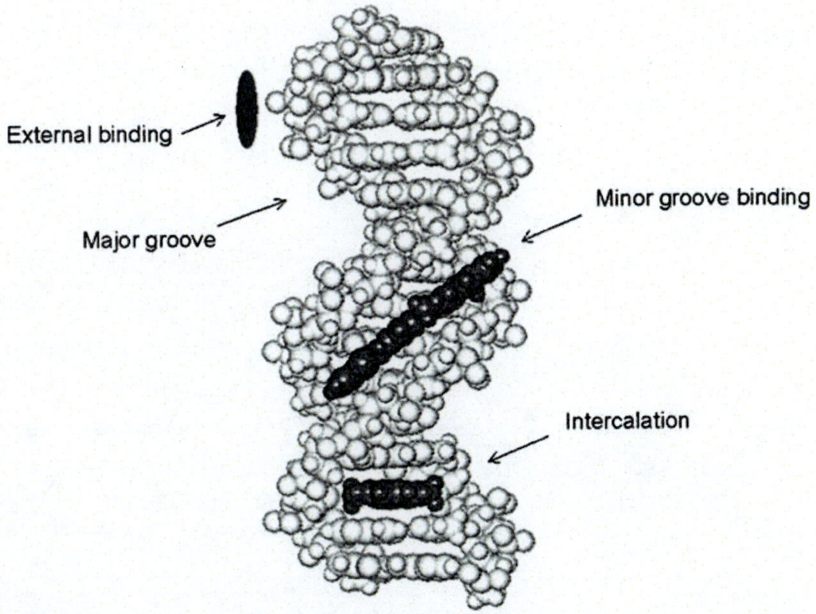

Fig. 1 Binding modes of small molecules to dsDNA

2.2 Cyanine Dyes as Nucleic Acids-Specific Fluorescent Probes

Cyanine dyes are a subclass of polymethines (Fig. 2), which contain odd number of methine groups and nitrogen-based heterocyclic end-groups. Widely used classification of dyes relates to the symmetry of the chromophore. Specifically, symmetrical dyes are composed of identical heterocycles linked at the same position, while unsymmetrical dyes consist either of two different heterocycles or two identical heterocycles linked at different positions. Examples of symmetrical and unsymmetrical cyanine dyes are shown in Fig. 2 [3–5].

Wide range of unsymmetrical polymethine cyanines with spectral maxima covering the visible region were developed as NA (nucleic acids) probes. The monomethine cyanine dyes of SYBR family are known for today as the most sensitive DNA and RNA probes having sensitivity level that is in orders of magnitude higher than that of classical DNA stain Ethidium Bromide. The most sensitive monomethine dye SYBR Gold enables the visualization of about 20 pg of dsDNA in gels and is dozens of times more sensitive than the traditionally used Ethidium Bromide [6].

Due to their unique properties, cyanine dyes are very efficient as probes for nucleic acids detection in homogeneous analysis [7].

Namely, cyanines used for this application have:

- High molar extinction values in visible spectral region (>50,000 M^{-1} cm^{-1})
- Low ($\leq$0.01) quantum yield of free dye (when unbound with NA)
- Sharply increased fluorescence upon binding with nucleic acids (in 2–3 orders)
- Valuable binding constants for their complexes with NA and low affinity to other biopolymers
- Absence of sequence specificity
- Wide linear dynamic detection range
- Modest photostability

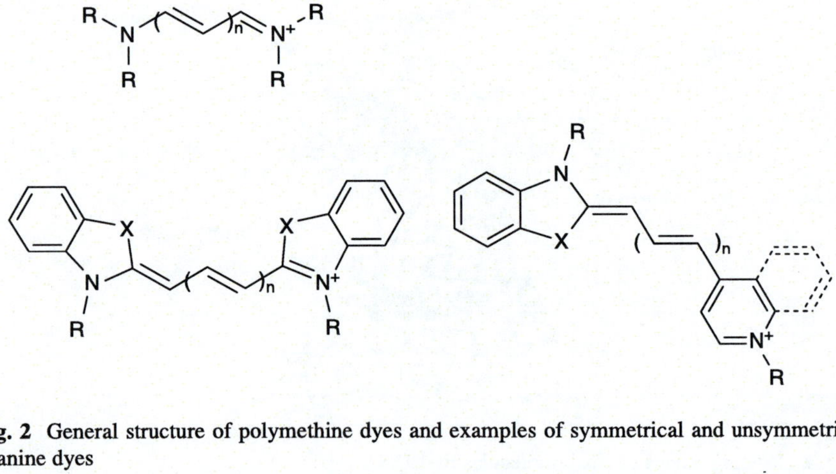

Fig. 2 General structure of polymethine dyes and examples of symmetrical and unsymmetrical cyanine dyes

In unbound state, the cyanine dyes show large amplitude motion around the methine bridge that brings the dye into the nonfluorescent conformation with perpendicular mutual orientation of heterocycles [8, 9], while when bound to DNA they are believed to increase their emission intensity due to the restriction of this internal motion and fixation of the planar conformation of the dye.

2.2.1 Unsymmetrical Cyanine Dyes

Since the 1990s, unsymmetrical cyanine dyes are applied for DNA visualization and quantification in such modern techniques as fluorescent microscopy, gel electrophoresis, and capillary electrophoresis. [9]. Spectral properties of these dyes are superior to other noncyanine dyes, such as Ethidium Bromide, DAPI, and Hoechst.

In 1986, Lee et al. had first shown [10] that monomethine cyanine dye Thiazole Orange (TO) could be applied for fluorescent detection of nucleic acids in solution (Fig. 3).

While the quantum yield of unbound TO is very insignificant (0.0002) upon complex formation with DNA, the quantum yield value of the dye increases thousand times and reaches 0.2. (Fig. 4) [10]. It was shown that the use of TO for DNA visualization on agarose gel allows to achieve sensitivity up to 50 times higher than for classic DNA stain Ethidium Bromide [11].

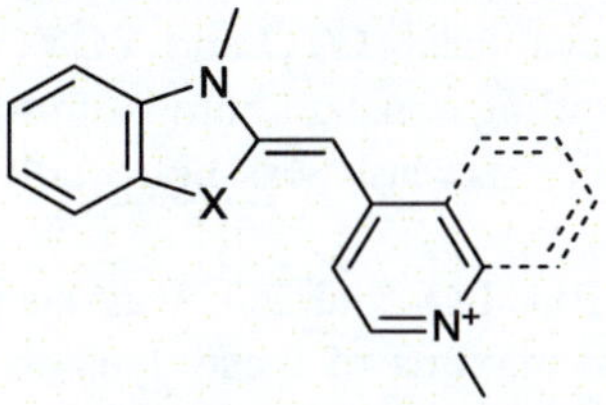

Dye	X	Heterocycle	λ_{ex}, nm
BO	S	pyridine	445
TO	S	quinoline	510
PO	O	pyridine	435
YO	O	quinoline	480

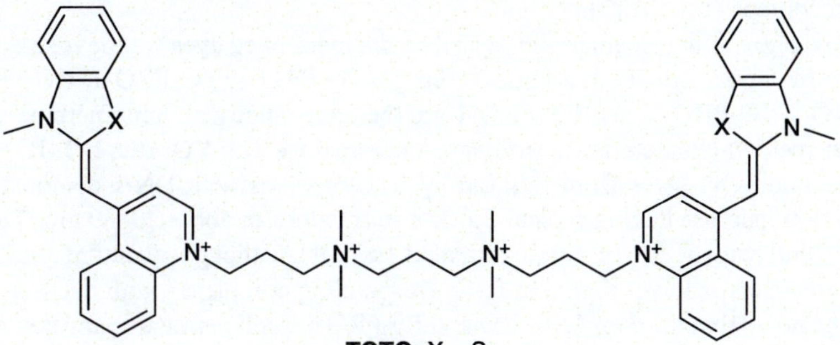

TOTO: X = S;
YOYO: X = O

Fig. 3 Monomeric and homodimeric monomethine cyanine dyes

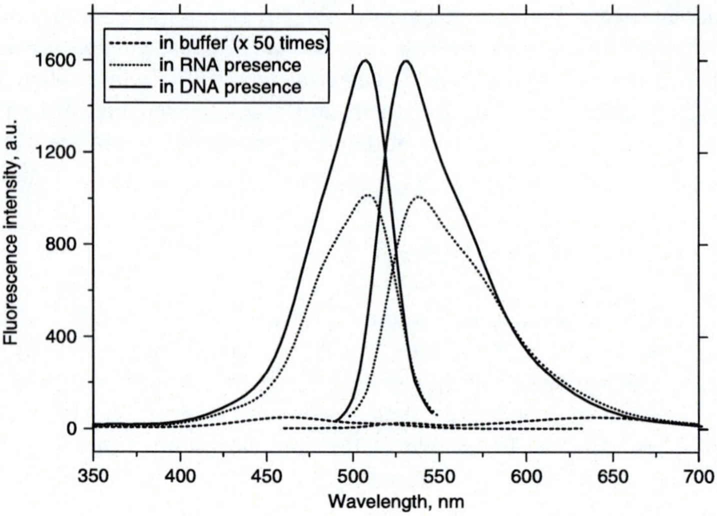

Fig. 4 Fluorescence excitation (*left*) and emission (*right*) spectra of the dye TO (5×10^{-6} M) in unbound state and in the presence of nucleic acids (6×10^{-5} M b.p. of DNA and 1.2×10^{-4} b. of RNA). The low-intensity spectra of free dye are multiplied in 50 times. Fluorescence excitation maxima were situated at 460 nm (free dye), 508 nm (dye + DNA), and 510 nm (dye + RNA)

Benzothiazole analog of TO, the dye Oxazole Yellow (YO), demonstrated similar spectral properties. To enhance the affinity of cyanine dyes to DNA Glazer and coworkers developed the dimeric unsymmetrical cyanines TOTO and YOYO (Fig. 3) [12]. For these dyes, a quadratic increase in binding constant compared to the monomer forms was expected, but observed affinity increase was noticeably lower than expected one.

A series of monomethine dyes, particularly YO, BO, LO, and PO, was also proposed for the detection of nucleic acids. Excitation maxima of these dyes are situated at various spectral regions, thus the dyes could be used with different excitation sources (Fig. 3) [7].

A wide range of fluorescent probes based on the mentioned dyes was developed by Invitrogen Inc. (now Life Technologies) (Fig. 5). TO-PRO-1, YO-PRO-1, BO-PRO-1, LO-PRO-1, JO-PRO-1, and PO-PRO-1 are the unsymmetrical monomethine cyanines with spectral characteristics similar to their analogs TO, YO, BO, LO, JO, and PO correspondingly. Dyes from this family in complexes with DNA demonstrate high emission increase having typical Stock's shift values of about 20–30 nm. These dyes also bind with RNA or single-stranded (ss) DNA, though quantum yield of formed complexes is lower than that for corresponding complexes with dsDNA [7]. It should be noted that the dyes LO-PRO and JO-PRO are cell permeable and could be applied for in vivo studies [13, 14].

Dyes of SYTO family [7] were proposed by Invitrogen Inc. for the visualization of nucleic acids in living cells. SYTO 11, SYTO 13, and SYTO 16 were shown

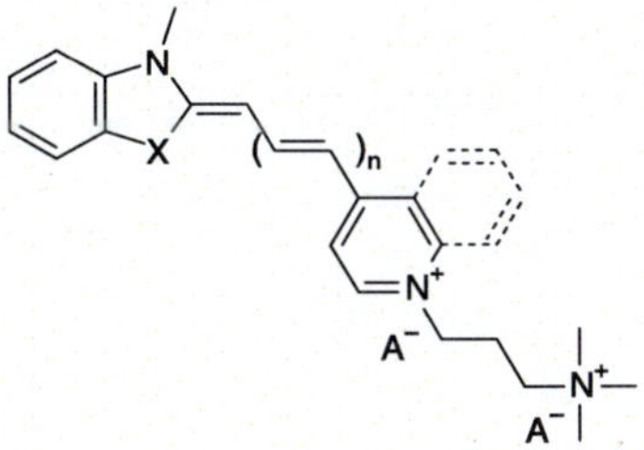

Dye	X	n	Heterocycle	$\lambda_{ex}/\lambda_{em}$, nm
PO-PRO	O	0	pyridine	435/455
BO-PRO	S	0	pyridine	462/481
YO-PRO	O	0	quinoline	491/509
TO-PRO	S	0	quinoline	515/531
PO-PRO-3	O	1	pyridine	539/567
BO-PRO-3	S	1	pyridine	575/599
YO-PRO-3	O	1	quinoline	612/631
TO-PRO-3	S	1	quinoline	642/661

Fig. 5 Mono- and trimethinecyanines developed by Invitrogen Inc. for DNA detection

Fig. 6 SYBR Green I [15]

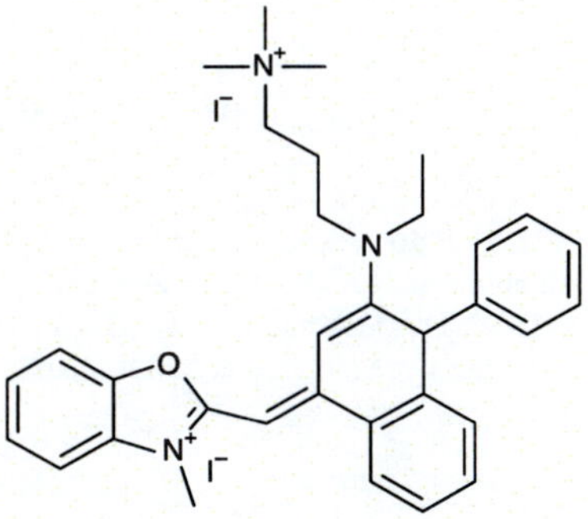

to have high efficiency in this application. Developed monomethine cyanine dye SYBR Green I (Fig. 6) significantly increases its fluorescence intensity in nucleic acids presence (Fig. 7) and allows visualizing about 20 pg of dsDNA per band [3].

SYBR Green I is also widely used for amplification products detection in real-time PCR method [16, 17]. Other SYBR dyes, such as SYBR Gold and SYBR Green II, are used correspondingly for ss/dsDNA and ssDNA/RNA detection. On the basis of cyanine dyes products, PicoGreen, OilGreen, and RiboGreen were developed for sensitive detection and quantification of dsDNA, ssDNA, and RNA in solution [7].

For unsymmetrical cyanine dyes, the binding mechanism may depend on both the dye and DNA nature as well as on the dye/DNA concentrations ratio. Thiazole Orange (TO) and Oxazole Yellow (YO) that were initially proposed as DNA probes appeared to interact with DNA via intercalation [18, 19]. For the SYBR Green I, it was shown that both intercalation and groove-binding modes of interaction with

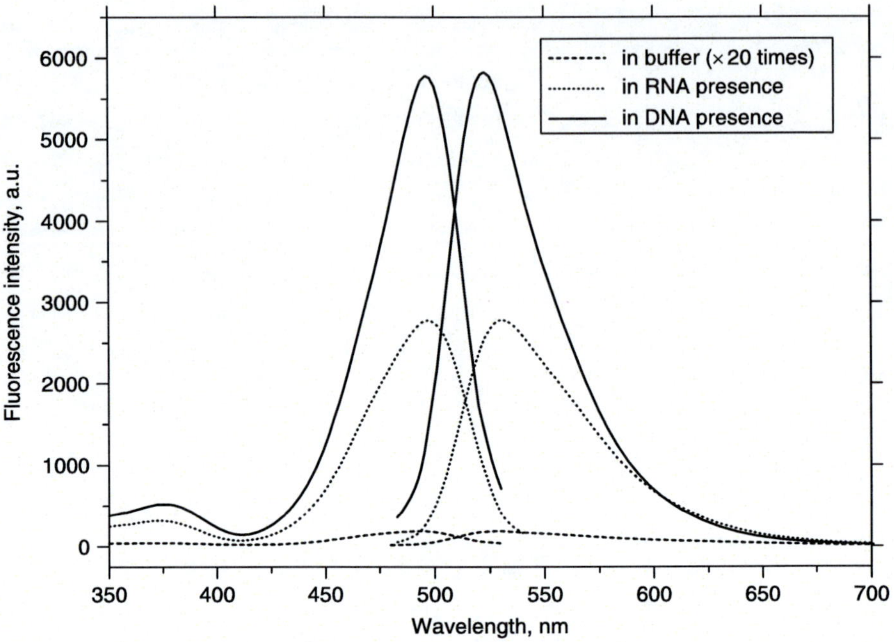

Fig. 7 Fluorescence excitation (*left*) and emission (*right*) spectra of the dye SYBR Green I (5×) in unbound state and in the presence of nucleic acids (6×10^{-5} M b.p. of DNA and 1.2×10^{-4} b. of RNA). The low-intensity spectra of free dye are multiplied 20 times. Fluorescence excitation maxima were situated at 494 nm (free dye), 496 nm (DNA), and 496 nm (RNA)

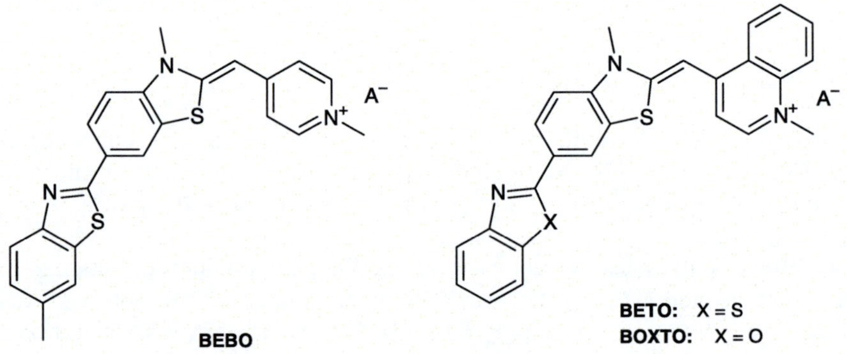

Fig. 8 Unsymmetrical cyanine dyes BEBO, BETO, and BOXTO

DNA could take place. The binding mode could depend on DNA base pairs (b.p.)/ dye ratio, namely for this ratio higher than 6 bp to 1 dye molecule the prevailing mode is intercalation, while at lower ratios the dye molecule binds with the dsDNA minor groove [15].

Unsymmetrical cyanine dye BEBO (Fig. 8), which is considered to be a minor groove binder, was obtained on the basis of the same chromophore as the intercalating dye BO, but was extended with a benzothiazole moiety. This gives BEBO a crescent shape, which is similar to that of other minor groove binders, e.g., dyes of Hoechst family [20, 21]. Later, two new crescent-shaped unsymmetrical cyanine dyes were synthesized – BETO and BOXTO (Fig. 8). The increase in fluorescence intensity upon binding to DNA is 300-fold for BOXTO and 130-fold for BETO.

2.2.2 Symmetrical Cyanine Dyes

Trimethinecyanines

As it was mentioned above, traditionally only unsymmetrical cyanine dyes were proposed for nucleic acids visualization and quantification. As a rule symmetrical benzothiazole or benzoxazole cyanine demonstrate insufficient emission enhancement in the presence of nucleic acids.

It was shown that incorporation of substituent into the polymethine chain leads to the loss of planarity of the dye molecule in an excited state that results in the destruction of the conjugation of the dye π-electron system and, consequently, to the decreasing of dye intrinsic fluorescence [22]. In this case, the fluorescent response of dye to the presence of nucleic acids becomes more pronounced.

Benzothiazole methyl-β-substituted trimethine dye Cyan 2 (Figs. 9 and 10) was the first symmetrical cyanine proposed as probe for DNA detection in gel and solution [23]. In opposite to Cyan 46 (Fig. 9), this dye has low emission in free state and interacts with DNA with hundred times emission increase having quantum yield about 0.33 (Table 1).

It was shown that methyl- and alkyl-meso-substituents were optimal for the designing of DNA-sensitive dyes on the basis of symmetrical trimethine cyanines. At the same time, the dyes containing bulky aryl or benzyl substituents in the β-position in presence of DNA demonstrate rather low emission intensity values [27]. It could be explained by the steric hindrances for DNA/dye complex formation caused by the bulky meso-substituents.

For symmetrical β-methyl-substituted benzoxazole dye Cyan 2-O (Fig. 11), intrinsic emission value and emission in complex with dsDNA is higher as compared to the corresponding values for its benzothiazole analog Cyan 2. For the

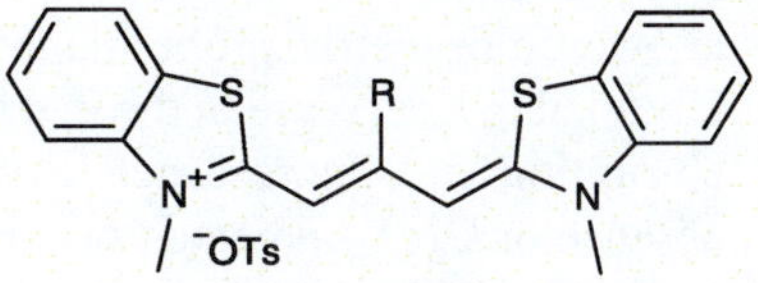

Cyan 46: R = H
Cyan 2: R = Me

Fig. 9 Symmetrical cyanine dyes Cyan 46 and Cyan 2

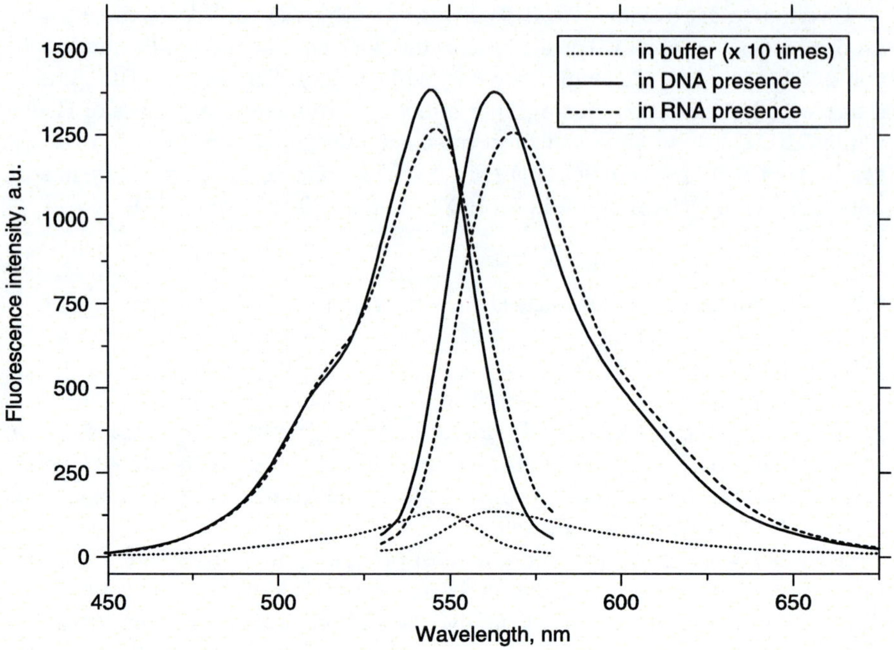

Fig. 10 Fluorescence excitation (*left*) and emission (*right*) spectra of the dye Cyan 2 (5×10^{-6} M) in unbound state and in the presence of nucleic acids (6×10^{-5} M b.p. of DNA and 1.2×10^{-4} b. of RNA). The low-intensity spectra of free dye are multiplied in 20 times. Fluorescence excitation maxima were situated at 535 nm (free dye), 546 nm (DNA), and 535 nm (RNA)

Table 1 Spectral properties of the trimethine cyanine dyes in aqueous buffer and in the presence of DNA and RNA [24–26]

Dye	Free dye in buffer			Dye + DNA		Dye + RNA	
	λ_{ex} (nm)	λ_{em} (nm)	I_0 (a.u.)	I^{DNA} (a.u.)	I^{DNA}/I_0	I^{RNA} (a.u.)	I^{RNA}/I_0
Cyan 2	535	564	13.5	1,380	102	1,268	93.0
Cyan 46	553	571	272.0	297	1.1	590	2.2
Cyan 2-O	485	504	94.0	2,600	27	2,250	23.0
Stains-All	645	656	24.0	19	0.79	112	4.6
CPentV	454	489	29.0	2,958	102	521	18.0
CPentE	590	619	46.0	2,000	43	440	9.5

λ_{ex} (λ_{em}) – wavelength of excitation (emission) maximum; I_0 (I^{DNA}, I^{RNA}) – fluorescence intensity of dye in aqueous buffer in free form (in the presence of DNA, RNA) in arbitrary units (a.u.)

known colorimetric probe, naphthathiazole dye Stains-All (Fig. 11), addition of DNA led to a decrease in the dye fluorescence intensity. It should be mentioned that absorption and fluorescence spectra for Stains-All, both in the presence and in absence of DNA, indicated occurrence of aggregation (Table 1).

For trimethine cyanine α,γ-bridged dyes (Fig. 12), it was found that the nature of the bridge group significantly influenced the spectral–luminescent properties of the dyes, namely, the absorption and fluorescence wavelength values (Table 1).

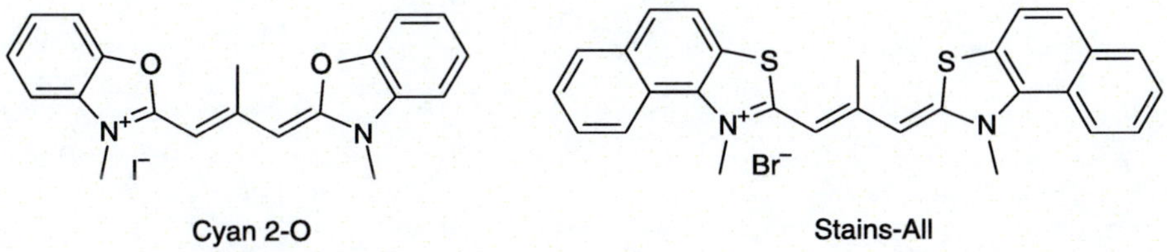

Fig. 11 Trimethine cyanine dyes Cyan 2-O and Stains-All

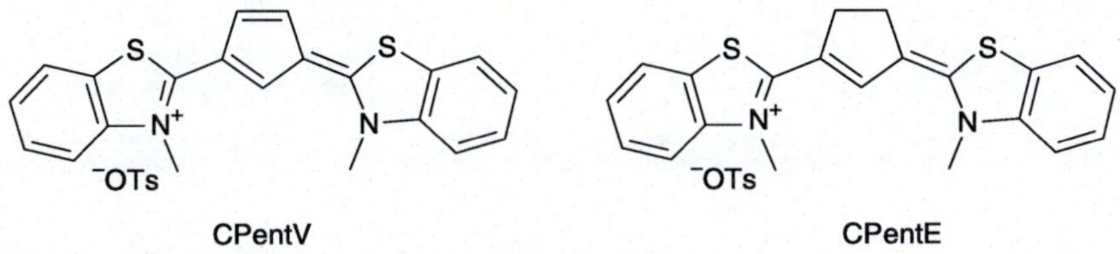

Fig. 12 Trimethine cyanine α,γ-bridged dyes

Fluorescence intensities of both dyes when free are not high. After binding with DNA, however, their emission increased by up to two orders of magnitude with high fluorescence intensity values. It should be noted that for both dyes, the emission intensity in DNA complexes was 3–5 times greater than it was observed for corresponding dye/RNA complexes (Table 1) [24, 28].

For trimethine dyes CPent V and CCyan 2-O, the ability to stain covalently closed and linear double-stranded DNA molecules in agarose gels was shown. Under standard conditions, UV illumination, green filter, and black-and-white photo film, these dyes visualize the bands containing 8 ng of dsDNA [29].

While it was shown that Cyan 2 predominantly interacted with dsDNA via intercalation [30], a groove-binding mechanism of interaction is proposed for both bridged dyes [28]. Existence of both groove-binding and intercalation mechanisms for trimethines has also been suggested [31–33].

Symmetrical Cyanines with Crescent Shape or Long Polymethine Chain as Potential Groove Binders

Designing of dyes that interact with DNA via groove-binding mode is of current importance for such actual and powerful technique as real-time PCR, which requires selective detection of double-stranded regions of DNA. Structurally close to the intercalator Cyan 2, trimethinecyanines with substituents in the positions 6 of benzothiazole moiety (Fig. 13) were studied as probes for NA detection. Such 6,6′-substituents imparted an overall crescent shape to dye molecules, thus possibly creating a preference for groove binding [34].

6,6′-Benzoyl-amino-disubstituted trimethine cyanines T-304, T-306, and T-307 demonstrated sufficient emission intensity and high (up to 200 times) emission

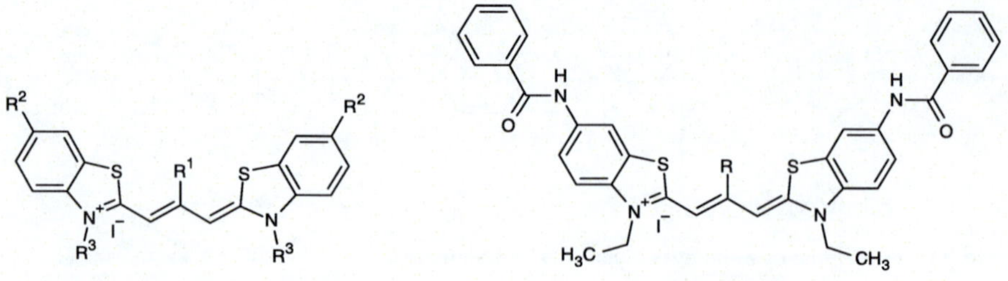

R1 = H; CH_3;
R2 = $NHCH_3$; $N(CH_3)_2$; $N(CH_2CH_3)_2$;
R3 = CH_3; CH_2CH_3

T-304: R = H;
T-306: R = CH_3;
T-305: R = CH_2CH_3;
T-307: R = CH_2OCH_3

Fig. 13 Trimethine cyanine dyes with substituents in the 6 position of the benzothiazole moieties

Table 2 Selected spectral–luminescent characteristics of 6,6′-benzoyl-amino-disubstituted trimethine dyes in aqueous buffer and in the presence of DNA and RNA [34]

Dye	Free dye in buffer			Dye + DNA		Dye + RNA	
	λ_{ex} (nm)	λ_{em} (nm)	I_0 (a.u.)	I^{DNA} (a.u.)	I^{DNA}/I_0	I^{RNA} (a.u.)	I^{RNA}/I_0
T-304	562	715	8.6	610	71	6.4	0.74
T-305	533	629	15.0	88	6	33.0	2.2
T-306	505	655	5.4	1120	207	42.0	7.7
T-307	503	672	4.1	890	222	29.0	7.25
		725	4.0				

λ_{ex} (λ_{em}) – wavelength of excitation (emission) maximum; I_0 (I^{DNA}, I^{RNA}) – fluorescence intensity of dye in aqueous buffer in free form (in the presence of DNA, RNA) in arbitrary units (a.u.)

enhancement in the presence of DNA (Fig. 13, Table 2). Incorporation of benzoyl-amino groups into the 6,6′-positions also permits construction of a dsDNA-sensitive fluorescent probes using a trimethine cyanine with an unsubstituted polymethine chain (T-304). Typically, unsubstituted dyes (e.g., Cyan 46; Fig. 9) have high intrinsic fluorescence intensity and weakly respond on nucleic acids presence. Contrariwise, the 6,6′-disubstituted trimethine cyanine T-304 demonstrates weak emission when unbound and upon interaction with DNA its fluorescence intensity increases up to 70 times.

It was shown that incorporation of methyl-, dimethyl-, diethyl-, and benzyl-amino substituents into the 6,6′-positions created substantial selectivity of the dyes for dsDNA compared to RNA. Such a preference in fluorescence complex formation with dsDNA over the mostly single-stranded RNA could be indicative of the groove-binding mode. Taking into account that Cyan 2 interacts with dsDNA via intercalation, it could be suggested that the incorporation of certain substituents into the 6,6′-positions of the trimethine cyanine dye changes the mode of its binding to dsDNA from intercalation to groove binding.

Pentamethine Cyanine Dyes

At the present time, diagnostic systems based on DNA fluorescent detection mostly involve intercalating monomethine cyanine dyes with excitation wavelengths of about 500 nm. Long-wavelength probes, however, could allow detection using a spectral region where no intrinsic fluorescence of biomolecules occurs. Moreover, long-wavelength dyes could be applied in multicolor detection experiments. Also it is known that groove binding becomes more prevalent for the dyes with more elongated polymethine chain. Pentamethine cyanines have been considered to form groove-binding complexes with DNA [35, 36]. Thus, using these dyes could permit to obtain long-wavelength fluorescent probes for the selective detection of double-stranded regions of DNA.

Pentamethine cyanines nonsubstituted in polymethine chain have rather high level of own fluorescence intensity and slightly change it in nucleic acids presence, the mostly known among them is indolenile dye Cy5, widely used fluorescent label. For dyes with γ substituent in the polymethine chain, the spectral response on DNA presence was also insignificant. Benzothiazole pentamethine cyanines dyes with a cyclohexene or cyclopentene group in the polymethine chain (Fig. 14), which were assumed to be DNA groove binders, were studied as fluorescent probes for detection of nucleic acids. The substituents in the cyclic group of the dye were found to influence significantly on its sensitivity to DNA [37].

These dyes demonstrated insignificant or moderate fluorescence intensity in free state. They form bright complexes with nucleic acids having emission increase up

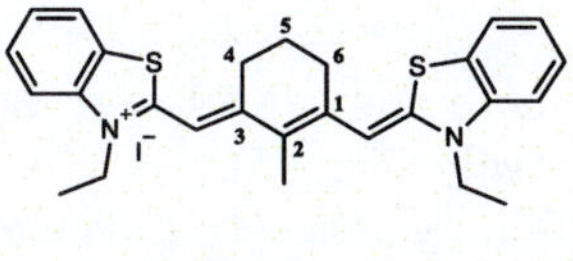

SL-251

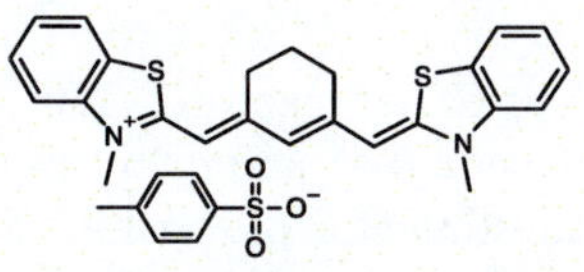

SL-370

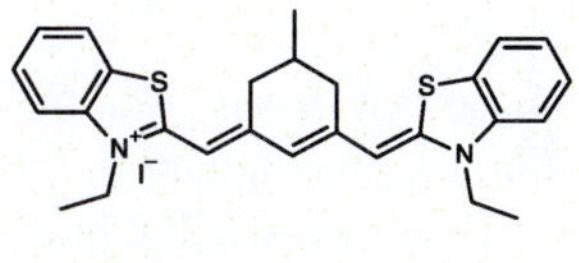

SL-372

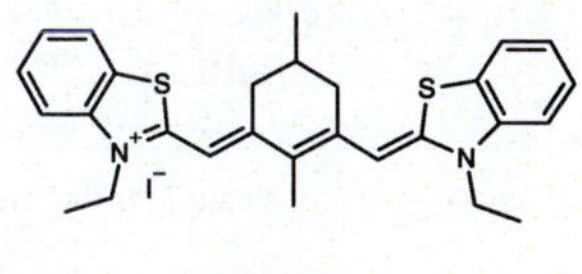

SL-377

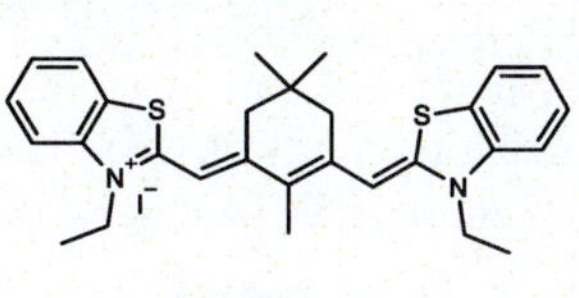

SL-957

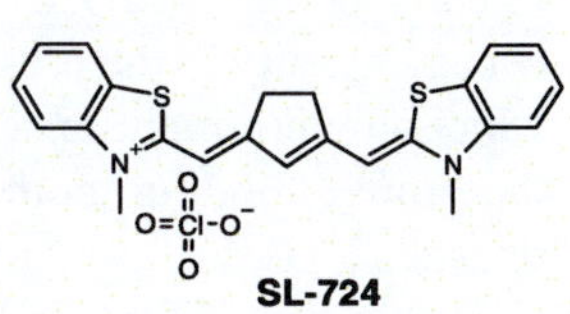

SL-724

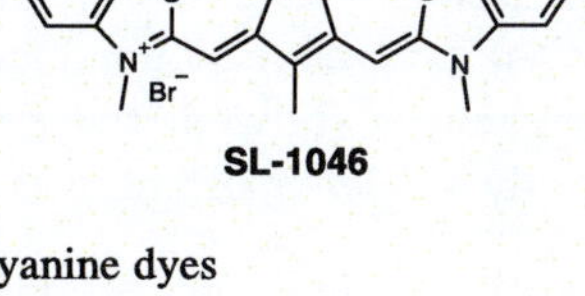

SL-1046

Fig. 14 Bridged pentamethine cyanine dyes

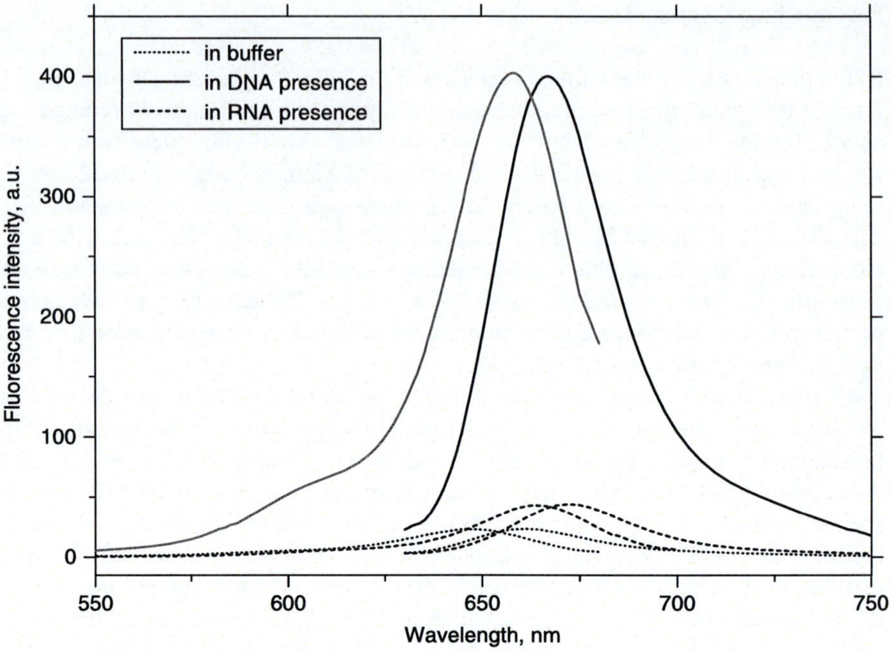

Fig. 15 Fluorescence excitation (*left*) and emission (*right*) spectra of the dye SL-251 (5×10^{-6} M) in aqueous buffer and in the presence of DNA (6×10^{-5} M b.p.) and RNA (1.2×10^{-4} b.). Fluorescence was excited at 620 nm; maxima of fluorescence emission were observed at 660 nm (in the presence of DNA) and at 670 nm (in the presence of RNA). Fluorescence excitation spectra are normalized to the corresponding emission spectra

to dozens times. For the dyes SL-251 and SL-370 in the presence of DNA, the fluorescence quantum yields were 0.41 and 0.66, respectively (Fig. 15).

It was shown that bulky substituent in the position 5 of the cyclohexene fragment hinders the dye/DNA interaction. On the other hand, the presence of methyl substituent in the position 2 of the cyclohexene or cyclopentene group is suggested to enhance the dye/DNA complex stability. Thus, the highest increase of fluorescence intensity in the presence of DNA was observed for dyes SL-251 and SL-1046, which contain mentioned substituents (Fig. 14). These results are consistent with the groove-binding mode of the dye/DNA interaction proposed for pentamethine dyes [37].

It should be noted that bridged pentamethines also give fluorescent response to the RNA presence, which is mostly pronounced for dyes SL-251 (Fig. 15) and SL-377 (emission intensity increased in 13.3 and 19.7 times, respectively) (Table 3).

The study of specificity of bridged pentamethine cyanine dyes for AT-rich and GC-rich nucleotide sequences showed that such dyes possess significant AT-preference that could be the evidence of interaction with DNA via the groove-binding mechanism [37].

Table 3 Selected spectral–luminescent characteristics of pentamethine dyes in aqueous buffer and in the presence of DNA and RNA [37]

Dye	Free dye in buffer			Dye + DNA		Dye + RNA	
	λ_{ex} (nm)	λ_{fl} (nm)	I_0 (a.u.)	I^{DNA} (a.u.)	I^{DNA}/I_0	I^{RNA} (a.u.)	I^{RNA}/I_0
SL-251	643	659	55	1,182	21.5	753	14
SL-1046	600	616	247	7,727	31.3	772	3.1
SL-370	636	654	285	2,455	8.6	818	2.9
SL-377	644	662	36	360	10	705	19.6

λ_{ex} (λ_{em}) – wavelength of excitation (emission) maximum; I_0 (I^{DNA}, I^{RNA}) – fluorescence intensity of dye in aqueous buffer in free form (in the presence of DNA, RNA) in arbitrary units (a.u.)

2.3 *Styrylcyanine Dyes as Two-Photon Excited Fluorescence Probes for DNA Detection and Imaging*

The using of two-photon excitation (TPE) of fluorescent probes in procedures of biological objects detection permits deeper penetration of exciting beam into the tissue; excitation of fluorescence in near infrared spectral region, where the biological objects are transparent; and decreased photodamage of the studied object. Such probes are required for two-photon laser scanning microscopy that is known as one of the most powerful tools for cell and tissue imaging. Monomethine cyanine dye Cyan 40 was used for the two-photon fluorescent visualization of KB (oral epidermoid carcinonoma) cells with fluorescence microscopy method [38].

Representatives of styryl dyes were proposed as the probes for the TPE fluorescent detection of DNA. In opposite to cyanine dyes, styryl dyes are known to have high two-photon absorption cross-section values [39, 40]. This property is demonstrated with the example of benzothiazolium styryl, for which the strong fluorescence upon TPE by 1,064-nm irradiation was observed [41]. On the other hand, significant fluorescent response on the dsDNA presence is observed for these dyes class [42, 43].

Studies of the series of monomer and homodimer benzothiazolium styrylcyanines have shown the efficiency of these dyes as TPE nucleic acids sensitive dyes. It was shown that the dyes modified with spermine-like linkage/tail group demonstrated increased sensitivity to DNA. They have low intrinsic emission and enhance their fluorescence intensity up to three orders of magnitude in the presence of DNA. Complexes of studied dyes with DNA also demonstrate intensive emission upon the single-photon excitation and two-photon excitation [44].

The monomer Bos-3 and homodimer DBos-21 styrylcyanine dyes (Fig. 16) increase their emission intensity by 2–3 orders of magnitude upon interaction with DNA, having rather high values of two-photon absorption (TPA) cross-section, which are comparable with the values of TPA cross-section of the rhodamine dyes [45].

The values of TPA cross-sections at 1,064 nm are close for both dyes in complexes with dsDNA (2.1×10^{-50} cm^4 s for Bos-3 and 1.7×10^{-50} cm^4 s for Dbos-21) (Fig. 17), but TPA cross-sections at 880 nm are considerably different (68×10^{-50} cm^4 s for Bos-3 and 24×10^{-50} cm^4 s for Dbos-21).

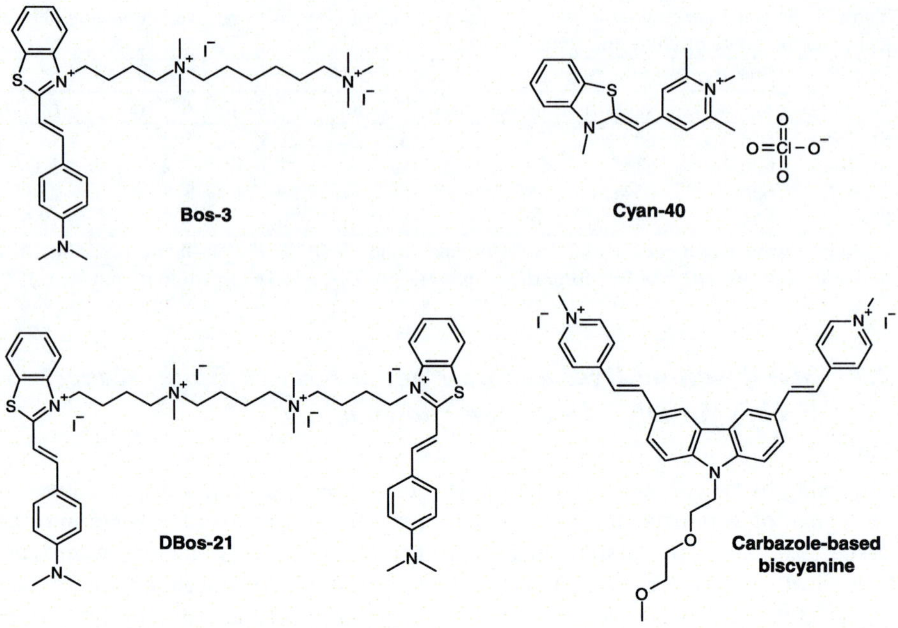

Fig. 16 Two-photon excitable dyes

It was shown that both dyes are cell permeable, but homodimer DBos-21 produces noticeably brighter staining of HeLa cells compared with monomer Bos-3. DBos-21 molecules initially bind to the nucleic acids containing cell organelles (presumable mitochondria) and are able to penetrate into the cell nucleus [45].

Recently, it was shown that carbazole-based biscyanine exhibits high sensitivity and efficiency as a fluorescent light-up probe for dsDNA, which shows selective binding toward the AT-rich regions. The synergetic effect of the bischromophoric skeleton gives a several-fold enhancement in a two-photon absorption cross-section as well as a 25- to 100-fold enhancement in TPE fluorescence upon dsDNA binding [46].

2.4 *Dyes for Detection of Noncanonical DNA Structures*

In living organisms, mostly genomic DNAs exist in a double-stranded helical form, but the occurrences of some other conformations also seems possible. The guanine-rich nucleic acid sequences are known to fold into four-stranded (G-quadruplexes, G4-DNA) or triple-stranded (triplex) structures in which the nucleotide bases are connected by means of Hoogsteen base-pairing bonding. Therefore, the development of a specific probe able to distinguish quadruplex or triplex conformations from a canonical dsDNA may have a significant scientific and practical importance.

Among the dyes proposed for the detection of G4-DNA regions, porphyrin dyes have been found to give noticeable fluorescent response in the presence of

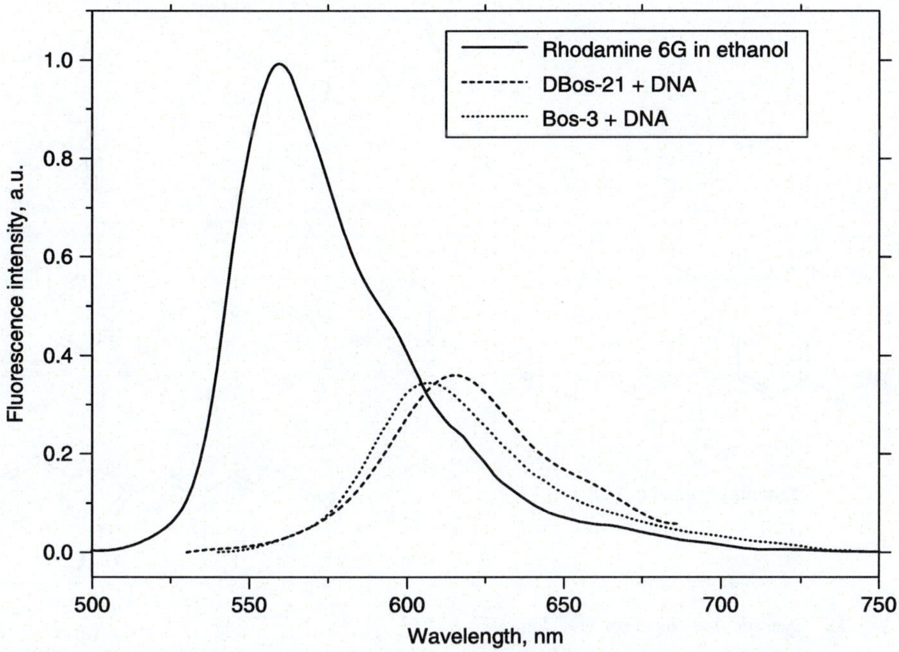

Fig. 17 TPE fluorescence spectra of Bos-3 and DBos-21 in DNA presence in 0.05 M Tris–HCl buffer, pH 8.0. Dye and DNA concentrations were 1.5×10^{-6} and 1.8×10^{-4} M b.p., respectively. Excitation with 1,064 nm from YAG:Nd^{3+} 20 ns pulsed laser. Spectrum of Rhodamine dye in ethanol is presented as a reference that allows to estimate the emission intensity of discussed styryl dyes

quadruplex but not duplex DNA [47–49]. Recently, triphenylmethane dyes were proposed as fluorescent probes for G-quadruplex recognition. These dyes have shown the ability to distinguish intramolecular G4 complex from ss and dsDNA, but they do not discriminate intermolecular G4 from ss or dsDNA [50].

Thiazole Orange (TO) (Fig. 3) was shown to strongly and selectively bind to G-quadruplexes and triplexes [51]. TO was used in the G4-fluorescent intercalator displacement method aimed on evaluating G4-DNA binding affinity and quadruplex-over-duplex DNA selectivity of putative ligands. Also on the base of TO dye the quadruplex-selective fluorescent probe was synthesized by modification of this dye with effective G-quadruplex ligand [52].

Monomethine dye Cyan 40 (Fig. 16) demonstrates strong preference to triplex DNA form over double stranded and quadrulex DNA. Meso-substituted trimethine Cyan 2 (Fig. 9) binds strongly and preferentially to triple- and quadruple-stranded DNA forms, as compared to dsDNA. Highly fluorescent complexes of Cyan 2 with DNA triplexes and G-quadruplexes and Cyan 40 with DNA triplexes are very stable and do not dissociate during gel electrophoresis [53].

The symmetrical benzothiazole trimethine cyanine DTDC and benzoxazole pentamethine cyanine dyes DODC (Fig. 18) were utilized in fluorescence method of

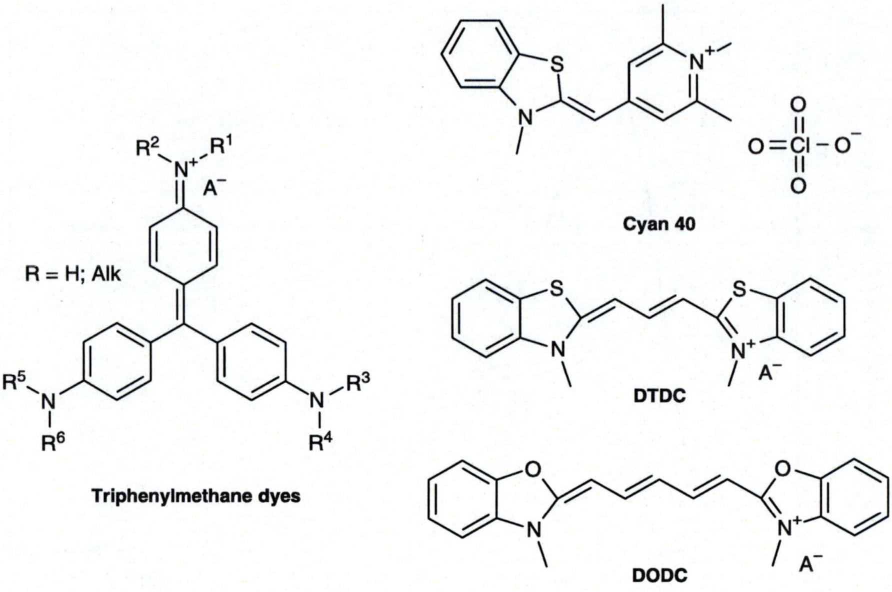

Fig. 18 Cyanine dyes used for G-4 detection

mix-and-measure screening of ligand binding to quadruplex DNAs of diverse sequence and structure. Since fluorescence of the dye molecules increases when the dye binds to quadruplex DNA, the addition of ligands can decrease the dye fluorescence [54].

3 Dyes for Proteins Detection

The field of proteomics requires new, highly quantitative electrophoresis techniques that can interface seamlessly with improved microanalytical methods and that can be performed in an increasingly high-throughput environment. As a consequence, a number of new dyes and staining systems recently have come to market including those discussed below. As proteomic analysis become integral to understanding disease states and normal metabolic pathways, it is increasingly important to design the techniques generating these data.

3.1 Fluorescent Probes for Proteins Detection on Polyacrylamide Gels

For decades, gel electrophoresis has been the most widely used technique for protein analysis. Traditionally, staining methods using silver or Coomassie brilliant blue (CBB) have been used for protein detection. With the rapid growth of proteomics, however, the limitations and experimental disadvantages of these routine detection methods have become obvious [7]. This has attracted the attention of companies providing analytical reagents to the global markets, resulting in production of new reagents with the claim of more efficient detection for protein analysis based on fluorescence. Fluorescent dyes, recently developed for protein visualization and the main areas of their application are described here. Table 4 lists all of the dyes discussed below.

3.1.1 SYPRO Orange, SYPRO Red, and SYPRO Tangerine

The merocyanine dyes SYPRO orange, SYPRO red (Fig. 19), and SYPRO tangerine developed by Invitrogen Inc. (now Life Technologies) as protein stains opened new opportunities for proteomics. These stains allowed a simple, rapid, and sensitive method of proteins detection following SDS or native polyacrylamide gel electrophoresis.

Although the exact mechanism of the interaction between proteins and SYPRO protein stains has not been fully characterized, the interaction is dependent upon an

Table 4 Spectral characteristics and considered class of the discussed dyes for denatured proteins detection

Stain	λ_{ex} (nm)	λ_{em} (nm)	Class
SYPRO red	300/550	630	Merocyanine
SYPRO orange	300/470	570	Merocyanine
SYPRO tangerine	300/490	640	Merocyanine
SYPRO ruby	280/450	610	Ru^{2+} complex
Deep purple stain	528	594	Sulfoindocarbocyanine
Flamingo	512	535	Unknown
Coomassie Fluor Orange	300/470	570	Merocyanine
Lucy 506	504	515	Carbocyanine
Lucy 565	565	584	Carbocyanine
Lucy 569	565	581	Carbocyanine

λ_{ex} (λ_{em}) – wavelength of excitation (emission) maximum of the dyes in presence of proteins

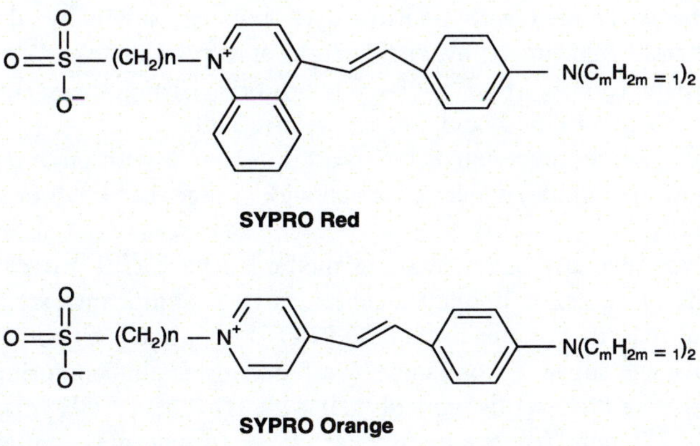

Fig. 19 Supposed structures of the SYPRO red and SYPRO Orange merocyanine dyes [55]

initial binding of SDS to the proteins. The dyes appear to bind to the SDS coat that surrounds proteins in SDS-polyacrylamide gels. The staining results observed with these dyes consequently exhibit relatively little protein-to-protein variation and are linearly related to protein mass.

It is claimed that the staining procedure takes from 30 to 60 min and does not involve destaining step, and that the staining is stable for many hours [7]. Staining also can be accomplished by including dye in the running buffer with a brief one-step destaining procedure following electrophoresis. A partially destained gel may be restained by addition of corresponding fresh SYPRO protein stain.

Using this group of SYPRO dyes allows detecting as little as 4–10 ng, equaling the sensitivity of silver staining techniques on commonly used configuration and surpassing the best colloidal CBB staining methods available. For cytochrome c oxidase subunits I and Va, mouse IgG light chain, protein G, NeutraLite avidin, streptavidin, pancreatic RNase A, β-bungarotoxin small subunit, and the histones staining with SYPRO Orange and SYPRO Red protein gel stains allowed detection of lower amounts of mentioned polypeptides than was possible with silver staining [56].

Because the SYPRO Orange and SYPRO Red dyes do not covalently bind to proteins, the stained proteins can be subsequently analyzed by microsequencing [57] or mass spectrometry [58, 59]. However, these dyes are not recommended for staining the gels prior to blotting, as there is a significant loss of sensitivity when proteins are stained with the SYPRO Orange or SYPRO Red dyes in typical Western blotting buffers.

SYPRO Tangerine gel stain is an extremely versatile stain for proteins in SDS gels. Because proteins stained with SYPRO Tangerine dye are not fixed, they can be used for zymography (in-gel enzyme activity) assays. Stained proteins can also be eluted from gels and used for further analysis. This stain does not alter protein structure and does not interfere with mass spectrometry. In addition, in contrast to SYPRO Red and SYPRO Orange, staining with SYPRO Tangerine does not interfere with the transfer of proteins to membranes, allowing visualization before proceeding with Western blotting [7].

Protein bands stained with SYPRO dyes can be excited by ultraviolet light at approximately 300 nm or by visible light at approximately 470 nm (SYPRO Orange), 550 nm (SYPRO Red), and 490 nm (SYPRO Tangerine). These three dyes emit maximally at 570, 630, and 640 nm, respectively.

In addition, fluorescence intensity is said to be linear with protein quantity over three orders of magnitude, which is a much wider range than CBB or silver staining can provide [7].

By contrast to many silver staining methods, the SYPRO dyes do not stain nucleic acids or bacterial lipopolysaccharides to a significant extent [60]. There are some limitations for the use of SYPRO dyes, such as poor sensitivity for detecting proteins separated on isoelectric focusing or nondenaturing gels. Moreover, although SYPRO stains worked well with 1D SDS-PAGE gels, their performance with 2D gels (especially thicker, large-format gels) failed to achieve sensitivity levels seen in 1D SDS-PAGE separations.

3.1.2 SYPRO Ruby

SYPRO Ruby was presented by Invitrogen Inc. as the new fluorescent stain of choice for 2D PAGE separations [7, 58] (Fig. 20). This dye is a luminescent metal chelate of a ruthenium ion with organic ligands that allows sensitive fluorescent detection of proteins separated by SDS-polyacrylamide and 2D gels. This stain provides improved sensitivity and is less affected by the presence of nonprotein components present in the gel. The SYPRO Ruby dye-based staining method is also compatible with downstream applications such as Edman-based protein sequencing and peptide mass profiling by mass spectrometry-based methods.

The protein binding mechanism is quite different from that of SYPRO Orange, Red, and Tangerine dyes in that it does not bind through intercalation into sodium dodecyl sulfate (SDS) micelles [61–64]. In such a way, SYPRO Ruby binds to proteins by electrostatic interaction of its sulfonate groups with basic amino acid residues on the proteins and coordination binding of the ruthenium ion with the polypeptide backbone [65].

Due to the simple staining procedure, this dye is ideal for high-throughput gel staining and large-scale proteomics applications. Because the dye is not present during electrophoresis, aberrant migration of proteins is avoided. The sensitivity (down to 1 ng of protein) exhibited by SYPRO Ruby is as good as or better than those of the best silver staining techniques and superior to them in terms of ease of use, linear dynamic range, and compatibility with downstream microchemical characterization techniques [66–68]. SYPRO Ruby has insignificant protein-to-protein variability and stains most classes of proteins including glycoproteins, phosphoproteins, lipoproteins, calcium binding proteins, fibrillar proteins, and other proteins that are difficult to stain by traditional methods. Moreover, heavily glycosylated proteins, which stain poorly with silver techniques, are detected

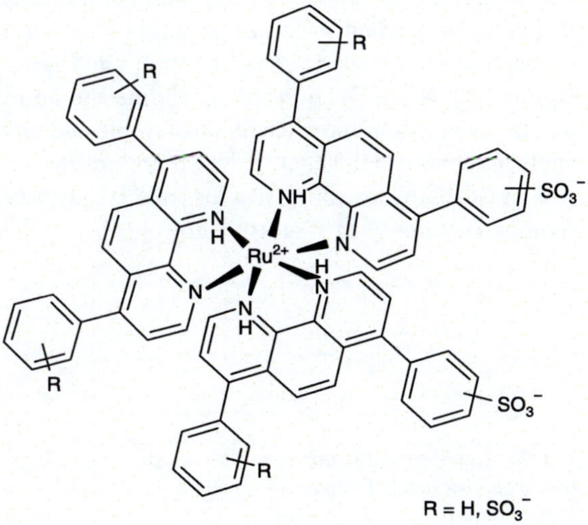

Fig. 20 Supposed structure of the SYPRO Ruby fluorescent stain [55]

successfully by SYPRO Ruby. Unlike silver staining, SYPRO Ruby does not stain extraneous nucleic acids, lipids, or carbohydrates in the sample [69, 70].

Because SYPRO Ruby contains the transition metal ruthenium as its fluorophore, it has exceptional photostability, allowing long exposure times for maximum sensitivity, and can be imaged on UV as well as laser-based gel scanners [71]. The dye is compatible with a wide range of light sources including a 302-nm UV transilluminator, 473-nm SHG Nd-YAG laser, 488-nm argon-ion laser, 532-nm SHG Nd-YAG laser, xenon arc lamp, blue fluorescent light bulb, and blue LED. The emission maximum for this dye is 610 nm.

Although SYPRO Ruby is more sensitive than SYPRO Orange, Red, and Tangerine, its staining is slower, requiring about 4 h to give optimal results.

3.1.3 Deep Purple Dye

This compound was isolated from the fungus *Epicoccum nigrum* by Bell and Karuso [72] and now proposed as a fluorescent stain for visualization of proteins in 1D and 2D polyacrylamide gels (Fig. 21). Formerly named Lightning Fast, it is now distributed by GE Healthcare under the trade name Deep Purple Total Protein Stain.

The reactive component of the Deep Purple stain is epicocconone, a nonfluorescent azophilone that becomes fluorescent when it interacts with protein [73]. The primary mechanism of epicocconone binding is thought to be through hydrophobic interaction with the lipophilic tail of SDS bound to protein [74]. The stain employs fixation with a low concentration of aqueous alcohol to remove SDS from the gel background, and the dye binds to both lysine residues and residual SDS present in the gel and bound to the protein [75, 76].

The dye appears to give low background fluorescence and have a large dynamic range (about four orders of magnitude) and a sensitivity even higher than SYPRO Ruby, at least when it comes to staining proteins electroblotted to nitrocellulose (Amersham's Hybond ECL) or polyvinylidene fluoride (PVDF, Amersham's Hybond-P). Also, Deep Purple is claimed to be less expensive than SYPRO Ruby and to give clearer, more easily discernible and more accurately quantifiable protein spots and bands. However, it was reported that its photostability is lower than that of SYPRO Ruby, resulting in a loss of 83% after 19 min under UV transillumination (compared to 44% of SYPRO Ruby) [76].

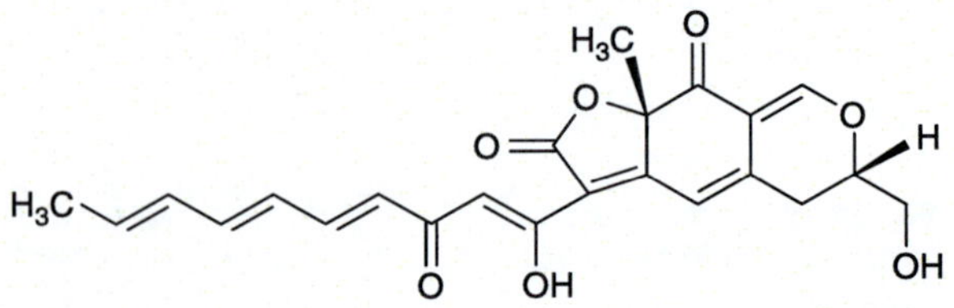

Fig. 21 Supposed structure of the epicocconone Deep Purple stain [55]

Using Deep Purple stain, less than 0.5-ng protein per band can be detected in 1D gels and less than 1-ng protein per band when blotted onto nitrocellulose or PVDF membranes. Deep Purple™ Total Protein Stain is used to stain both 1D and 2D gels in both backed and unbacked format. The stain can be used in every step of protein analysis from sample preparation to mass spectrometry, and it is compatible with MALDI-TOF MS and Edman sequencing (http://www.gelifesciences.com/aptrix/upp01077.nsf/content/2d_electrophoresis~2delectrophoresis_detection_reagents~staining_reagents~deep_purple?OpenDocument&parentid=25900083&moduleid=165652).

Deep Purple can be detected using a wide variety of imaging devices including industry standard instrumentation such as the Typhoon fluorescent scanner, CCD cameras, UV transilluminators, and some light boxes. The dye can be excited using standard laser sources, i.e., violet (395 nm), blue (488 nm), and green (532 nm), and it possesses red fluorescence with an emission maximum at 605 nm. The dye gives poor staining of the glycoprotein, glucose oxidase, and the phosphoprotein, pepsin. Deep Purple tends to have lower sensitivity on 1D and 2D Bis–Tris gels compared to Tris– glycine gels (http://www.invitrogen.com/downloads/Comparative_Perf_-Fluoresc_Protein_Stains.pdf) [77].

3.1.4 Flamingo

Flamingo is a fluorescent dye from Bio-Rad Laboratories Inc. It undergoes dramatic fluorescence enhancement in the presence of denatured proteins, thus allowing sensitive visualization of proteins that have been separated by 1D or 2D SDS-PAGE. Flamingo exhibits insignificant protein-to-protein variability, making results more consistent and reproducible compared to some other staining methods. Gels stained with Flamingo have low background, making results easier to see [78]. Besides, Flamingo staining is reported to be fully compatible with peptide mass fingerprinting by MALDI-MS and to result in a lower incidence of oxidative protein modification than SYPRO Ruby [79].

Gels stained with Flamingo may be assessed using a variety of fluorescence imaging systems. In the presence of proteins, Flamingo has an excitation maximum at 512 nm and an emission maximum at 535 nm. Thus, the optimal imaging systems are laser-based fluorescence scanners capable of exciting and detecting at 510 and 535 nm, respectively. The manufacturer claims that, when using 532-nm laser excitation, the in-gel protein detection is possible down to 0.25 ng. Because this dye has a secondary excitation at 271 nm, it can be excited also with UV light; however, the detection limit is then two times higher (0.5 ng).

It has been claimed that Flamingo visualized 5–10 ng of protein on Tris–glycine and Bis–Tris 1D gels using an excitation wavelength of 488 nm with a 555-nm LP emission filter. The Flamingo stain was said to show higher sensitivity in 1D Bis–Tris gels compared to Tris–glycine gels, while the opposite was the case for 2D gels, where Flamingo was more sensitive with Tris–glycine gels.

When SYPRO Ruby, Deep Purple, and Flamingo stains were compared directly using the same protein sample and the same emission filter set, it was found that with UV excitation, SYPRO Ruby was nine times brighter than Deep Purple and 83 times brighter than Flamingo; with 488-nm excitation, SYPRO Ruby was four times brighter than Deep Purple and eight times brighter than Flamingo; and with 532-nm excitation, Deep Purple was 1.3 times brighter than SYPRO Ruby and three times brighter than Flamingo.

Fluorescence enhancement of Flamingo is also related linearly to protein concentration and allows linear quantitation over four orders of magnitude. A disadvantage of Flamingo is the time-consuming protocol. It requires a 2-h fixation step and 3 h for staining. This dye is also less sensitive with 2D gels compared to the SYPRO Ruby and Deep Purple dyes.

3.1.5 Coomassie Fluor Orange

Invitrogen Inc. (now Life Technologies) proprietary Coomassie Fluor Orange protein gel stain provides fast, simple, sensitive staining of proteins in electrophoretic gels and allows detection of as little as 8 ng of protein per minigel band, which is comparable to Coomassie® Brilliant Blue (CBB) stains. Among the advantages of Coomassie Fluor Orange over conventional colorimetric stains the easy-to-use staining procedure should be admitted. Using this dye, the fixation and staining procedure could be performed in one step with no prewashing or destaining steps, and staining is complete in as little as 30 min. Stained proteins can be visualized using a standard 300-nm UV transilluminator or a laser-based scanner.

Coomassie Fluore Orange is reported to be chemically similar to merocyanine fluorescent dyes SYPRO Red and SYPRO Orange.

3.1.6 Lucy Dyes

Three fluorescent protein gel stains (Lucy 506, Lucy 569, and Lucy 565) for nonspecific visualization of proteins have been developed by OTAVA lab in cooperation with Sigma-Aldrich [80] (Fig. 22).

The main advantages of these stains are their high sensitivity, low protein-to-protein variability, wide concentration range, and simple and rapid staining protocols.

The Lucy stains are the carbocyanine dyes. They bind to proteins via hydrophobic interactions with the lipophilic tail of SDS molecules bound to proteins. These dyes have high molar extinction coefficients and high fluorescence quantum yields in the presence of protein/SDS mixture, and fluorescence intensity increase by two orders of magnitude upon addition to protein/SDS system. The

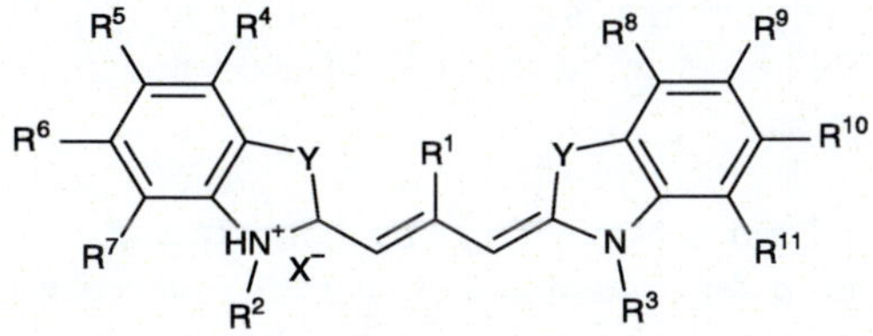

Fig. 22 General structure of Lucy dyes

excitation maxima of Lucy 506, Lucy 565, and Lucy 569 are situated at 506, 565, and 569 nm [81].

According to literature data, each Lucy dye has its own benefits. In such a way, Lucy 565 allows neutral gel staining (e.g., before Western blotting), Lucy 569 excels by a linear response over an extraordinary broad linear dynamic range, and Lucy 506 shows highest sensitivity – using this stain under the blue light excitation, it is possible to visualize the band containing 2 ng of BSA.

The dye Lucy 565 demonstrates the same sensitivity with UV as well as with blue light excitation source and allows detection of about 8–10 ng of protein per minigel band. The Lucy 569 dye can be successfully applied with standard 313-nm UV transilluminator and visualized 5–10 ng of protein per band.

All Lucy dyes showed poor staining of the α1 acid glycoprotein and reduced staining of glucose oxidase. This may be explained by the SDS-meditated mechanism by which these dyes bind to protein. While the Lucy dyes are less sensitive than ruthenium-containing SYPRO Ruby on 1D gels and tend to have lower sensitivity with Bis–Tris gels, they exhibit the same sensitivity as the widely used merocyanine dyes SYPRO Orange and SYPRO Tangerine, which bind with protein by a mechanism similar to that of Lucy dyes.

3.2 *Fluorescent Probes for Proteins In-Solution Detection*

The traditional methods of proteins detection and quantification in solution are Lowry [82] and Bradford methods [83]. However, these methods have some limitations, namely narrow linear range or slow reaction time, which restrict their practical applications. Up-to-date fluorometric methods are widely used for investigation and detection of proteins having the advantage over other methods for their high sensitivity, selectivity, and convenience [84].

3.2.1 NanoOrange

The NanoOrange product of Invitrogen Inc. provides an ultrasensitive assay for measuring the concentration of proteins in solution and is ideal for quantitating protein samples before gel electrophoresis and western blot analysis [7].

NanoOrange is a merocyanine dye [85, 86] that produces a large increase in fluorescence quantum yield upon noncovalent interaction with lipid-coated proteins.

SYPRO dyes (SYPRO Red, SYPRO Orange, and SYPRO Tangerine) are closely related to NanoOrange by their structure. The latter group of dyes is designed to be used for protein detection in the presence of SDS [87, 88], while SDS is not a component of the commercially available NanoOrange Protein Quantification Kit. According to the technical note [86], the bound dye is efficiently excited at 488 nm and possesses a large Stokes' shift with maximum emission at about 580 nm. Therefore, the NanoOrange assay is well suited for use with standard fluorescence microplate readers, fluorometers, and some laser scanners [85, 89]. The dye allowed the detection of 10 ng/mL to 10 μg/mL of protein with a standard fluorometer, offering a broad, dynamic quantitation range and improved sensitivity relative to absorption-based protein solution assays [90].

The protein-to-protein variability of the NanoOrange assay was comparable to those of standard assays, including Lowry, bicinchoninic acid, and Bradford procedures. It was also shown that the NanoOrange assay is useful for detecting relatively small proteins or large peptides, such as aprotinin and insulin.

Despite of the insensitiveness of NanoOrange to the presence of reducing agents, nucleic acids, and free amino acids, using this assay was somewhat sensitive to the presence of several common contaminants found in protein preparations such as salts and detergents.

3.3 Squaraine Dyes as Albumin-Sensitive Fluorescent Probes

Up-to-date fluorometric methods are proposed for detection and studies of albumins having the advantage over other methods for their high sensitivity, selectivity, and convenience [84, 91]. The quantitative determination of albumins in biological fluids is very important for clinical diagnosis.

The interest in squaraine compounds has been recently renewed due to their potential usefulness in a large number of technologically relevant fields such as NIR-emitting fluorescent probes. In such a way, squaraines were reported as efficient noncovalent labels for albumins [92], exhibiting high quantum yields when bound to these proteins [93].

Recently, benzothiazole and benzoselenazole squaraines were proposed as fluorescent probes for albumins detection, demonstrating bright fluorescence in the presence of both human serum albumin (HSA) and bovine serum albumin (BSA). For squaraine dyes with *N*-hexyl pendent groups (Fig. 23), about a 100- to 540-fold fluorescence intensity increase upon albumins addition was observed (Table 5, Fig. 24). It was shown that generally, squaraines with long *N*-hexyl pendent groups demonstrated higher emission increase values upon proteins addition compared with their analogs with short *N*-ethyl tails. Benzothiazole

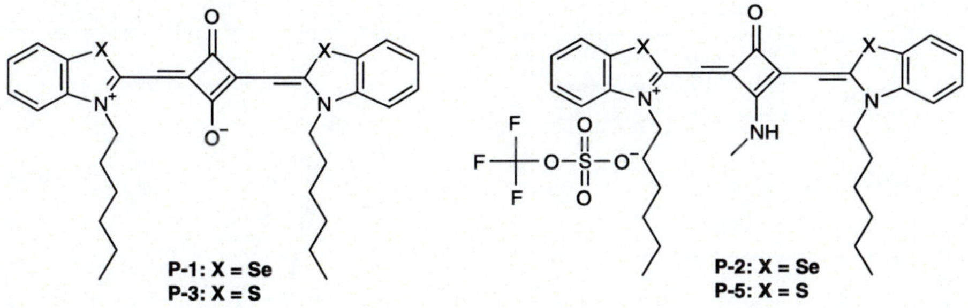

Fig. 23 Albumin-sensitive squaraine dyes containing *N*-hexyl groups

Table 5 Selected spectral–luminescent characteristics of squaraine dyes in aqueous buffer and in the presence of HSA and BSA

Dye	Free dye in buffer			Dye + HSA		Dye + BSA	
	λ_{ex} (nm)	λ_{fl} (nm)	I_0 (a.u.)	I^{HSA} (a.u.)	I^{HSA}/I_0	I^{BSA} (a.u.)	I^{BSA}/I_0
P-1	663	686	1.0	306	306	540	540
P-2	690	700	2.2	232	105	520	236
P-3	671	690	3.6	502	139	1,241	345
P-5	648	668	2.5	261	104	943	377
P-6	652	658	12.0	623	52	387	32
P-16	651	661	17.0	559	33	631	37

λ_{ex} (λ_{em}) – wavelength of excitation (fluorescence) maximum; I_0 (I^{HSA}, I^{BSA}) – fluorescence intensity of dye in aqueous buffer in free form (in the presence of HSA, BSA) in arbitrary units (a.u.)

dye P-3 (Fig. 23) allows quantification of HSA in the range from 0.2 to 500 μg/ml that is comparable with commercially used dyes such as CBB and Pyrogallol Red Protein [94, 95].

Also, benzothiazole dyes with ethyldiamino and *N,N*-dimethylhydrazino subsituents demonstrated good emission intensities in complexes with albumins and sufficient emission increasing in up to 52 times [96] (Fig. 25, Table 5).

In such a way, squaraines seem to be a promising tool for fluorescent detection of proteins and could be proposed for further studies as albumin-specific probes.

3.4 Fluorescent Dyes for Amyloid Structures Detection

The deposition of insoluble protein aggregates known as amyloid fibrils in brain tissues is associated with a number of neurodegenerative diseases, including Alzheimer's and Parkinson's diseases.

Despite the lack of amino acid sequence homology among the amyloidogenic proteins, the common morphology of amyloid fibrils was revealed as a cross β-sheet. The fluorescent molecules benzothiazole dye Thioflavin T and symmetrical

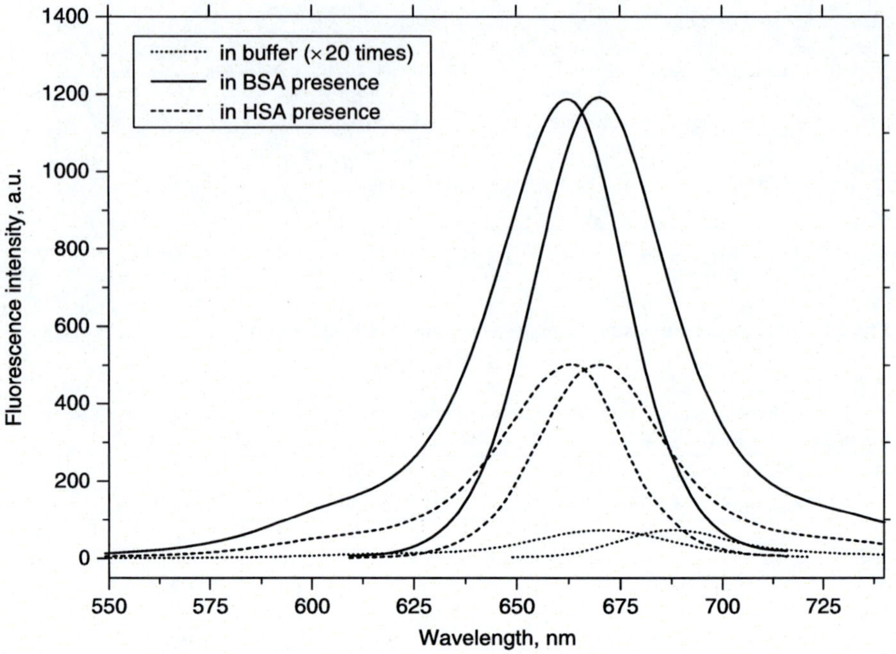

Fig. 24 Fluorescence excitation (*left*) and emission (*right*) spectra of the dye P-3 (5×10^{-6} M) in unbound state and in the presence of 3 μM solution albumins. The low-intensity spectra of free dye are multiplied 20 times. Excitation maxima are situated at 671 nm (free dye) and at 663 nm (dye in BSA or HSA presence)

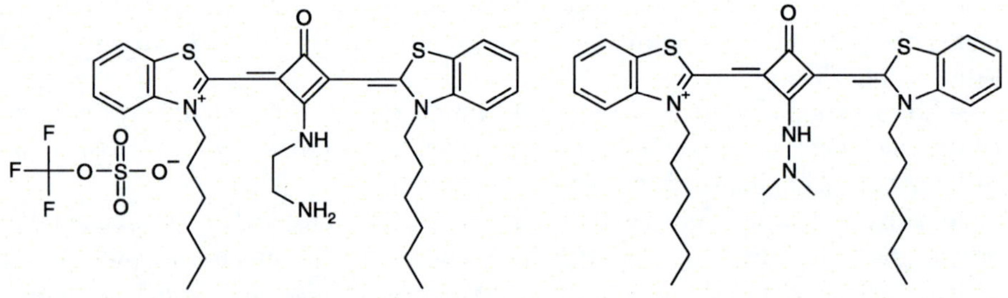

Fig. 25 Aza-substituted squaraine dyes for albumin detection

sulfonated azo dye Congo Red are the most frequently used dyes to detect the presence of amyloid deposits [97–99] (Fig. 26). These compounds upon binding to the amyloid nano-fibrils demonstrate selective fluorescence intensity increase (Thioflavin T) and undergo spectral shifts (Congo Red, Thioflavin T) that do not occur in the presence of other forms of the protein, such as monomer or amorphous aggregates.

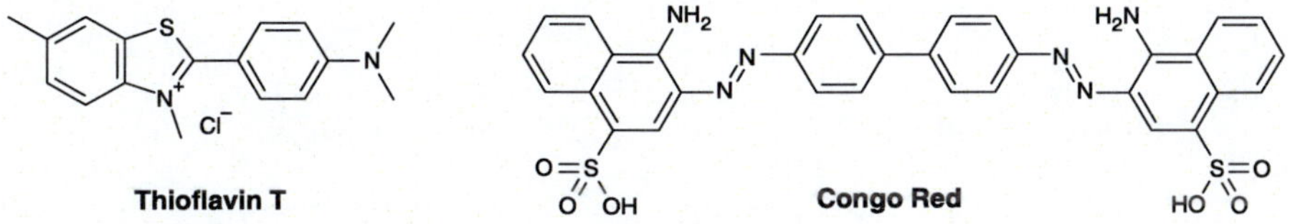

Fig. 26 Amyloid-specific probes Thioflavin T and Congo Red

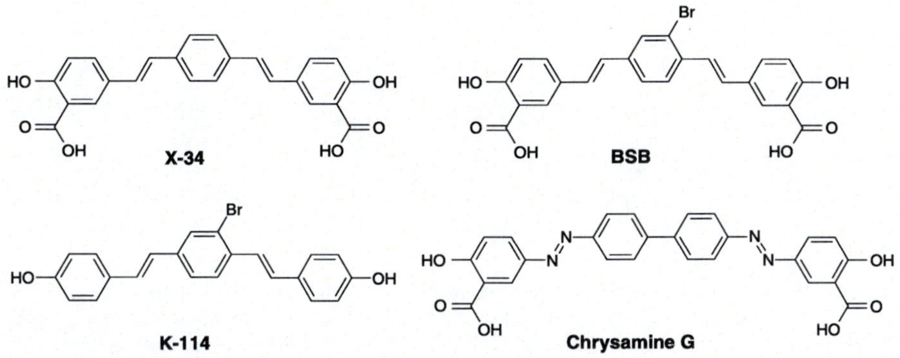

Fig. 27 Chemical structures of amyloid-specific derivatives of Congo Red – Chrysamine-G, X-34, K114, and BSB

Besides these dyes, very few new compounds have been used extensively to detect amyloid inclusions. Most of these new dyes, including Chrysamine-G [100], X-34 [101], K114 [99], and BSB [102] (Fig. 27), are derived from the structure of Congo Red.

However, despite the widespread use of existing probes for amyloid detection, they can give inconsistent and possibly inaccurate results [103].

Main disadvantages of these dyes are poor reproducibility of results observed for Thioflavin T and limitations in use in some of experimental systems that occurs for both Congo Red and Thioflavin T.

Also, Thioflavin T is not always a quantitative predictor of fibrillization degree because its fluorescence can vary depending on the structure and morphology of the fibrils. Congo Red was shown to bind also to native α-helical proteins such as citrate synthase and interleukin-2. Furthermore, Congo Red has been established as an inhibitor of fibril formation for a number of proteins, which makes impossible to use this dye for aggregation kinetics studies. As a consequence of its poor optical properties, the Congo Red derivative Chrysamine-G only weakly stains neuritic plaques and cerebrovascular amyloid in postmortem tissue [104].

Recently, cyanine dyes were proposed as specific fluorescent probes for aggregated protein structures [105, 106]. The trimethine cyanines (Figs. 28 and 29) demonstrated a significant fluorescence intensity increase upon interaction with aggregated into amyloid fibrils proteins β-lactoglobulin (BLG) [105, 107] and insulin [107] and

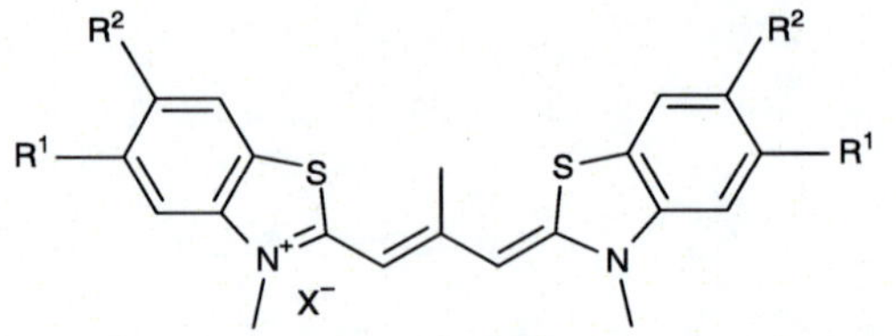

7519 R^1 = OMe; R^2 = H; X = ClO_4
7515 R^1= OH; R^2 = H; X = OTs
7523 R^1 = H; R^2 = OMe; X = OTs

Fig. 28 General structure of fluorescent probes for specific detection of fibrillar insulin

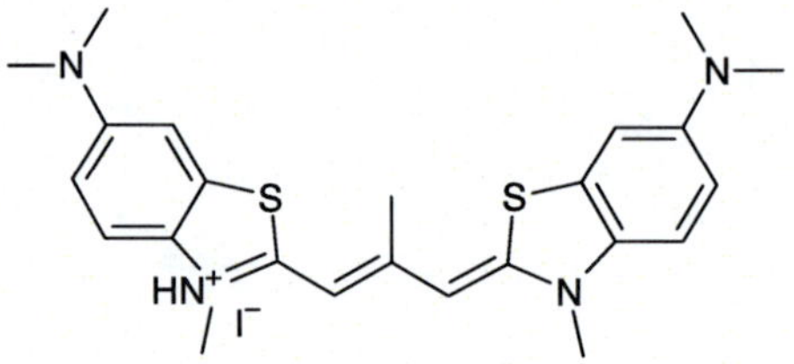

T-49

I_0 = 1.2 a.u.
I_M = 1.8 a.u.
I_F = 26 a.u.

Fig. 29 Structure of amyloid-sensitive cyanine dye T-49 and its spectral–luminescent properties in (5 × 10^{-6} M) in Tris–HCl buffer, pH = 7.9 and in presence of monomeric and aggregated β-lactoglobulin (both in concentration of 10 μm). Excitation and fluorescence maxima for this dye in free state are situated at 580 and 633 nm, respectively. I_0 – fluorescence intensity of dye in buffer in free form; I_M (I_F) – emission intensity of dye in presence of 10^{-6} M of monomeric (fibrillar) BLG

Table 6 Selected spectral–luminescent characteristics of trimethine dyes in aqueous buffer and in the presence of monomeric and fibrillar insulin

Dye	Free dye			In presence of insulin			
	λ_{ex} (nm)	λ_{em} (nm)	I_0 (a.u.)	[F]			[M]
				λ_{ex} (nm)	λ_{em} (nm)	I_f (a.u.)	I_m (a.u.)
7515	567	578	2	585	595	86	2.2
7519	562	581	5	580	590	129	3.5
7523	558	574	3	583	593	27.4	4.1

λ_{ex} (λ_{em}) – wavelength of excitation (emission) maximum; I_0 (I_M, I_F) – fluorescence intensity of dye in aqueous buffer in free form (in the presence of monomeric and fibrillar insulin) in arbitrary units (a.u.)

showed a relatively weak response in the presence of corresponding native proteins (Table 6, Fig. 30).

Also cyanine dyes appeared to be sensitive to fibrillar forms of Parkinson disease associated protein α-synuclein. Monomethine and trimethine cyanines based on benzothiazole, pyridine, and quinoline heterocycle end groups (Fig. 31) were shown to be efficient fluorescent probes for fibrillar α-synuclein detection, demonstrating comparable or higher emission intensity enhancements than the classic amyloid stain Thioflavin T [106].

Based upon the dye/amyloid fibril binding model proposed by Krebs for Thioflavin T [108] and on the molecular dimensions of series of cyanine dyes, it

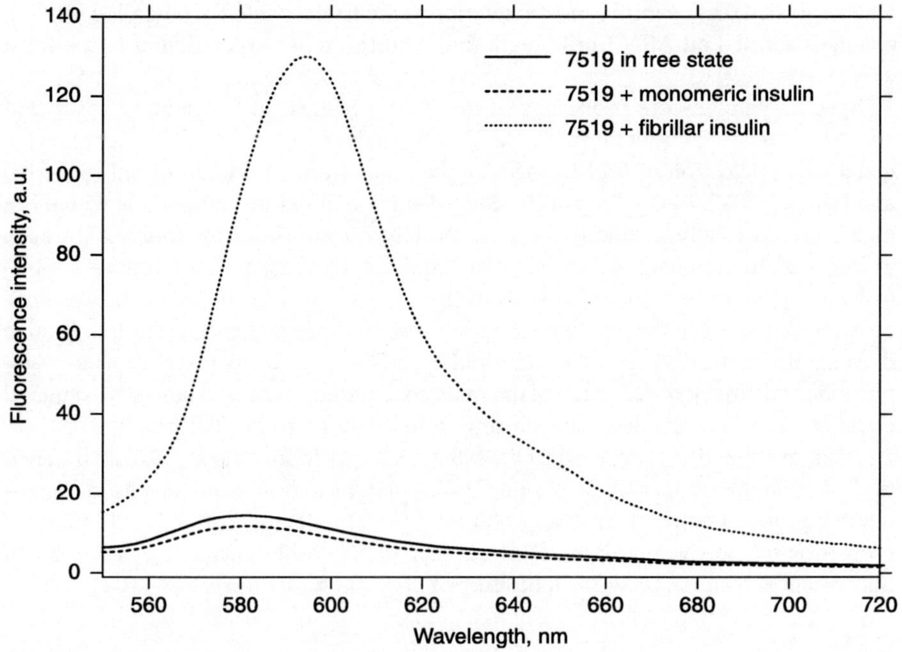

Fig. 30 Emission spectra of cyanine dye 7519 (5×10^{-6} M) in Tris–HCl buffer, pH = 7.9 and in presence of monomeric and aggregated insulin (both in concentration of 10 μm). Fluorescence was excited at 562 nm (free dye and dye in monomeric insulin presence) and at 580 nm (in fibrillar insulin presence)

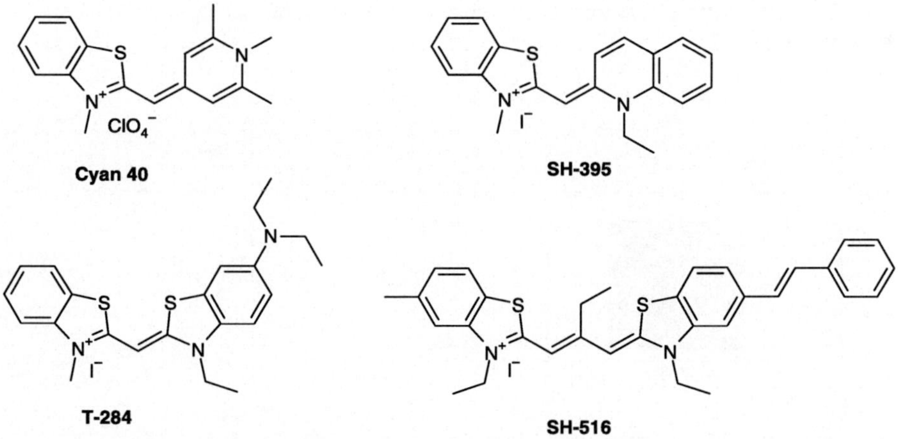

Fig. 31 Monomethine cyanine dyes based on benzothiazole (T-284), pyridine (Cyan 40), and quinoline (SH-395) heterocycle end groups and benzothiazole trimethine (SH-516) cyanine dye for sensitive detection of fibrillar ASN

was suggested that cyanines most likely insert themselves into the so-called binding channels formed on ASN fibrils, with their shortest axis perpendicular to the fibril axis [106].

Spectral–luminescent properties of the most ASN-sensitive cyanine dyes T-284 and SH-516 are presented in Table 7. These dyes appeared to be suitable for quantitative detection of fibrillar ASN in the range from ~1.5 to 28 μg/ml (SH-516) and 19 μg/ml (T-284) (Fig. 31) [109]. Such detection limits are comparable with those of the commercially available dyes for protein quantification in solution. In such a way, the fluorescence of the Congo red analog (trans, trans)-1-bromo-2,5-bis-(4-hydroxy) styrylbenzene (K114) increases linearly up to 70 μg/ml in the presence of ASN fibrils, while the dye concentration was 10 times higher than for the cyanine dyes studied here [99]. Another amyloid-specific agent, *trans*-resveratrol (a major phenolic constituent of red wine) at the same concentration as the cyanine dyes studied could be used for ASN detection starting from 20 μg/ml of fibrillar protein [110].

Also, cyanine dyes appeared to have the ability to follow the step-by-step transition of monomeric α-synuclein into fibrils, demonstrating good results of reproducibility, much better than it was observed for Thioflavin T (Fig. 32). Mentioned properties of the cyanine dyes allowed successful development of cyanine-based fluorescent screening assay for inhibitors of amyloid fibrils formation [107].

Table 7 Selected spectral characteristics of the cyanine dyes in aqueous buffer and in the presence of monomeric and aggregated into fibrils α-synuclein

Dye	Free dye			In α-synuclein presence		
	λ_{ex} (nm)	λ_{em} (nm)	I_0 (a.u.)	λ_{ex} (nm)	λ_{em} (nm)	I_F/I_M
T-284	441	563	3.0	470	560	27.4
SH-516	558	569	2.1	574	584	16.5
Thio-T	442	486	1.5	449	480	13.3

λ_{ex} (λ_{em}) – wavelength of excitation (emission) maximum; I_0 (I_M, I_F) – fluorescence intensity of dye in aqueous buffer in free form (in the presence of monomeric and fibrillar α-synuclein) in arbitrary units (a.u.)

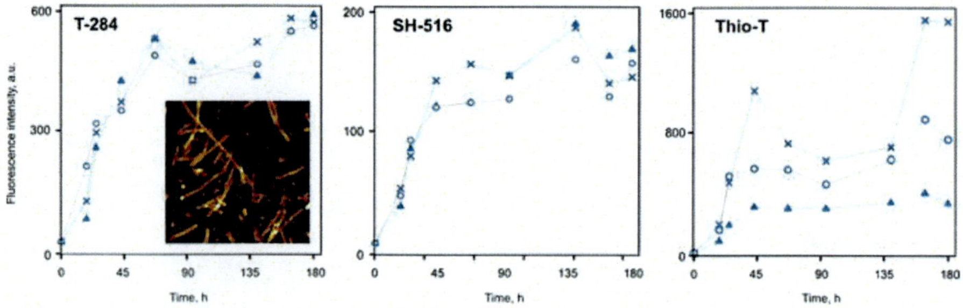

Fig. 32 Kinetics of wild-type human ASN aggregation assessed with SH-516, T-284, and Thioflavin T fluorescence emission using a 2.5-μM dye concentration in working solutions. Experiments were performed in triplicate (*crosses*, *circles*, and *triangles* in each of the graphs). The *inset* shows an atomic force micrograph of fibrillar aggregates of wild type ASN at time point 120 h

4 Conclusions

A wide range of fluorescent reagents for commonly used techniques are presented for today on imaging biological market. Thus, extremely sensitive unsymmetrical cyanine dyes with various spectral–luminescent characteristics are developed for nucleic acids visualization on gel, in living cells and detection of amplification products in real-time PCR. Using monomethine dye SYBR Green I allows detection of as little as 20 pg of dsDNA in postelectrophoresis staining procedure.

For nonspecific detection of proteins on SDS-PAGE gels, a series of metal complexes of Ru^{2+}, merocyanine dyes, and carbocyanine dyes was applied as fluorescent stains. Using these dyes allows the detection of proteins in concentrations ranging from 1 ng.

We could suggest that existed fluorescent dyes applied in routine techniques satisfy the majority of requirements of these techniques; therefore, the widening of the range of such dyes does not seem to be urgent.

At the same time there exists a deficiency of the dyes for the more specific tasks, such as the determination of certain conformations of nucleic acids or the detection of proteins of certain families or folding types.

Thus, for the detection of double-stranded regions of DNA, symmetrical and asymmetrical cyanine dyes with crescent-shaped molecule could be applied. Among the dyes that are able to interact with noncanonical triplex and quadruplex DNA structures are porphyrin, triphenylmethane, and cyanine dyes.

Mono and trimethine cyanine dyes have shown the ability to distinguish β-pleated amyloid formations from parent monomeric proteins. For squaraine dyes, the high sensitivity for albumin proteins was demonstrated.

References

1. Wolfbeis OS, Böhmer M, Dürkop A, Enderlein J, Gruber M, Klimant I, Krause C, Kürner J, Liebsch G, Lin Z, Oswald B, Wu M (2002) Advanced luminescent labels, probes and beads and their application to luminescence bioassay and imaging. In: Fluorescence spectroscopy, imaging and probes; Eds. Kraayenhof R, Visser AJWG, Gerritsen HC, Springer, Berlin/New York, 2002, pp. 3–42
2. Blackburn G, Gait M (1990) Nucleic acids in chemistry and biology. IRL, Oxford
3. Yarmoluk SM, Kovalska VB, Losytskyy MYu (2008) Symmetric cyanine dyes for detecting nucleic acids. Biotech Histochem 83:131–145
4. Hammer FM (1964) Cyanine dyes and related compounds. Wiley, New York
5. Armitage BA (2005) Cyanine dye-DNA interactions: intercalation, groove binding, and aggregation. DNA binders and related subjects. Top Curr Chem 253:55–76
6. Tuma RS, Beaudet MP, Jin X, Jones LJ, Cheung C-Y, Yue S, Singer VL (1999) Characterization of SYBR Gold nucleic acid gel stain: a dye optimized for use with 300-nm ultraviolet transilluminators. Anal Biochem 268:278–288
7. Haugland RP (2002) Handbook of fluorescent probes and research products, 9th edn. Molecular Probes Inc, Eugene
8. Sahyun MRV, Serpone N (1997) Photophysics of thiacarbocyanine dyes: relaxation dynamics in a homologous series of thiacarbocyanines. J Phys Chem 101:9877–9883

9. Murphy S, Schuster GB (1995) Electronic relaxation in a series of cyanine dyes: evidence for electronic and steric control of the rotational rate. J Phys Chem B 99:8516–8518
10. Lee GL, Chen C-H, Chiu LA (1986) Thiazole Orange: a new dye for reticulocyte analysis. Cytometry 7:508–517
11. Rye HS, Quesada MA, Peck K, Mathies RA, Glazer AN (1990) High sensivity two-color detection of double-stranded DNA with confocal fluorescence gel scanner using ethidium homodimer and thiazole orange. Nucleic Acids Res 19:327–333
12. Rye HS, Yue S, Wemmer DE, Quesada MA, Haughland RP, Mathies RA, Glazer A (1992) Stable fluorescent complexes of double-stranded DNA with bis-intercalating asymmetric cyanine dyes: properties and applications. Nucleic Acids Res 20:2803–2812
13. Frey T (1995) Nucleic acid dyes for detection of apoptosis in live cells. Cytometry 21: 265–274
14. Ankarcrona M, Dypbukt JM, Bonfoco E, Zhivotovsky B, Orrenius S, Lipton SA, Nicotera P (1995) Glutamate-induced neuronal death: a succession of necrosis or apoptosis depending on mitochondrial function. Neuron 15:961–973
15. Zipper H, Brunner H, Bernhagen J, Vitzthum F (2004) Investigations on DNA intercalation and surface binding by SYBR Green I, its structure determination and methodological implications. Nucleic Acids Res 32:e103
16. Karlsen F, Steen HB, Nesland JM (1995) SYBR Green I staining increases the detection sensitivity of viruses by polymerase chain reaction. J Virol Methods 55:153–156
17. Ishiguro T, Saitoh J, Yawata H, Yamagishi H, Iwasaki S, Mitoma Y (1995) Homogeneous quantitative assay of hepatitis C virus RNA by polymerase chain reaction in the presence of fluorescent intercalater. Anal Biochem 229:207–213
18. Nygren J, Svanvik N, Kubista M (1998) The interactions between the fluorescent dye thiazole orange and DNA. Biopolymers 46:39–51
19. Larsson A, Carlsson C, Jonsson M, Albinsson B (1994) Characterization of the binding of the fluorescent dyes YO and YOYO by polarized light spectroscopy. J Am Chem Soc 116:8459–8465
20. Karlsson J, Lincoln P, Westman G (2003) Synthesis and DNA binding studies of a new asymmetric cyanine dye binding in the minor groove of [poly(dA-dT)]$_2$. Bioorg Med Chem 11:1035–1040
21. Karlsson HJ, Eriksson M, Perzon E, Åkerman B, Lincoln P, Westman G (2003) Groove-binding unsymmetrical cyanine dyes for staining of DNA: syntheses and characterization of DNA-binding. Nucleic Acids Res 31:6227–6234
22. Tyutyulkov N, Fabian J, Mehlhorn A, Dietz F, Tadjer A (1991) Polymethine dyes: structure and properties. St. Kliment Ohridsky University Press, Sofia
23. Yarmoluk SM, Kovalska VB, Lukashov SS, Slominskii YuL (1999) Interaction of cyanine dyes with nucleic acids. XII. β-substituted carbocyanines as possible fluorescent probes for nucleic acids detection. Bioorg Med Chem Lett 9:1677–1678
24. Valyukh IV, Kovalska VB, Slominskii YL, Yarmoluk SM (2002) Spectroscopic studies of α, γ-disubstituted trimethine cyanine: new fluorescent dye for nucleic acids. J Fluoresc 12:105–107
25. Lukashov SS, Makovenko IE, Losytskyy MYu, Slominskii YuL, Yarmoluk SM (2001) The interaction of cyanine dyes with nucleic acids. XXXIII. Meso-methylsubstituted trimethin-cyanines, as possible probes for fluorescent nucleic acid detection. Biopolym Cell 17:448–454 (in Ukrainian)
26. Kovalska VB, Valyukh IV, Lukashov SS, Slominskii YuL, Yarmoluk SM (2002) An investigation of tricarbocyanines “Stains-All” and “iso-Stains-All” as fluorescent nucleic acids probes. J Fluoresc 12:209–212
27. Lukashov SS, Losytskyy MYu, Kovtun YuP, Yarmoluk SM (2002) Interaction of cyanine dyes with nucleic acids. Meso-substituted trimethincyanines for fluorescent detection of nucleic acids. Biopolym Cell 18:243–252 (in Ukrainian)

28. Valyukh I, Slobodyanyuk A, Kovalska V, Grangan A, Slominskii YL, Yarmoluk SM (2002) Spectral properties of symmetrical trimethine cyanines with α-, γ-substituted polymethine chain in the presence of nucleic acids. J Phys Stud 6:236–242
29. Matselyukh BP, Yarmoluk SM, Matselyukh AB, Kovalska VB, Kocheshev IO, Kryvorotenko DV, Lukashov SS (2003) Interaction of cyanine dyes with nucleic acids: XXXI. Using of polymethine cyanine dyes for the visualization of DNA in agarose gels. J Biochem Biophys Methods 57:35–43
30. Yarmoluk SM, Lukashov SS, Losytskyy MYu, Akerman B, Kornyushyna OS (2002) Interaction of cyanine dyes with nucleic acids: XXVI. Intercalation of the trimethine cyanine dye Cyan 2 into double-stranded DNA: study by spectral luminescence methods. Spectrochim Acta A Mol Biomol Spectrosc 58:3223–3232
31. Mikheikin AL, Zhuze AL, Zasedatelev AS (2000) Binding of symmetrical cyanine dyes into the DNA minor groove. J Biomol Struct Dyn 18:59–72
32. Milanovich N, Suh M, Jankowiak R, Small GJ, Hayes JM (1996) Binding of TO-PRO-3 and TOTO-3 to DNA: fluorescence and hole-burning studies. J Phys Chem 100:9181–9186
33. Rampal S, Liu MS, Chen FTA (1997) Characterization of TO-PRO-3 as an intercalator for double-stranded DNA analysis with red diode laser-induced fluorescence detection. J Chromatogr A 781:357–365
34. Kovalska VB, Volkova KD, Losytskyy MYu, Tolmachev OI, Balanda AO, Yarmoluk SM (2006) Studies of 6, 6′-disubstituted benzothiazole trimethine cyanine dyes as fluorescent dyes for DNA detection. Spectrochim Acta A 65:271–277
35. Smith JO, Olson DA, Armitage BA (1999) Molecular recognition of PNA-containing hybrids: spontaneous assembly of helical cyanine dye aggregates on PNA templates. J Am Chem Soc 121:2686–2695
36. Seifert JL, Connor RE, Kushon SA, Wang M, Armitage BA (1999) Spontaneous assembly of helical cyanine dye aggregates on DNA nanotemplates. J Am Chem Soc 121:2987–2995
37. Losytskyy MYu, Volkova KD, Kovalska VB, Makovenko IE, Slominskii Yu L, Tolmachev OI, Yarmoluk SM (2005) Fluorescent properties of pentamethine cyanine dyes with cyclopentene and cyclohexene group in presence of biological molecules. J Fluoresc 15:849–857
38. Ohulchanskyy TYu, Pudavar HE, Yarmoluk SM, Yashchuk VM, Bergey EG, Prasad PN (2003) A monomethine cyanine dye Cyan 40 for two-photon-excited fluorescence detection of nucleic acids and their visualization in live cells. Photochem Photobiol 77:138–145
39. Albota M, Beljonne D, Brédas J-L, Ehrlich JE, Fu J-Y, Heikal AA, Hess SE, Kogej T, Levin MD, Marder SR, McCord-Maughon D, Perry JW, Röckel H, Rumi M, Subramaniam G, Webb WW, Wu X-L, Xu C (1998) Design of organic molecules with large two-photon absorption cross sections. Science 281:1653–1656
40. Fujita H, Nakano M, Takahata M, Yamaguchi K (2002) A new strategy of enhancing two-photon absorption in conjugated molecules: introduction of charged defects. Chem Phys Lett 358:435–441
41. Wu L-Z, Tang X-J, Jiang M-H, Tung C-H (1999) Two-photon induced fluorescence of novel dyes. Chem Phys Lett 315:379–382
42. Kovalska VB, Kryvorotenko DV, Balanda AO, Losytskyy MYu, Tokar VP, Yarmoluk SM (2005) Fluorescent homodimer styrylcyanines: synthesis and spectral-luminescent studies in nucleic acids and protein complex. Dyes Pigments 67:47–54
43. Balanda AO, Volkova KD, Kovalska VB, Losytskyy MYu, Tokar VP, Prokopets VM, Yarmoluk SM (2007) Synthesis and spectral-luminescent studies of novel 4-oxo-4, 6, 7, 8-tetrahydropyrrolo[1, 2-alpha]thieno[2, 3-d]pyrimidinium styryls as fluorescent dyes for biomolecules detection. Dyes Pigments 75:25–31
44. Tokar VP, Losytskyy MYu, Kovalska VB, Kryvorotenko DV, Balanda AO, Prokopets VM, Galak MP, Dmytruk IM, Yashchuk VM, Yarmoluk SM (2006) Fluorescence of styryl dyes-DNA complexes induced by single- and two-photon excitation. J Fluoresc 16:783–791
45. Tokar VP, Losytskyy MYu, Ohulchanskyy TY, Kryvorotenko DV, Kovalska VB, Balanda AO, Dmytruk IM, Prokopets VM, Yarmoluk SM, Yashchuk VM (2010) Styryl dyes as two-photon

excited fluorescent probes for DNA detection and two-photon laser scanning fluorescence microscopy of living cells. J Fluoresc 20:865–872
46. Andrade CD, Yanez CO, Rodriguez L, Belfield KD (2010) A series of fluorene-based two-photon absorbing molecules: synthesis, linear and nonlinear characterization, and bioimaging. J Org Chem 75:3975–3982
47. Lubitz I, Borovok N, Kotlyar AB (2007) Interaction of monomolecular G4-DNA nanowires with TMPyP: evidence for intercalation. Biochemistry 46:12925–12929
48. Wei C, Jia G, Yuan J, Feng Z, Li C (2006) A spectroscopic study on the interaction of porphyrin with G-quadruplex DNAs. Biochemistry 45:6681–6691
49. Arthanari H, Basu S, Kawano TL, Bolton PH (1998) Fluorescent dyes specific for quadruplex DNA. Nucleic Acids Res 26:3724–3728
50. Guo J-H, Zhu L-N, Kong D-M, Shen H-X (2009) Triphenylmethane dyes as fluorescent probes for G-quadruplex recognition. Talanta 80:607–613
51. Lubitz I, Zikich D, Kotlyar A (2010) Specific high-affinity binding of Thiazole Orange to triplex and G-quadruplex DNA. Biochemistry 49:3567–3574
52. Yang P, De Cian A, Teulade-Fichou M-P, Mergny J-L, Monchaud D (2009) Engineering bisquinolinium/thiazole orange conjugates for fluorescent sensing of G-Quadruplex DNA. Angew Chemie Int Ed 48:2188–2191
53. Kovalska VB, Losytskyy MYu, Yarmoluk SM, Lubitz I, Kotlyar AB (2011) Mono and trimethine cyanines Cyan 40 and Cyan 2 as probes for highly selective fluorescent detection of non-canonical DNA structures. J Fluoresc 21:223–230.
54. Paramasivan S, Bolton PH (2008) Mix and measure fluorescence screening for selective quadruplex binders. Nucleic Acids Res 36:e106
55. Volkova KD, Kovalska VB, Yarmoluk SM (2007) Modern techniques for protein detection on polyacrylamide gels: problems arising from the use of dyes of undisclosed structures for scientific purposes. Biotech Histochem 82:201–208
56. Steinberg T, Jones L, Haugland R, Singer V (1996) SYPRO orange and SYPRO red protein gel stains: one-step fluorescent staining of denaturing gels for detection of nanogram levels of protein. Anal Biochem 239:223–237
57. Neubert R, Schiewe J (1996) Vitamin analysis using capillary zone electrophoresis. J Am Biotechnol Lab 14:12–14
58. Lauber WM, Carroll JA, Dufield DR, Kiesel JR, Radabaugh MR, Malone JP (2001) Mass spectrometry compatibility of two-dimensional gel protein stains. Electrophoresis 22: 906–918
59. Patton WF (1998) Proteomics – IBC's Second International Conference. Practical applications of protein based tools for drug discovery. IDrugs 1:299–301
60. Steinberg TH, Lauber WM, Berggren K, Kemper C, Yue S, Patton WF (2000) Fluorescence detection of proteins in sodium dodecyl sulfate-polyacrylamide gels using environmentally benign, nonfixative, saline solution. Electrophoresis 21:497–508
61. Fazekas de St GS, Webster R, Datyner A (1963) Two new staining procedures for quantitative estimation of proteins on electrophoretic strips. Biochim Biophys Acta 71:377–391
62. Merril CR (1990) Gel-staining techniques. Methods Enzymol 182:477–488
63. Chrambach A, Reisfeld RA, Wyckoff M, Zaccari J (1967) A procedure for rapid and sensitive staining of protein fractionated by polyacrylamide gel electrophoresis. Anal Biochem 20:150–154
64. Patton WF (2002) Detection technologies in proteome analysis. J Chromatogr B Anal Technol Biomed Life Sci 771:3–31
65. Berggren KN, Schulenberg B, Lopez MF, Steinberg TH, Bogdanova A, Smejkal G, Wang A, Patton WF (2002) An improved formulation of SYPRO ruby protein gel stain: comparison with the original formulation and with a ruthenium II tris (bathophenanthroline disulfonate) formulation. Proteomics 2:486–498
66. Lopez M, Berggren K, Chernokalskaya E, Lazarev A, Robinson M, Patton W (2000) A comparison of silver stain and SYPRO Ruby Protein Gel Stain with respect to protein

detection in two-dimensional gels and identification by peptide mass profiling. Electrophoresis 21:3673–3683
67. Nishihara J, Champion K (2002) Quantitative evaluation of proteins in one-and two-dimensional polyacrylamide gels using a fluorescent stain. Electrophoresis 23:2203–2215
68. Gerner C, Vejda S, Gelbmann D, Bayer E, Gotzmann J, Schulte-Hermann R, Mikulits W (2002) Concomitant determination of absolute values of cellular protein amounts, synthesis rates, and turnover rates by quantitative proteome profiling. Mol Cell Proteomics 1: 528–537
69. Berggren K, Chernokalskaya E, Steinberg TH, Kemper C, Lopez MF, Diwu Z, Haugland RP, Patton WF (2000) Background-free, high sensitivity staining of proteins in one- and two-dimensional sodium dodecyl sulfate-polyacrylamide gels using a luminescent ruthenium complex. Electrophoresis 21:2509–2521
70. Candiano G, Bruschi M, Musante L, Santucci L, Ghiggeri GM, Carnemolla B, Orecchia P, Zardi L, Righetti PG (2004) Blue silver: A very sensitive colloidal Coomassie G-250 staining for proteome analysis. Electrophoresis 25:1327–1333
71. Walker JM (2002) The protein protocols handbook, 2nd edn. Humana, Totowa, NJ
72. Bell PJ, Karuso P (2003) Epicocconone, a novel fluorescent compound from the fungus Epicoccum nigrum. J Am Chem Soc 125:9304–9305
73. Ferrari BC, Attfield PV, Veal DA, Bell PJ (2003) Application of a novel fluorescent dye to the detection of Giardia cysts. J Microbiol Methods 52:133–135
74. Mackintosh JA, Choi H-Y, Bae S-H, Veal DA, Bell PJ, Ferrari BC, van Dyk D, Verrills NM, Paik Y-K, Karuso P (2003) A fluorescent natural product for ultra sensitive detection of proteins in 1-D and 2-D gel electrophoresis. Proteomics 3:2273–2288
75. Smejkal GB (2004) The Coomassie chronicles: past, present and future perspectives in polyacrylamide gel staining. Expert Rev Proteomics 1:381–387
76. Smejkal GB, Robinson MH, Lazarev A (2004) Comparison of fluorescent stains: relative photostability and differential staining of proteins in two-dimensional gels. Electrophoresis 25:2511–2519
77. Miller I, Crawford J, Gianazza E (2006) Protein stains for proteomic applications: Which, when, why? Proteomics 6:5385–5408
78. Bermudez A, Daban J, Garcia J, Mendez E (1994) Direct blotting, sequencing and immunodetection of proteins after five-minute staining of SDS and SDS-treated IEF gels with Nile red. Biotechniques 16:621–624
79. Tom Berkelman (2007) Development Of Flamingo™, a novel fluorescent dye for non-specific detection and quantitation of proteins in gels. In: The 2007 annual meeting of the american electrophoresis society (AES), Salt Lake City, UT, 152d
80. Kovalska VB, Kryvorotenko DV, Losytskyy MYu, Nording P, Rueck A, Schoenenberger B, Yarmoluk SM, Wah P (2006) Detection of polyamino acids using trimethincyanine dyes. US Patent 2006207881
81. Schonenberger B, Nording P, Yarmoluk S, Ruck A, Baumle M, Weber M (2006) New fluorescence dyes for protein gel stains. In: 7th Siena meeting "From genome to proteome: back to the future", Siena, Italy
82. Lowry OH, Rosebrough NJ, Randall RJ (1951) Protein measurement with the Folin phenol reagent. J Biol Chem 193:265–275
83. Bradford MM (1976) A rapid and sensitive method for the quantitation of microgram quantities of protein utilizing the principle of protein-dye binding. Anal Biochem 7:248–254
84. Chen Y, Yang J, Wang Z, Wu X, Wang F (2006) Scopoletine as fluorescence probe for determination of protein. Spectrochim Acta A Mol Biomol Spectrosc 66:686–690
85. Jones LJ, Haugland RP, Singer VL (2003) Development and characterization of the NanoOrange protein quantitation assay: a fluorescence-based assay of proteins in solution. Biotechniques 34:850–859
86. NanoOrange™ Protein Quantitation Kit (N-6666), Product Information, Invitrogen Inc. (probes.invitrogen.com/media/pis/mp06666.pdf)

87. SyproOrange™ and SyproRed™ Protein Gel Stains. Product Information, Invitrogen Inc. (probes.invitrogen.com/media/pis/mp06650.pdf)
88. Chiu TC, Lin YW, Huang CC, Chrambach A, Chang HT (2003) A simple, rapid, and sensitive method for analysis of SYPRO Red labeled sodium dodecyl sulfate-protein complexes by capillary electrophoresis with laser-induced fluorescence. Electrophoresis 24:1730–1736
89. Harvey MD, Bablekis V, Banks PR, Skinner CD (2001) Utilization of the non-covalent fluorescent dye, NanoOrange, as a potential clinical diagnostic tool: nanomolar human serum albumin quantitation. J Chromatogr B Biomed Sci Appl 754:345–356
90. Stoyanov AV, Fan ZH, Das C, Ahmadzadeh H, Mei Q, Mohammed S (2006) On the possibility of applying noncovalent dyes for protein labeling in isoelectric focusing. Anal Biochem 350:263–267
91. Kessler M, Wolfbeis OS (1993) Method for detection and determination of human serum albumin. US Patent 5182214
92. Jisha VS, Arun KT, Hariharan M, Ramaiah D (2006) Site-selective binding and dual mode recognition of serum albumin by a squaraine dye. J Am Chem Soc 128:6024–6025
93. Nakazumi H, Colyer ChL, Kaihara K, Yagi S, Hyodo Y (2003) Red luminescent squarylium dyes for noncovalent HSA labeling. Chem Lett 32:804–805
94. Volkova KD, Kovalska VB, Tatarets AL, Patsenker LD, Kryvorotenko DV, Yarmoluk SM (2007) Spectroscopic study of squaraines as protein-sensitive fluorescent dyes. Dyes Pigments 72:285–292
95. Volkova KD, Kovalska VB, Losytskyy MY, Bento A, Reis LV, Santos PF, Almeida P, Yarmoluk SM (2008) Studies of benzothiazole and benzoselenazole squaraines as fluorescent probes for albumins detection. J Fluoresc 18:877–882
96. Volkova KD, Kovalska VB, Losytskyy MYu, Reis LV, Santos PF, Almeida P, Lynch DE, Yarmoluk SM (2011) Aza-substituted squaraines for the fluorescent detection of albumins Dyes Pigments 90:41–47
97. Puchtler H, Sweat F, Levine M (1962) On the binding of Congo Red by amyloid. J Histochem Cytochem 10:355–364
98. Naiki H, Higuchi K, Matsushima K, Shimada A, Chen W-H, Hosokawa M, Takeda T (1990) Fluorometric examination of tissue amyloid fibrils in murine senile amyloidosis: use of the fluorescent indicator, Thioflavin T. Lab Invest 62:768–773
99. Crystal AS, Giasson BI, Crowe A, Kung M-P, Zhuang Z-P, Trojanowski JQ, Lee VM-Y (2003) A comparison of amyloid fibrillogenesis using the novel fluorescent compound K114. J Neurochem 86:1359–1368
100. Klunk WE, Debnath ML, Pettegrew JW (1995) Chrysamine-G binding to Alzheimer and control brain: autopsy study of a new amyloid probe. Neurobiol Aging 16:541–548
101. Styren SD, Hamilton RL, Styren GC, Klunk WE (2000) X-34, a fluorescent derivative of Congo Red: a novel histochemical stain for Alzheimer's disease pathology. J Histochem Cytochem 48:1223–1232
102. Schmidt ML, Schuck T, Sheridan S, Kung MP, Kung H, Zhuang ZP, Bergeron C, Lamarche JS, Skovronsky D, Giasson BI, Lee VM, Trojanowski JQ (2001) The fluorescent Congo red derivative, (trans, trans)-1-bromo-2, 5-bis-(3-hydroxycarbonyl-4-hydroxy)styrylbenzene (BSB), labels diverse beta-pleated sheet structures in postmortem human neurodegenerative disease brains. Am J Pathol 159:937–943
103. Murakami K, Irie K, Morimoto A, Ohigashi H, Shindo M, Nagao M, Shimizu T, Shirasawa T (2003) Neurotoxicity and physicochemical properties of Aβ mutant peptides from cerebral amyloid angiopathy implication for the pathogenesis of cerebral amyloid angiopathy and Alzheimer's disease. J Biol Chem 278:46179–46187
104. Khurana R, Uversky VN, Nielsen L, Fink AL (2001) Is Congo Red an amyloid-specific dye. J Biol Chem 276:22715–22721
105. Volkova KD, Kovalska VB, Balanda AO, Vermeij RJ, Subramaniam V, Slominskii YuL, Yarmoluk SM (2007) Cyanine dye–protein interactions: looking for fluorescent probes for amyloid structures. J Biochem Biophys Methods 70:727–733

106. Volkova KD, Kovalska VB, Balanda AO, Losytskyy MYu, Golub AG, Vermeij RJ, Subramaniam V, Tolmachev OI, Yarmoluk SM (2008) Specific fluorescent detection of fibrillar alpha-synuclein using mono- and trimethine cyanine dyes. Bioorg Med Chem 16:1452–1459
107. Volkova KD, Kovalska VB, Inshin D, Slominskii YL, Tolmachev OI, Yarmoluk SM (2010) Novel fluorescent trimethine cyanine dye 7519 for amyloid fibril inhibition assay. Biotech Histochem, Early Online 1–4
108. Krebs MRH, Bromley EHC, Donald AM (2005) The binding of thioflavin-T to amyloid fibrils: localisation and implications. J Struct Biol 149:30–37
109. Volkova KD, Kovalska VB, Segers-Nolten GM, Veldhuis G, Subramaniam V, Yarmoluk SM (2009) Explorations of the application of cyanine dyes for quantitative alpha-synuclein detection. Biotech Histochem 84:55–61
110. Ahn JS, Lee JH, Kim JH, Paik SR (2007) Novel method for quantitative determination of amyloid fibrils of a-synuclein and amyloid b/A4 protein by using resveratrol. Anal Biochem 367:259–265

Functional Nucleic Acids for Fluorescence-Based Biosensing Applications

Jennifer Lee, Lawrence Lin, and Yingfu Li

Abstract Functional nucleic acids (FNAs) are single-stranded DNA or RNA molecules that are capable of carrying out the function of ligand binding (by aptamers), catalysis (by nucleic acid enzymes), or both (by aptazymes). Many FNAs have been shown to be suitable molecular recognition elements for many molecular targets, including small molecules and proteins, and have been examined for a variety of biosensing applications. In this chapter, we present a focused discussion on the use of FNAs for the development of fluorescence-based biosensors or bioassays. First, we briefly discuss the technique of "in vitro selection" by which artificial FNAs can be isolated from random-sequence DNA or RNA pools. This is followed by a survey of various strategies in employing aptamers for fluorescence assay development. Finally, we review emerging applications to explore nucleic acid enzymes (ribozymes, DNAzymes, and aptazymes) as fluorescent biosensing probes.

Keywords Aptamers · Aptazymes · Biosensing · DNAzymes · Fluorescence · Functional nucleic acids · In vitro selection · Molecular recognition elements · Nucleic acid enzymes · Ribozymes

---

J. Lee and L. Lin
Department of Biochemistry and Biomedical Sciences, McMaster University, 1200 Main Street West, Hamilton, ON, Canada L8N 3Z5

Y. Li (✉)
Department of Biochemistry and Biomedical Sciences, McMaster University, 1200 Main Street West, Hamilton, ON, Canada L8N 3Z5
and
Department of Chemistry and Chemical Biology, McMaster University, 1200 Main Street West, Hamilton, ON, Canada L8N 3Z5
e-mail: liying@mcmaster.ca

A.P. Demchenko (ed.), *Advanced Fluorescence Reporters in Chemistry and Biology III: Applications in Sensing and Imaging*, Springer Ser Fluoresc (2011) 10: 201–222,
DOI 10.1007/978-3-642-18035-4_6,

Contents

1 Introduction

Nucleic acids – DNA and RNA – have long been known to be very important molecules in cells: DNA carries genetic information and RNA actively participates in many essential cellular processes, such as protein synthesis and gene regulation. Single-stranded (ss) nucleic acids have also been shown to possess the ability to perform two functions once thought exclusive to protein: catalysis and molecular recognition. ssDNA or RNA molecules with such functions are generally termed "functional nucleic acids (FNAs)" [1–3]. These include aptamers (ligand-binding ssDNAs or RNAs), ribozymes (RNA-based enzymes), DNAzymes or deoxyribozymes (ssDNA-based enzymes), and allosteric nucleic acid enzymes or simply aptazymes (ssDNAs or RNAs that contain both ligand-binding and catalytic elements which are specially arranged such that the catalytic activity is controlled by the target binding).

The Nobel Prize winning discovery of ribozymes in cellular systems by Thomas Cech and Sidney Altman in early 1980s [4, 5] provided the first evidence that RNA is able to execute very complex functions associated with biological catalysis. The invention of the "in vitro selection" or SELEX (which stands for "Systematic Evolution of Ligands by EXponential Enrichment") technique in the laboratories of Larry Gold, Jack Szostak, and Gerald Joyce in 1990 [6–8] has led to the development of many artificial FNAs [9–18].

In vitro selection is an experimental process to derive nucleic acid sequences with binding and/or catalytic activity from a man-made DNA or RNA library. A typical DNA library consists of 10^{13}–10^{16} sequences produced by chemical synthesis; an RNA library is further made from a DNA library by in vitro transcription. The sequence of a library contains a random-sequence domain flanked by two constant regions as primer binding sites for PCR (polymerase chain reaction) or RT-PCR (RT: reverse transcription). The library is subjected to repetitive cycles of selection and amplification: the selection step is to fractionate active sequences from inactive ones; the enriched fraction is then amplified by PCR (for DNA-based selection) or RT-PCR (for RNA-based selection). The amplified pool is used for the next round of selection and amplification and this process is repeated many times until the selected population exhibits a significant activity. Individual FNAs in the pool are identified by cloning and sequencing. By in vitro selection, FNA sequences are allowed to compete with each other for survival so that only the most competent species are progressively enriched and dominate in the end. Because the selection and amplification process is conducted entirely in a test tube, a researcher can set up experimental conditions to derive FNAs to meet a defined need. The ability to simultaneously screen as many as 10^{16} different molecules brings about a high probability of success.

Many naturally occurring RNA aptamers have recently been discovered in various biological systems as part of riboswitches. A riboswitch is a regulatory element typically residing in an untranslated region of a messenger RNA and provides metabolite-responsive controls over the expression of relevant genes in cells. A riboswitch is typically composed of a well conserved aptamer domain that recognizes a specific metabolite and an "expression platform" that relays the aptamer–metabolite interaction into a change in the level of expression of genes it helps to control [19, 20]. The metabolites for which a riboswitch has been discovered include vitamins [21–23], amino acids [24–27] and their derivatives [28–34], nucleobases [35–37] and their derivatives [38–47], sugar derivatives [48], and metal ions [49].

For the past three decades, FNAs have been widely examined as molecular tools for diverse applications in wide-ranging fields including chemistry, biology, biomedical sciences, and nanotechnology [3]. Engineering and application of FNAs have been extensively discussed in several books [50–52] and many review articles (e.g., [1, 3, 11, 53]). In this chapter, we will focus on reviewing fluorescence-based detection methods where an FNA is used as a molecular recognition element (MRE).

2 Fluorescent Aptamer Biosensors

A biosensor can be regarded as the marriage of two components – an MRE that interacts with a target of interest and a sensor element that transduces a ligand-binding event to a detectable signal. Over the years, substantial research efforts have been focused on the development of biosensors or bioassays that are highly sensitive and specific, versatile, stable, inexpensive, and convenient to use. FNAs are an emerging class of MRE that fulfills these criteria. This section reviews the progress made in the design and application of fluorescent aptamers as biosensors.

2.1 Label-Free Approach

One simple, label-free fluorescent detection method utilizes DNA intercalating dyes or cationic conjugated polymers to probe changes in DNA conformations. Certain organic dyes, such as TOTO (the chemical structures of organic dyes, including fluorophores and quenchers, that are discussed in this chapter are provided in Fig. 1), are able to intercalate into double-stranded DNA and the intercalated dyes often exhibit significantly different fluorescence properties. Thus, organic dyes can be used to monitor the formation or dissociation of double-helical elements of an aptamer as a result of ligand binding.

Fang and coworkers used TOTO to detect platelet-derived growth factor BB (PDGF-BB) (Fig. 2a) [54]. A fluorescence signal is present in the absence of PDGF-BB because the dye is able to intercalate into the double-stranded region in the PDGF-binding aptamer. When PDGF-BB is added to the sample, the binding between PDGF-BB and the aptamer promotes a conformational change that displaces bound dyes, causing a decrease in fluorescence signal. This method provides a simple and convenient way to monitor protein and nucleic acid interactions without the need of covalent labeling of either the analyte or the aptamer.

Leclerc and colleagues used polythiophene, a cationic conjugated polymer, to probe the binding between an antithrombin aptamer and human thrombin (Fig. 2b) [55]. The mechanism for detection is based on electrostatic interactions between the positively charged polymer and the negatively charged phosphate backbone found

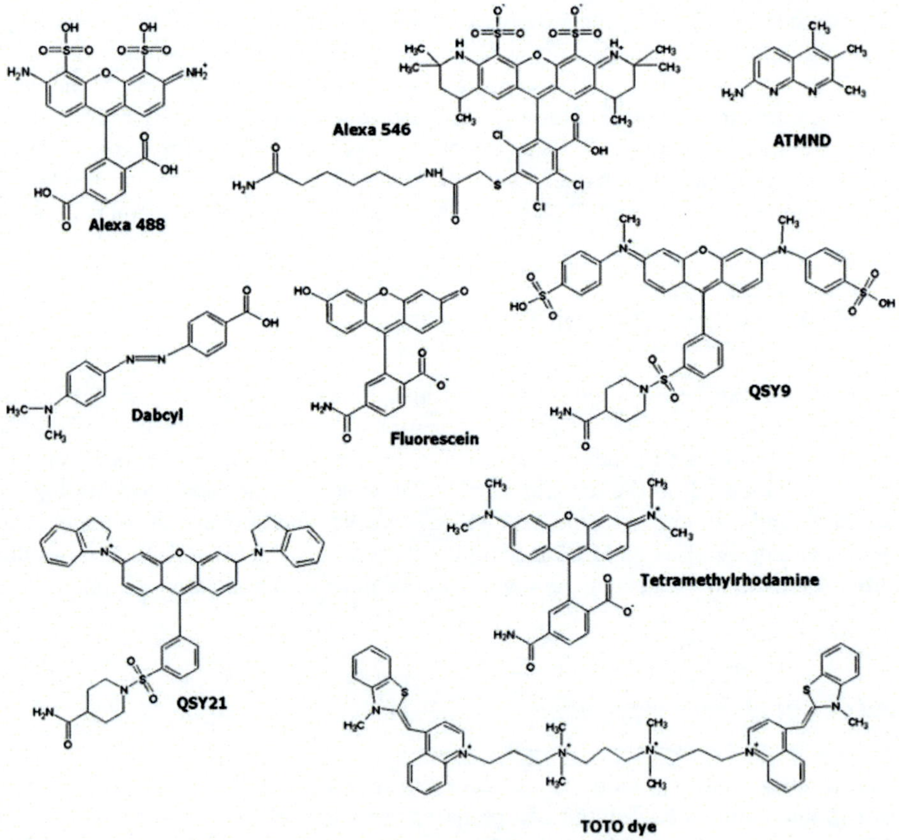

Fig. 1 Chemical structures of representative organic dyes

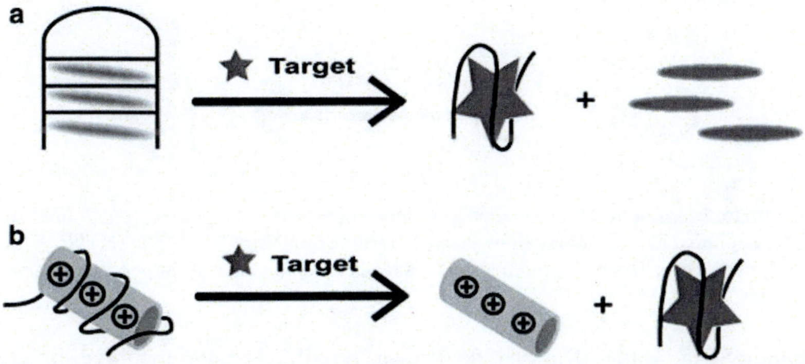

Fig. 2 Label-free approach for the design of fluorescence aptamer reporters. (**a**) Displacement of a fluorescent dye. In the absence of target, the binding between the aptamer (*black line*) and the DNA-binding dye (*round disks*) produces a fluorescence signal (*left*). The presence of the target causes the dissociation of the dye and as a result, the fluorescence signal is decreased (*right*). (**b**) Displacement of a fluorescent polymer. In the absence of target, the binding between the aptamer (*black line*) and the cationic conjugated polymer (*rod shape*) quenches the fluorescence signal (*left*). The target binding to the aptamer displaces the cationic conjugated polymer and the fluorescence signal is restored

in DNA and RNA. Polythiophene is a water-soluble polymer and fluoresces strongly as a free polymer. When it interacts with the DNA aptamer, the electrostatic interaction between the polymer and the aptamer promotes polymer–aptamer aggregation with the reduction in fluorescence. When thrombin is introduced, the aptamer folds into a guanine quadruplex structure to bind the protein; the resulting complex is only weakly associated with the polymer allowing the fluorescence signal to be restored.

The above label-free methods utilize an external reagent (organic dye or cationic conjugated polymer) to achieve target detection. The key advantage is the simplicity. However, label-free assays are susceptible to interference by other components in a sample [3] and certain organic dyes, such as ethidium bromide, are mutagenic and therefore cause health concerns. Furthermore, label-free methods cannot precisely probe a local conformation change in an aptamer and consequently, the fluorescence enhancement becomes difficult to optimize under different conditions [3].

2.2 *Single-Labeled Signaling Aptamers*

Fluorescence enhancement is governed by changes in the fluorophore's spectroscopic properties and therefore, susceptible to an aptamer's conformational change which can cause significant perturbations in a fluorophore's local electronic environment [56].

Ellington's group illustrated this concept of single-labeled signaling aptamers with modified constructs of an ATP-binding aptamer that are labeled with a single

Fig. 3 Single fluorophore labeling approach. Detection using an aptamer (*black line*) modified with a fluorophore (F). The interactions between target (*star shape*) and aptamer (*black line*) alter the fluorophore's electronic environment, which leads to an increase of fluorescence signal (*right*)

fluorophore (Fig. 3) [57]. Rational design was possible because the three-dimensional structure of this aptamer was available [58]. However, the signaling probes developed in this study only showed moderate fluorescence enhancement upon ATP binding. Ellington and coworkers then went on to perform an in vitro selection experiment to isolate ATP-binding aptamers containing fluorophore-modified nucleotide analogs [59]. The isolated signaling aptamers exhibit a 200-fold increase in binding affinity and a 2-fold fluorescence enhancement at the saturating ATP concentration.

Early attempts to convert aptamers into signaling probes required optimization steps based on the structure of the aptamer and significant fluorescence enhancement is not always observed in the presence of target [56]. It can be difficult to achieve a substantial change in fluorescence intensity with the single fluorophore-labeled aptamer approach. The site of attachment for a fluorophore on an aptamer is critical for both target recognition and fluorescence enhancement [3]. If a fluorophore is labeled at a location where no conformational change takes place during ligand binding, it will not produce detectable fluorescence enhancement when the cognate target is added. On the other hand, if a fluorophore is introduced at a conserved nucleotide, it may abolish the aptamer's binding affinity. The major limitations of the single-labeled approach include high background signal and the requirement of the availability of structural information for a specific aptamer. Therefore, it can be difficult to generalize this strategy for different aptamers.

2.3 Double-Labeled Signaling Aptamers

The double-labeled signaling aptamer approach involves labeling an aptamer with either a fluorophore and quencher pair for fluorescence quenching or a pair of fluorophores for fluorescence resonance energy transfer (FRET). This strategy often exploits conformational changes of aptamers to perturb spatial arrangements between the fluorophore/quencher or fluorophore/fluorophore pair; consequently, this allows the synchronization of the ligand-binding event with signal transduction.

One common double-labeled method is "aptamer beacon", based on the molecular beacon design that was originally developed for the detection of nucleic acid sequences through hybridizations [60, 61]. An aptamer beacon consists of a

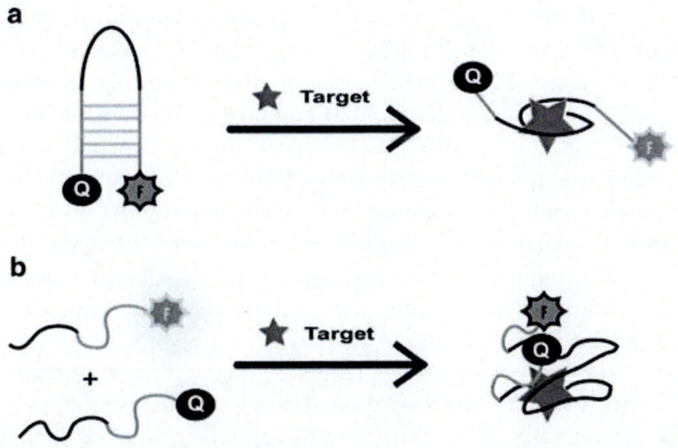

Fig. 4 Double-labeled approach. (**a**) Aptamer beacon design. Fluorescence is used to probe a hairpin structure (*left*) and a complex with target (*right*). Aptamer (*black line*) is labeled with a fluorophore (F) and a quencher (Q) appended on the terminal ends (*gray lines*). In the absence of target (*star shape*), fluorescence is low because the fluorophore and the quencher are oriented in close proximity (*left*). In presence of target, the binding between the aptamer and target disrupts the hairpin structure and a fluorescence signal results (*right*). (**b**) Self-assemble aptamer design. An aptamer is split into two single-strand oligonucleotides, one labeled a fluorophore (F) and the other with a quencher (Q). High fluorescence background results in the absence of target (*left*). These oligonucleotides assemble into a complex with the target (*star shape*) and the fluorescence background is decreased (*right*)

target recognition region flanked by complementary ends that are labeled with a fluorophore and a quencher (Fig. 4a). In the absence of target, the aptamer beacon folds into a hairpin structure so that the fluorophore and quencher are forced into close proximity for fluorescence quenching. In the presence of target, the binding between an aptamer and its target promotes a conformational change that disrupts the hairpin structure allowing the fluorescence signal to be restored. Hamaguchi et al. adapted the aptamer beacon design to construct a double-labeled aptamer that binds thrombin [62]. Complementary sequences were added at the terminal ends, which facilitate the formation of a hairpin structure in the absence of thrombin. Under this conformation, the fluorophore and quencher located on the terminal ends are closely positioned to achieve efficient fluorescence quenching. When thrombin is introduced, this aptamer undergoes a transition from the hairpin structure to the guanine quadruplex structure so that the fluorophore and quencher are separated and this spatial rearrangement restores the fluorescence signal. This fluorescent reporter shows a 2.5-fold increase in fluorescence when thrombin is present.

The self-assemble design requires an aptamer to be split into two oligonucleotides and each oligonucleotide is either labeled with a fluorophore or a quencher (Fig. 4b). Signal transduction is coupled to the ligand-dependent formation of a stable target-bound complex. Stojanovic and colleagues made biosensors that detect cocaine and ATP using the self-assemble design [63]. In the absence of cognate target, the fluorophore-labeled and quencher-labeled oligonucleotides do not assemble and therefore, a fluorescence signal is produced. In the presence of cognate target, these oligonucleotides undergo ligand-dependent self-assembly, causing the fluorophore and quencher to be in close proximity; consequently, the fluorescence signal is reduced. Yamamoto and coworkers constructed an aptamer beacon for the Tat protein of HIV that undergoes ligand-dependent self-assembly [64]. The Tat-binding aptamer was separated into two oligonucleotides, one strand was double-labeled with a fluorophore and a quencher and the other was unmodified. In the absence of Tat protein, the labeled strand forms a hairpin structure (like an aptamer beacon) and low fluorescence signal is produced. In the presence of Tat, the unmodified strand forms a complex with the labeled strand producing significant fluorescence enhancement (up to 14-fold).

The above examples illustrate that aptamers labeled with a fluorophore– quencher pair can effectively reduce the background level of fluorescence. However, the compatibility between the binding structure of an aptamer and the structural restrictions imposed by a given design can greatly affect the performance of the aptamer. For example, aptamers that cannot be split are not compatible with the self-assemble approach.

2.4 *Structure-Switching Signaling Aptamers*

Significant fluorescence enhancement from the aforementioned design strategies can be difficult to achieve because specific structural information of an aptamer is required to produce signaling aptamers with optimal settings. Our group came up with a more generalizable rational design strategy for making "structure-switching" fluorescent aptamer reporters (Fig. 5) [65]. This method utilizes fluorescence-labeled DNA aptamers and quencher-labeled complementary DNA strands. The production of a fluorescence signal is dependent on structural differences between the bound and unbound states of an aptamer. In the absence of the ligand for the aptamer, a weak DNA duplex is formed between the fluorophore-labeled aptamer and the quencher-labeled oligonucleotide, and this conformation allows the fluorophore and quencher to be in close proximity and therefore, the fluorescence from the fluorophore is quenched. In the presence of the ligand, the complex is formed

Fig. 5 Structure-switching approach. Target (*star shape*) promotes structure-switching of an aptamer from a DNA–DNA duplex (*left*) to a DNA–target complex (*right*). The dissociation of quencher-labeled DNA strand (QDNA) from the aptamer is accompanied with an increase of fluorescence signal. F and Q denote fluorophore and quencher, respectively

between the aptamer and its ligand causing the dissociation of the quencher-labeled oligonucleotide from the fluorophore-labeled aptamer with a concomitant increase in the fluorescence signal.

To demonstrate the structure-switching concept, the ATP- and thrombin-binding aptamers were converted into structure-switching reporters [65]. These fluorescent probes were able to provide real-time detection and delivered over 10-fold fluorescence enhancement when the cognate target was introduced. Furthermore, the rational design of structure-switching biosensors can be accomplished with either a tripartite (as shown in Fig. 5) or bipartite system, both of which have been shown to be able to signal the presence of target. Ellington's group has also developed structure-switching aptamers using quantum dots [66]. These studies demonstrate that the structure-switching design does not impose structure limitations that an aptamer needs to adapt in order to be signaling competent.

Similar to other rational design methods, structure-switching also has an inherent caveat, that is, the native binding affinity of an aptamer can be affected when the aptamer is confined in a duplex structure. Strategies to overcome this limitation have been reported in greater detail elsewhere [67]. In general, the number of base pairs occupied by the complementary DNA can be optimized to suit the temperature at which the biosensor will operate or mismatch base pairs can be employed to form a flawed duplex to facilitate the structure-switching process. Alternatively, the structure-switching concept can be implemented with in vitro selection to isolate aptamers with the "trained" structure-switching ability, as illustrated in a study by our group [68]. The DNA library used in this work contained a fixed region flanked by two variable domains that are further flanked by primer-binding sites. The fixed region is complementary to a short oligonucleotide modified with biotin, which allowed the library to be immobilized onto avidin beads. Upon ligand-binding, the variable domain would interact with the target and trigger a conformational change so that the aptamer became dissociated from the short oligonucleotide. Under this condition, the structure-switching mechanism is linked to the isolation of aptamers. The aptamer sequences are required not only to be able to recognize its cognate target but also respond in a structure-switching manner. In this way, the isolated aptamers possess the inherent ability to undergo conformational change via structure-switching. Consequently, the effect of duplex structure on an aptamer's binding affinity can be reduced. Fluorescent reporters for ATP and GTP were isolated by this approach. The obtained structure-switching fluorescent probes showed signaling capability and target specificity without additional optimization steps.

Tan and coworkers showed an approach using fluorophore-labeled aptamers that are covalently attached with their quencher-labeled complementary DNA through a PEG linker [69]. Their study showed that the number of base pairs between the complementary DNA and the aptamer can be reduced and potentially make the structure-switching process easier. Aptamers that bind ATP and thrombin were chosen to demonstrate this working principle. The ATP and the thrombin fluorescent probes showed a 30-fold and 16-fold fluorescence enhancement in the presence of cognate target, respectively.

By exploiting predictable conformation changes of aptamers, structure-switching provides a simple and general solution for the design of fluorescent aptamer biosensors. Our group further exploited the structure-switching aptamers for monitoring enzymatic reactions for alkaline phosphatase [70], screening of inhibitors for adenine deaminase in a high-throughput screening (HTS) format [71] and developing a nanoengine that utilizes structure-switching as a precise delivery mechanism [72]. Structure-switching signaling aptamers have also been incorporated into sol-gel derived silica to allow detection on solid support [73]. The implementation of the structure-switching concept has considerably simplified the operational procedures in these applications.

3 Fluorescent Nucleic Acid Enzyme Biosensors

As discussed earlier, deoxyribozymes (DNAzymes) and ribozymes, collectively known as nucleic acid enzymes, are single-stranded DNA or RNA molecules that exhibit catalytic activity [74]. Many nucleic acid enzymes exist in nature or have been isolated to catalyze different types of chemical reactions, such as DNA and RNA cleavage, ligation and DNA phosphorylation [75, 76]. Nucleic acid enzymes can function in two forms: *cis* and *trans*. In *cis*, the enzyme and the substrate are linked together, while in *trans* they are separated [77].

Many nucleic acid enzymes require certain divalent metal ions as cofactors to function properly, and thus, they can be used as sensors for these metal ions [78]. In addition, a nucleic acid enzyme can be coupled to an aptamer to produce an allosteric nucleic acid enzyme (also known as an aptazyme). In this case, a change in the aptamer's ligand-binding state (bound or unbound) is translated into a change in the catalytic activity of the nucleic acid enzyme. Therefore, an aptazyme can function as a sensor for a specific target, expanding the nucleic acid enzyme's sensing utility. Nucleic acid enzymes and aptazymes can also be easily developed into fluorescent sensors, making them useful for biosensing and related applications such as molecular computation [79].

3.1 Nucleic Acid Enzyme Sensors

Synthetic nucleic acid enzymes are typically developed through the process of in vitro selection (see the discussion in the Introduction), which starts with a pool of random DNA or RNA sequences. Sequences that possess the desired catalytic activity are separated from the noncatalytic sequences, and amplified. After several iterations of selection and amplification, catalytic sequences will prevail in the selected population [74]. These sequences can then be cloned, sequenced, and characterized. Some nucleic acid enzymes can then be further developed into different types of sensors, including fluorescent ones.

One common reaction catalyzed by DNAzymes is the cleavage of the embedded RNA linkage within a DNA substrate (chimeric DNA/RNA substrate). Lu and colleagues have found that one particular DNAzyme named "8-17" can highly efficiently cleave a chimeric DNA/RNA substrate in the presence of Pb^{2+}; this finding has allowed them to turn a DNAzyme into a Pb^{2+} sensor. For example, they have made 8-17 into a fluorescent sensor by attaching a fluorophore (F; specifically tetramethylrhodamine or TMR) to the 5′ end of the substrate and a quencher (Q; specifically 4-((4-(dimethylamino) phenyl) azo) benzoic acid or DABCYL) to the 3′ end of the DNAzyme (Fig. 6a). When the substrate and DNAzyme strands are hybridized, DABCYL quenches the fluorescence of TMR. In the presence of lead, the DNAzyme cleaves the substrate, generating two short DNA fragments that can no longer hybridize to the DNAzyme strand. The dissociation of these two

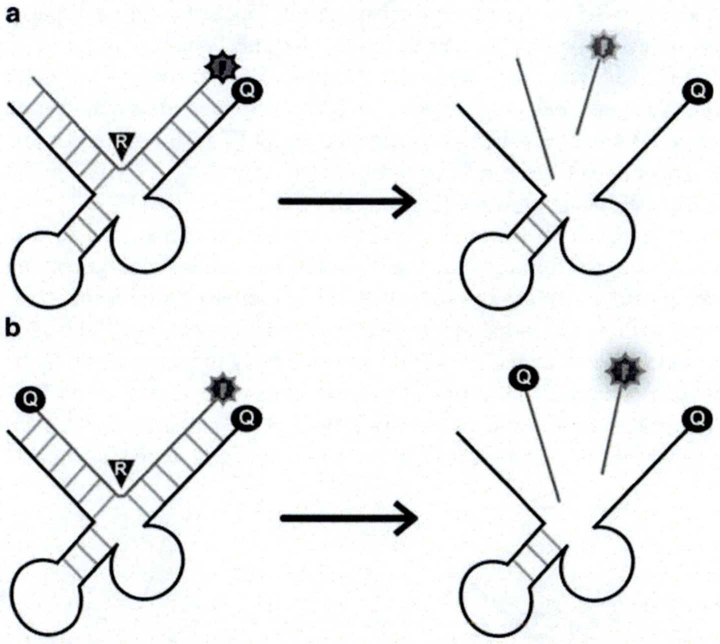

Fig. 6 Pb^{2+}-sensing 8-17 DNAzyme. (**a**) One fluorophore–one quencher design. F is TMR attached to one end of the substrate strand (*gray line*); Q is DABCYL linked to one end of the DNAzyme (*black line*); R denotes the RNA linkage to act as the cleavage site. The substrate strand hybridizes with the DNAzyme strand through Watson–Crick base-pair formation (*short gray dashes*), which leads to a low level of fluorescence due to the quenching by Q towards F. In the presence of lead ion, the substrate is cleaved by the DNAzyme at the location of R into two short fragments, which can no longer anneal with the DNAzyme strand. The separation of F and Q produces a high level of fluorescence. (**b**) One fluorophore–two-quencher design. This specific design places a fluorophore and quencher at the two opposite ends of the substrate which itself can form an intramolecular fold-back structure to bring F and Q close to each other. Thus, any free substrate that is not hybridized to the DNAzyme strand will stay in fluorescence quenching state to keep the background fluorescence low

fragments from the DNAzyme is accompanied with a datable fluorescence signal (Fig. 6a) [78].

A limitation of the above design is that the substrate molecules are not completely hybridized with the DNAzyme strands and thus, the presence of unquenched F can result in high background fluorescence. In an effort to overcome this issue, the Lu group employed a clever strategy: they redesigned the substrate strand with two important features. First, they attached a Q at one end of the substrate and an F to the other. Second, they reconfigured the sequence of the substrate to allow the formation of intramolecular base pairs to bring the two ends into a fold-back structure. Thus any free substrate molecules that are not annealed will adopt this internal structure to reduce the background fluorescence (Fig. 6b) [78].

The Lu group has also reported a different method for designing the 8-17 DNAzyme into a fluorescent lead sensor without having to label the DNA strands with fluorophores and quenchers (Fig. 7). In this case, they replaced one of the bases on one of the two 8-17 DNAzyme's substrate binding arms with a dSpacer to create a baseless (abasic) nucleotide unit in the DNAzyme. When the rest of the binding arm is hybridized with the substrate, 2-amino-5,6,7-trimethyl-1,8-naphthyridine (ATMND, a fluorescent compound) binds the dSpacer unit, which quenches its fluorescence. When the substrate is cleaved by 8-17 in the presence of Pb^{2+}, the substrate dissociates from the DNAzyme strand, releasing ATMND and producing a fluorescence enhancement [80].

Another approach to lower the background fluorescence was to use the 8-17 DNAzyme in the *cis* format where the 3′ end of the substrate and the 5′ end of 8-17 were linked with a 10-T base sequence. This increases the amount of substrate hybridizing with 8-17, reducing the background fluorescence. The length of the substrate was also optimized so that it can efficiently hybridize with 8-17 when the substrate is in one piece but efficiently dissociates once it is cleaved [81].

Our group has found that internally labeling the substrate with a fluorophore and quencher on opposite sides of the cleavage site can significantly reduce

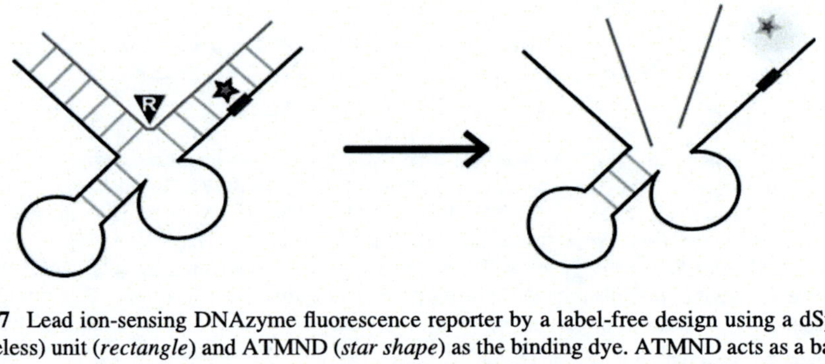

Fig. 7 Lead ion-sensing DNAzyme fluorescence reporter by a label-free design using a dSpacer (baseless) unit (*rectangle*) and ATMND (*star shape*) as the binding dye. ATMND acts as a base to occupy the abasic site and in doing so, the intrinsic fluorescence of ATMND is significantly reduced. In the presence of lead ion, the DNAzyme cleaves the substrate. This action dissolves the duplex structure between the DNAzyme and substrate and releases ATMND, leading to an increased level of fluorescence

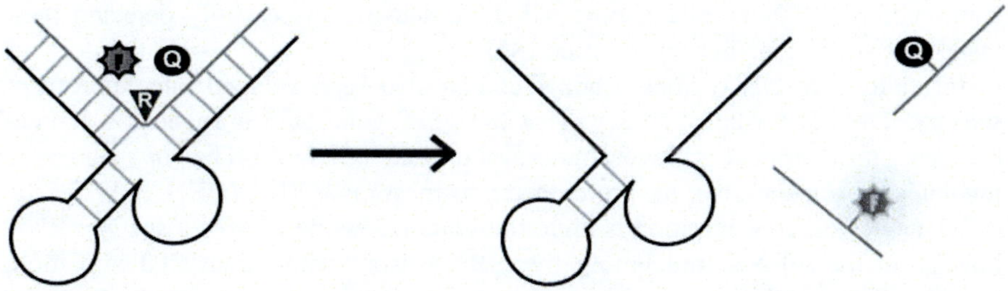

Fig. 8 Fluorescent reporter DNAzymes with internal labels. F: fluorophore; Q: quencher; R: cleavage site

background fluorescence (Fig. 8). This arrangement would produce low background fluorescence when the substrate was intact, and a large fluorescence enhancement when the RNA linkage was cleaved. The substrate was internally doubly labeled with one of three Alexa Fluor fluorophores (Alexa488, Alexa546, and Alexa647) and one of two quenchers (QSY9 and QSY21) at different locations along the substrate. It was found that the closer the fluorophore and quencher were to each other, the higher the fluorescence enhancement became once the substrate was cleaved. But if they were located too close to the cleavage site, the catalytic efficiency of 8-17 decreased [82].

Our group has also developed a series of fluorescence-generating DNAzymes by in vitro selection. These DNAzymes were selected to cleave an RNA linkage sandwiched between a fluorescein-labeled nucleotide on one side and a DABCYL quencher-labeled nucleotide on the other (Fig. 8). In their first report, they isolated a DNAzyme named "DEC22-18", which could efficiently cleave the embedded RNA linkage sandwiched between the fluorophore and quencher, producing a 10-fold fluorescence increase in 5 min [83]. Since the substrate was already internally labeled with fluorophore and quencher during in vitro selection and not just labeled afterwards, their presence did not interfere with DEC22-18's catalytic efficiency.

The same approach was also used by this group to develop a set of signaling DNAzymes that function under different pH conditions. For example, pH3DZ1, which was selected under a solution acidity of pH 3.0, had an optimal pH of 3.0. Similarly, pH6DZ1, which was selected at pH 6.0, exhibited an optimal pH of 6.0 [84].

Other fluorescent non-8-17-based DNAzyme sensors have been reported. A DNAzyme with high sensitivity and selectivity for uranyl (UO_2^{2+}) was developed into a fluorescent sensor using the one-fluorophore–two-quencher design. It was found to have a detection limit of 45 pM and had a greater than a million-fold selectivity over other metal ions [85]. The same one-fluorophore–two-quencher design was used to develop a Cu^{2+}-specific fluorescent sensor with a detection limit of 35 nM. It was developed by rational design using a DNA-cleaving DNAzyme that uses Cu^{2+} as a cofactor. Using the fluorescent DNAzyme sensors for Cu^{2+}, Pb^{2+}, and UO_2^{2+} in tandem, the Lu group tested different solutions with different combinations of these metal ions. All three sensors successfully detected their respective metal ions in each solution [85].

In addition to DNAzymes, ribozymes can also be developed into fluorescent sensors. The *glmS* ribozyme is part of the *glmS* gene, which encodes a protein enzyme whose product is glucosamine-6-phosphate (GlcN6P). The *glmS* ribozyme regulates the translation of the protein from its messenger RNA (mRNA). When the cellular GlcN6P concentration is high, the *glmS* ribozyme cleaves a specific RNA linkage in the mRNA, preventing the mRNA from being translated into more proteins. The *glmS* ribozyme is found in many Gram-positive bacteria, with the version in *Bacillus subtilis* having a detection limit of 200 nM of GlcN6P [48]. It cleaves in the presence of certain analogs of GlcN6P, but the cleavage rates are always lower compared to the rate in the presence of GlcN6P itself [86].

The *glmS* ribozyme of *Staphylococcus aureus* has been developed into a fluorescent sensor by converting it to *trans* and attaching 5/6-FAM (the FRET donor) to the 3′ end and Cy3 (the FRET acceptor) to the 5′ end of the substrate. When the substrate is cleaved, FRET between the two fluorophores is abolished. Using this fluorescent sensor, a high-throughput screen of a chemical library was conducted, searching for compounds that could stimulate *glmS* ribozyme cleavage, which would inhibit the bacteria's growth. Five compounds were found but they were determined to be false positives that did not increase *glmS* ribozyme cleavage. However, this study does demonstrate the possibility of using natural ribozymes as new drug targets [87].

Most fluorescent nucleic acid enzyme sensors have been developed for use in the analytical chemistry and biochemistry fields but the following fluorescent DNAzyme sensors have been developed for computational purposes [88]. A NOT gate was developed using the E6 DNAzyme. The presence of the oligonucleotide (as the input) prevents cleavage (0) and the absence of it allows cleavage (1). An AND gate using the 8-17 DNAzyme (requires two different oligonucleotide inputs for cleavage to occur) and an eXclusive OR (XOR) gate using a set of two E6 DNAzymes (cleavage occurs in the presence of one, and only one oligonucleotide input) were also developed. All substrates of the DNAzymes were labeled on the 3′ end with fluorescein and on the 5′ end with a tetramethylrhodamine quencher so the gates' outputs (amount of cleavage) could be monitored fluorescently. The gates, and various combinations thereof, could be adapted for use in potential diagnostic applications, such as whether a set of multiple molecular markers of disease are present or not [88].

3.2 Aptazyme Sensors

Aptazymes are typically developed by replacing a portion of a nucleic acid enzyme with an aptamer such that the activity of the nucleic acid enzyme is modulated by the appended aptamer. This can be accomplished by rational design, in vitro selection, or a combination of both [75].

The Breaker group developed the first group of aptazymes using several RNA aptamers and the well-known hammerhead ribozyme. For example, they replaced one variable stem (Stem II) of the hammerhead ribozyme with an RNA aptamer that binds flavin mononucleotide (FMN), and more importantly, they devised a four random-sequence base-pair bridge to link the aptamer to the ribozyme. This approach created a library of molecules from which a specific bridge was selected to provide two functions: (1) in the absence of FMN, it prevents the hammerhead ribozyme from cleaving; (2) in the presence of FMN, it registers the conformation change of the FMN aptamer when bound to its target. The bridge then allows the hammerhead ribozyme to cleave, indicating the presence of FMN (Fig. 9a) [89].

Other aptazymes were developed by replacing the FMN aptamer with other aptamers (such as the theophylline-binding aptamer), while leaving the bridge

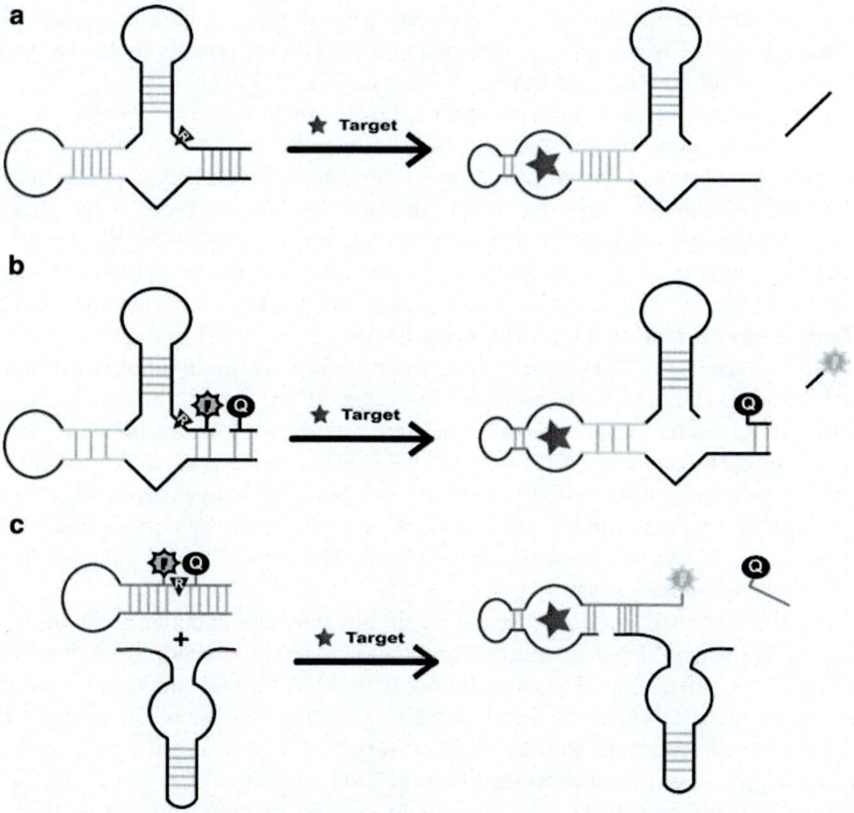

Fig. 9 Several aptazymes demonstrating different ways of an aptamer's allosteric control over the nucleic acid enzyme. Drawing in *dark gray* is the aptamer, that in *light gray* is the base-pair bridge, *black line* is the nucleic acid enzyme, the *triangle* with an imbedded "R" represents the cleavage site, and the *star shape* denotes the target. (**a**) The FMN aptamer linked to the hammerhead ribozyme via a base-pair bridge. (**b**) The caffeine-responsive aptazyme fluorescent reporter. (**c**) An ATP-binding aptamer used to regulate the availability of the substrate (*gray*) to a DNAzyme (MgZ)

intact [89]. The theophylline aptazyme was later developed into a fluorescent sensor by attaching Cy3 (the FRET donor) to the 5′ end and Cy5 (the FRET acceptor) to the 3′ end of the substrate. FRET between Cy3 and Cy5 was abolished when the substrate was cleaved and dissociated in the presence of theophylline [90].

A fluorescent aptazyme sensor for Hg^{2+} was also developed by rational design using the fluorescent UO_2^{2+} sensor seen earlier. The sequence of its hairpin loop was modified so that it would only hybridize into a hairpin loop when Hg^{2+} ions were present. The sensor had a detection limit of 2.4 nM Hg^{2+} and had a selectivity of greater than 100,000-fold over other metal ions [91].

An aptazyme can also be developed through in vitro selection using a random-sequence library containing a known nucleic acid enzyme; the use of such a library is to facilitate the selection of de novo aptamers that can regulate the enzyme activity through target binding. For example, aptazymes for caffeine and aspartame in the setting of the hammerhead ribozyme were developed by this method [92]. These aptazymes were further developed into fluorescent sensors by further modification with fluorescein and DABCYL system (Fig. 9b) [92].

Although nucleic acid enzymes that catalyze RNA cleavage reactions are the most popular, they are not the only allosteric reporters used. Aptazymes were developed by rational design using the L1 ribozyme, which catalyzes the ligation of two RNA strands. A specific stem in the ribozyme was replaced with either the ATP or theophylline aptamer so that RNA ligation activity of the ribozyme was controlled by the addition of ATP or theophylline. An FMN aptazyme was also developed by in vitro selection for a bridge linking the FMN aptamer and the aforementioned stem in the L1 ribozyme [93].

Other fluorescent ATP-sensing aptazymes employ alternate rationally designed arrangements. One of these arrangements uses EC56, the minimized version of pH6DZ1. In *trans*, EC56 was developed into a fluorescent aptazyme by replacing one of its stems with the ATP aptamer [77]. When there is no ATP present, the DNAzyme cannot fold properly into its catalytically active structure, so little cleavage occurs. But when ATP is present, the ATP aptamer portion binds ATP, allowing the DNAzyme portion to fold properly and cleave its substrate, producing a fluorescence enhancement [77].

Another rationally designed arrangement has the ATP aptamer modulating the activity of the MgZ DNAzyme by a completely different method. MgZ cleaves the single RNA linkage in the same fluorophore-RNA linkage-quencher sandwich design as other DNAzymes developed by our group. In *trans*, an aptamer was attached to the substrate so that when no target was present, the aptamer would hybridize with the substrate, preventing MgZ from cleaving the substrate. But when the target was present, the aptamer would change its conformation to bind the target, allowing MgZ to cleave the substrate, producing a fluorescence enhancement (Fig. 9c) [94].

Fluorescent nucleic acid enzyme sensors and their allosteric counterparts have tremendous potential in many fields of study including diagnostics, biosensing, molecular computation, and drug screening [79]. Some of their many uses have already been demonstrated. As seen above, the fluorescent *glmS* ribozyme was used in a drug screen [87]. As well, the UO_2^{2+} fluorescent sensor accurately determined the concentration of UO_2^{2+} in contaminated soil samples [85]. The Pb^{2+} fluorescent sensor accurately determined the concentration of Pb^{2+} in lake water samples [78]. Considering the significant advancements taken by the field of nucleic acid enzymes in its short history, it will be interesting to see their future applications, especially outside of academic institutions. For more information on these types of fluorescent sensors, please see [79].

4 Concluding Remarks

The discovery and application of FNAs has been an extremely exciting and productive research arena for the past 30 years. Various biological systems have invented many ribozymes and riboswitches and use them as important players for precision controls of various biological processes such as RNA splicing, protein synthesis, and gene expression. The human invention of the in vitro selection technique has provided researchers an extremely powerful yet simple test-tube technique to isolate man-made DNA and RNA aptamers, ribozymes and deoxyribozymes, and allosteric ribozymes and deoxyribozymes from synthetic DNA or RNA libraries. It is certain that continuing applications of this technique will generate more and better FNAs. With the build-up of a large reservoir of FNAs, more and more of them will be explored as building blocks to make useful molecular tools for research and real-world applications.

FNAs have found, and will continue to find, useful applications in many fields, such as bioanalysis, therapeutics, diagnostics, chemical biology, and nanotechnology. This chapter has intended to serve the purpose of providing a focused review of only a particular research field where FNAs have been explored for the design of fluorescent biosensors and bioassays. As discussed in this chapter, different dye moieties (fluorophores and quenchers) can be used to provide aptamers and aptazymes with many choices of fluorescence signaling set-ups such that a molecular recognition event can be reported through fluorescence signals. The availability of both a wide selection of fluorophores and quenchers and a large collection of FNAs will continue to provide a great opportunity for the design of highly effective fluorescent reporters for many important applications such as drug screening, metabolite profiling, and any other sensing application in need. With the increasing demand of molecular reporters that can make a quick response to the presence of the target of interest with high specificity, a large detection dynamic range and a fluorescence output, we hope our chapter will motivate more researchers to consider fluorescent FNA sensors as a possible choice when they seek to design a biosensor for their applications of interest.

References

1. Navani NK, Li Y (2006) Nucleic acid aptamers and enzymes as sensors. Curr Opin Chem Biol 10:272–281
2. Silverman SK (2009) Artificial functional nucleic acids: aptamers, ribozymes, and deoxyribozymes identified by in vitro selection. In: Li Y, Lu Y (eds) Functional nucleic acids for analytical applications. Springer, New York, pp 47–108
3. Liu J, Cao Z, Lu Y (2009) Functional nucleic acid sensors. Chem Rev 109:1948–1998
4. Kruger K, Grabowski PJ, Zaug AJ et al (1982) Self-splicing RNA: autoexcision and autocyclization of the ribosomal RNA intervening sequence of Tetrahymena. Cell 31:147–157
5. Guerrier-Takada C, Gardiner K, Marsh T et al (1983) The RNA moiety of ribonuclease P is the catalytic subunit of the enzyme. Cell 35:849–857
6. Ellington AD, Szostak JW (1990) In vitro selection of RNA molecules that bind specific ligands. Nature 346:818–822
7. Tuerk C, Gold L (1990) Systematic evolution of ligands by exponential enrichment: RNA ligands to bacteriophage T4 DNA polymerase. Science 249:505–510
8. Robertson DL, Joyce GF (1990) Selection in vitro of an RNA enzyme that specifically cleaves single-stranded DNA. Nature 344:467–468
9. Chapman KB, Szostak JW (1994) In vitro selection of catalytic RNAs. Curr Opin Struct Biol 4:618–622
10. Breaker RR (1997) DNA aptamers and DNA enzymes. Curr Opin Chem Biol 1:26–31
11. Wilson DS, Szostak JW (1999) In vitro selection of functional nucleic acids. Annu Rev Biochem 68:611–647
12. Li Y, Breaker RR (1999) Deoxyribozymes: new players in the ancient game of biocatalysis. Curr Opin Struct Biol 9:315–323
13. Famulok M, Mayer G, Blind M (2000) Nucleic acid aptamers-from selection in vitro to applications in vivo. Acc Chem Res 33:591–599
14. Cech TR (2002) Ribozymes, the first 20 years. Biochem Soc Trans 30:1162–1166
15. Emilsson GM, Breaker RR (2002) Deoxyribozymes: new activities and new applications. Cell Mol Life Sci 59:596–607
16. Achenbach JC, Chiuman W, Cruz RP, Li Y (2004) DNAzymes: from creation in vitro to application in vivo. Curr Pharm Biotechnol 5:321–336
17. Hobartner C, Silverman SK (2007) Recent advances in DNA catalysis. Biopolymers 87:279–292
18. Schlosser K, Li Y (2009) Biologically inspired synthetic enzymes made from DNA. Chem Biol 16:311–322
19. Stoddard CD, Batey RT (2006) Mix-and-match riboswitches. ACS Chem Biol 1:751–754
20. Winkler WC, Breaker RR (2003) Genetic control by metabolite-binding riboswitches. Chembiochem 4:1024–1032
21. Borovok I, Gorovitz B, Schreiber R et al (2006) Coenzyme B12 controls transcription of the Streptomyces class Ia ribonucleotide reductase nrdABS operon via a riboswitch mechanism. J Bacteriol 188:2512–2520
22. Nahvi A, Sudarsan N, Ebert MS et al (2002) Genetic control by a metabolite binding mRNA. Chem Biol 9:1043
23. Warner DF, Savvi S, Mizrahi V, Dawes SS (2007) A riboswitch regulates expression of the coenzyme B12-independent methionine synthase in Mycobacterium tuberculosis: implications for differential methionine synthase function in strains H37Rv and CDC1551. J Bacteriol 189:3655–3659
24. Mandal M, Lee M, Barrick JE et al (2004) A glycine-dependent riboswitch that uses cooperative binding to control gene expression. Science 306:275–279
25. Grundy FJ, Lehman SC, Henkin TM (2003) The L box regulon: lysine sensing by leader RNAs of bacterial lysine biosynthesis genes. Proc Natl Acad Sci USA 100:12057–12062

26. Rodionov DA, Vitreschak AG, Mironov AA, Gelfand MS (2003) Regulation of lysine biosynthesis and transport genes in bacteria: yet another RNA riboswitch? Nucleic Acids Res 31:6748–6757
27. Sudarsan N, Wickiser JK, Nakamura S et al (2003) An mRNA structure in bacteria that controls gene expression by binding lysine. Genes Dev 17:2688–2697
28. Gilbert SD, Rambo RP, Van Tyne D, Batey RT (2008) Structure of the SAM-II riboswitch bound to S-adenosylmethionine. Nat Struct Mol Biol 15:177–182
29. Corbino KA, Barrick JE, Lim J et al (2005) Evidence for a second class of S-adenosylmethionine riboswitches and other regulatory RNA motifs in alpha-proteobacteria. Genome Biol 6:R70
30. Epshtein V, Mironov AS, Nudler E (2003) The riboswitch-mediated control of sulfur metabolism in bacteria. Proc Natl Acad Sci USA 100:5052–5056
31. Fuchs RT, Grundy FJ, Henkin TM (2006) The S(MK) box is a new SAM-binding RNA for translational regulation of SAM synthetase. Nat Struct Mol Biol 13:226–233
32. McDaniel BA, Grundy FJ, Artsimovitch I, Henkin TM (2003) Transcription termination control of the S box system: direct measurement of S-adenosylmethionine by the leader RNA. Proc Natl Acad Sci USA 100:3083–3088
33. Winkler WC, Nahvi A, Sudarsan N et al (2003) An mRNA structure that controls gene expression by binding S-adenosylmethionine. Nat Struct Biol 10:701–707
34. Wang JX, Lee ER, Morales DR et al (2008) Riboswitches that sense S-adenosylhomocysteine and activate genes involved in coenzyme recycling. Mol Cell 29:691–702
35. Johansen LE, Nygaard P, Lassen C et al (2003) Definition of a second Bacillus subtilis pur regulon comprising the pur and xpt-pbuX operons plus pbuG, nupG (yxjA), and pbuE (ydhL). J Bacteriol 185:5200–5209
36. Mandal M, Breaker RR (2004) Adenine riboswitches and gene activation by disruption of a transcription terminator. Nat Struct Mol Biol 11:29–35
37. Mandal M, Boese B, Barrick JE et al (2003) Riboswitches control fundamental biochemical pathways in Bacillus subtilis and other bacteria. Cell 113:577–586
38. Roth A, Winkler WC, Regulski EE et al (2007) A riboswitch selective for the queuosine precursor preQ1 contains an unusually small aptamer domain. Nat Struct Mol Biol 14:308–317
39. Meyer MM, Roth A, Chervin SM et al (2008) Confirmation of a second natural preQ1 aptamer class in Streptococcaceae bacteria. RNA 14:685–695
40. Mironov AS, Gusarov I, Rafikov R et al (2002) Sensing small molecules by nascent RNA: a mechanism to control transcription in bacteria. Cell 111:747–756
41. Rodionov DA, Vitreschak AG, Mironov AA, Gelfand MS (2002) Comparative genomics of thiamin biosynthesis in procaryotes. New genes and regulatory. J Biol Chem 277:48949–48959
42. Cheah MT, Wachter A, Sudarsan N, Breaker RR (2007) Control of alternative RNA splicing and gene expression by eukaryotic riboswitches. Nature 447:497–500
43. Wachter A, Tunc-Ozdemir M, Grove BC et al (2007) Riboswitch control of gene expression in plants by splicing and alternative 3′ end processing of mRNAs. Plant Cell 19:3437–3450
44. Croft MT, Moulin M, Webb ME, Smith AG (2007) Thiamine biosynthesis in algae is regulated by riboswitches. Proc Natl Acad Sci USA 104:20770–20775
45. Bocobza S, Adato A, Mandel T et al (2007) Riboswitch-dependent gene regulation and its evolution in the plant kingdom. Genes Dev 21:2874–2879
46. Winkler WC, Cohen-Chalamish S, Breaker RR (2002) An mRNA structure that controls gene expression by binding FMN. Proc Natl Acad Sci USA 99:15908–15913
47. Sudarsan N, Lee ER, Weinberg Z et al (2008) Riboswitches in eubacteria sense the second messenger cyclic di-GMP. Science 321:411–413
48. Winkler WC, Nahvi A, Roth A et al (2004) Control of gene expression by a natural metabolite-responsive ribozyme. Nature 428:281–286
49. Cromie MJ, Shi Y, Latifi T, Groisman EA (2006) An RNA sensor for intracellular Mg(2+). Cell 125:71–84

50. Klussmann S (2006) The aptamer handbook: functional oligonucleotides and their applications. Wiley-VCH, Weinheim
51. Mascini M (2009) Aptamers in bioanalysis. Wiley, New York
52. Li Y, Lu Y (2009) Functional nucleic acids for analytical applications. Springer, New York
53. Lu Y, Liu J (2006) Functional DNA nanotechnology: emerging applications of DNAzymes and aptamers. Curr Opin Biotechnol 17:580–588
54. Zhou C, Jiang Y, Hou S et al (2006) Detection of oncoprotein platelet-derived growth factor using a fluorescent signaling complex of an aptamer and TOTO. Anal Bioanal Chem 384:1175–1180
55. Ho HA, Leclerc M (2004) Optical sensors based on hybrid aptamer/conjugated polymer complexes. J Am Chem Soc 126:1384–1387
56. Nutiu R, Li Y (2004) Structure-switching signaling aptamers: transducing molecular recognition into fluorescence signaling. Chemistry 10:1868–1876
57. Jhaveri SD, Kirby R, Conrad R et al (2000) Designed signaling aptamers that transduce molecular recognition to changes in fluorescence intensity. J Am Chem Soc 122: 2469–2473
58. Hermann T, Patel DJ (2000) Adaptive recognition by nucleic acid aptamers. Science 287:820–825
59. Jhaveri S, Rajendran M, Ellington AD (2000) In vitro selection of signaling aptamers. Nat Biotechnol 18:1293–1297
60. Tyagi S, Kramer FR (1996) Molecular beacons: probes that fluoresce upon hybridization. Nat Biotechnol 14:303–308
61. Wang K, Tang Z, Yang CJ et al (2009) Molecular engineering of DNA: molecular beacons. Angew Chem Int Ed Engl 48:856–870
62. Hamaguchi N, Ellington A, Stanton M (2001) Aptamer beacons for the direct detection of proteins. Anal Biochem 294:126–131
63. Stojanovic MN, de Prada P, Landry DW (2000) Fluorescent sensors based on aptamer self-assembly. J Am Chem Soc 122:11547–11548
64. Yamamoto R, Baba T, Kumar PK (2000) Molecular beacon aptamer fluoresces in the presence of Tat protein of HIV-1. Genes Cells 5:389–396
65. Nutiu R, Li Y (2003) Structure-switching signaling aptamers. J Am Chem Soc 125:4771–4778
66. Levy M, Cater SF, Ellington AD (2005) Quantum-dot aptamer beacons for the detection of proteins. Chembiochem 6:2163–2166
67. Nutiu R, Li Y (2005) Aptamers with fluorescence-signaling properties. Methods 37:16–25
68. Nutiu R, Li Y (2005) In vitro selection of structure-switching signaling aptamers. Angew Chem Int Ed Engl 44:1061–1065
69. Tang Z, Mallikaratchy P, Yang R et al (2008) Aptamer switch probe based on intramolecular displacement. J Am Chem Soc 130:11268–11269
70. Nutiu R, Yu JM, Li Y (2004) Signaling aptamers for monitoring enzymatic activity and for inhibitor screening. Chembiochem 5:1139–1144
71. Elowe NH, Nutiu R, Allali-Hassani A et al (2006) Small-molecule screening made simple for a difficult target with a signaling nucleic acid aptamer that reports on deaminase activity. Angew Chem Int Ed Engl 45:5648–5652
72. Nutiu R, Li Y (2005) A DNA-protein nanoengine for "on-demand" release and precise delivery of molecules. Angew Chem Int Ed Engl 44:5464–5467
73. Rupcich N, Nutiu R, Li Y, Brennan JD (2005) Entrapment of fluorescent signaling DNA aptamers in sol-gel-derived silica. Anal Chem 77:4300–4307
74. Silverman SK (2005) In vitro selection, characterization, and application of deoxyribozymes that cleave RNA. Nucleic Acids Res 33:6151–6163
75. Breaker RR (2004) Natural and engineered nucleic acids as tools to explore biology. Nature 432:838–845
76. Li Y, Breaker RR (2001) In vitro selection of kinase and ligase deoxyribozymes. Methods 23:179–190

77. Shen Y, Chiuman W, Brennan JD, Li Y (2006) Catalysis and rational engineering of trans-acting pH6DZ1, an RNA-cleaving and fluorescence-signaling deoxyribozyme with a four-way junction structure. Chembiochem 7:1343–1348
78. Liu J, Lu Y (2003) Improving fluorescent DNAzyme biosensors by combining inter- and intramolecular quenchers. Anal Chem 75:6666–6672
79. Chiuman W, Li Y (2009) Fluorescent Ribozyme and Deoxyribozyme Sensors. In: Li Y, Lu Y (eds) Functional nucleic acids for analytical applications. Springer Science + Business Media, New York, NY, pp 131–153
80. Xiang Y, Tong A, Lu Y (2009) Abasic site-containing DNAzyme and aptamer for label-free fluorescent detection of Pb(2+) and adenosine with high sensitivity, selectivity, and tunable dynamic range. J Am Chem Soc 131:15352–15357
81. Wang H, Kim Y, Liu H et al (2009) Engineering a unimolecular DNA-catalytic probe for single lead ion monitoring. J Am Chem Soc 131:8221–8226
82. Chiuman W, Li Y (2007) Efficient signaling platforms built from a small catalytic DNA and doubly labeled fluorogenic substrates. Nucleic Acids Res 35:401–405
83. Mei SH, Liu Z, Brennan JD, Li Y (2003) An efficient RNA-cleaving DNA enzyme that synchronizes catalysis with fluorescence signaling. J Am Chem Soc 125:412–420
84. Liu Z, Mei SH, Brennan JD, Li Y (2003) Assemblage of signaling DNA enzymes with intriguing metal-ion specificities and pH dependences. J Am Chem Soc 125:7539–7545
85. Liu J, Brown AK, Meng X et al (2007) A catalytic beacon sensor for uranium with parts-per-trillion sensitivity and millionfold selectivity. Proc Natl Acad Sci USA 104:2056–2061
86. Lim J, Grove BC, Roth A, Breaker RR (2006) Characteristics of ligand recognition by a glmS self-cleaving ribozyme. Angew Chem Int Ed Engl 45:6689–6693
87. Blount K, Puskarz I, Penchovsky R, Breaker R (2006) Development and application of a high-throughput assay for glmS riboswitch activators. RNA Biol 3:77–81
88. Stojanovic MN, Mitchell TE, Stefanovic D (2002) Deoxyribozyme-based logic gates. J Am Chem Soc 124:3555–3561
89. Soukup GA, Breaker RR (1999) Engineering precision RNA molecular switches. Proc Natl Acad Sci USA 96:3584–3589
90. Sekella PT, Rueda D, Walter NG (2002) A biosensor for theophylline based on fluorescence detection of ligand-induced hammerhead ribozyme cleavage. RNA 8:1242–1252
91. Liu J, Lu Y (2007) Rational design of "turn-on" allosteric DNAzyme catalytic beacons for aqueous mercury ions with ultrahigh sensitivity and selectivity. Angew Chem Int Ed Engl 46:7587–7590
92. Ferguson A, Boomer RM, Kurz M et al (2004) A novel strategy for selection of allosteric ribozymes yields RiboReporter sensors for caffeine and aspartame. Nucleic Acids Res 32:1756–1766
93. Robertson MP, Ellington AD (2000) Design and optimization of effector-activated ribozyme ligases. Nucleic Acids Res 28:1751–1759
94. Chiuman W, Li Y (2007) Simple fluorescent sensors engineered with catalytic DNA 'MgZ' based on a non-classic allosteric design. PLoS One 2:e1224

Part IV
Cell Imaging with Organic Dyes

Covalent Labeling of Biomolecules in Living Cells

Tilman Plass and Carsten Schultz

Abstract Putting fluorescent and other labels on proteins and small molecules in living cells is an important tool for studying cell biology and developing drugs. The required chemical reactions need to be bioorthogonal to produce selectivity. Here, we summarize currently used bioorthogonal reactions that have been successfully applied for labeling biomolecules in cells. Further, we discuss the various methods available to include orthogonally reactive groups into proteins.

Keywords Activity-based probes · Bioorthogonal reactions · Cycloaddition reactions · Expanded genetic code · Fusion proteins

Contents

T. Plass and C. Schultz (✉)
European Molecular Biology Laboratory, Meyerhofstraße 1, 69117 Heidelberg, Germany
e-mail: carsten.schultz@embl.de

A.P. Demchenko (ed.), *Advanced Fluorescence Reporters in Chemistry and Biology III: Applications in Sensing and Imaging*, Springer Ser Fluoresc (2011) 10: 225–262,
DOI 10.1007/978-3-642-18035-4_7,

1 Introduction

Arguably the most relevant and the most common instrument in biology is the microscope. There is hardly another experiment where we learn more about cells and organisms then when we are able to observe the object of interest in real time and with as much resolution as possible. What we are able to observe, however, depends in part on the probes we are able to introduce to cells. Surely, fluorescent and luminescent probes are the most effective ones and an entire industry is developing to make optical imaging available for biology. While stains and fluorescent dyes for organelles and fluorescently labeled antibodies have a long tradition in histochemistry and cell biology, the introduction of fluorescent protein fusion proteins has revolutionized our ability to follow protein location and protein function in living cells. Yet, it is not a trivial task to stain a protein of interest (POI) specifically without manipulating its expression level, metabolism, and molecular interaction potential. Even less accessible are other cellular molecules that are only indirectly depending on the genetic code, such as carbohydrates, nucleotides, and lipids. In addition, ex vivo tagging and subsequent administration to cells often results in mislocation of the biomolecules. Therefore, specific labeling techniques that attach a fluorophore or any other label at a distinct location inside the living or fixed cell are highly desirable for all biologically relevant molecules. There is now a growing branch of chemical biology addressing this issue. In this chapter, we will discuss the current state-of-the-art in performing bioorthogonal labeling reactions on biomolecules in cells.

2 The Chemical Bioorthogonality Approach

Chemical reactions, even when performed under controlled conditions in a glass flask, are not always producing the desired results. Even more difficult to predict are unnatural reactions in cells. Physiological conditions include moderate temperatures, mostly neutral pH, a large variety of competing functional groups, high ion concentrations, and last but not least water as the solvent. These conditions restrict the scope of many chemical reactions commonly used in synthetic organic chemistry. Therefore, a limited set of useful chemical transformations is currently available.

About a decade ago, Sharpless and coworkers established the so-called click chemistry concept that included a number of very reliable chemical reactions. This concept influenced both chemical synthesis and chemical biology strongly in the following years [1]. The idea is not to copy the exceptional synthetic power of nature, e.g., by trying to mimic enzymes, but instead to consider nearly perfect reactions known to chemists. Amongst those, transformations resulting in new carbon–heteroatom links are particularly promising because nature's ability to form new carbon–carbon (C–C) bonds in its width is still unattainable. A process or reaction to earn click chemistry status has to fulfill certain stringent criteria: (1) high yield (high thermodynamic driving force, usually greater than 20 kcal mol^{-1}) and fast reaction rates at relatively low biomolecule concentrations; (2) simple reaction conditions (such as insensitivity to oxygen and water or neutral pH); (3) simple product isolation; (4) benign solvent (such as water); (5) produce no or only inoffensive byproducts; (6) wide in scope and modular; (7) stereospecific; and (8) the product must be stable under physiological conditions.

The most common categories for these often called spring-loaded transformations belonging to the click chemistry toolbox are [1]: (1) cycloadditions of unsaturated species, particularly 1,3-diploar [3+2] and [4+2] cycloadditions; (2) additions to C–C multiple bonds, such as Michael additions of nucleophilic reactants, but also especially oxidative reaction types such as dihydroxylation and epoxidation; (3) carbonyl chemistry ('nonaldol' type), such as the formation of amides, hydrazones, oxime ethers, (thio)ureas, and aromatic heterocycles; and (4) nucleophilic substitutions, for instance ring-opening reactions of strained heterocyclic electrophiles, such as aziridines and epoxides.

Even though many of these reactions are very reliable for organic synthesis, chemical reactions suitable for chemical biology – that explicitly includes reaction types which also work under physiological conditions – have to satisfy another requirement: bioorthogonality. The property of bioorthogonality implies per definition that reactants must not cross-react with abundant nucleophiles and electrophiles inside cells. Instead, they should only react with the artificially introduced reaction partners [2].

For applications to biological problems, as will be addressed in this chapter, the emphasis lies on making use of bioorthogonal functionalities for labeling or modifying different kinds of biomolecules covalently in vitro as well as in vivo. Furthermore, ways to combine two or more bioorthogonal reactions which also have to be chemically orthogonal to each other in a multiparameter experiment will be discussed.

The latest techniques and achievements in covalent labeling of biomolecules can be roughly divided into two categories: on the one hand, there are several genetically encodable tags dedicated to study protein function and structure. Fusion protein tags usually consist of at least a few amino acids up to sequences of tens of kDa. As described in Sect. 5, these tags are employed by a diverse set of small molecule probes including fluorophores and antigens in a time- and space-dependent manner. In this context, the direct fluorescent labeling with genetically encoded autofluorescent proteins (AFPs) is explicitly not mentioned here because this method works outside the principle of bioorthogonal labeling.

On the other hand, there are many other very important and intriguing classes of molecules and compounds occurring in cells that are not directly dependent on the genetic code. Nonetheless, they are as important for all life forms as proteins are.

Among those, carbohydrates, nucleotides, and lipids are among the most prominent building blocks of living organisms and very specific labeling techniques are necessary to gain insights into their intracellular distribution and function. To study nonproteinogenic biomolecules in their native environment, techniques such as harnessing metabolic pathways for introducing modified molecules and supplying cells with an external source of modified molecules received a lot of attention in the past years. Novel advanced labeling and imaging technologies for small molecule probes contributed eminently to a better understanding of processes of life on the molecular level. Again, bioorthogonality is the key feature.

3 Suitable Bioorthogonal Reactions

Bioorthogonal ligation reactions afford the covalent fusion of two molecules by utilizing functional groups which are distinct from those present in biomolecules. Amongst those, the Staudinger ligation, the 1,3-diploar [3+2] and other cycloaddition reactions, oxime ligation, and the hydrazone coupling are the most prominent representatives. All mentioned reaction types have been demonstrated to be valuable tools for labeling biomolecules in vitro as well as in vivo. The scope of each bioorthogonal reaction depends on the delivery of the reactants to cells, competing endogenous functional groups that might cross-react with the target molecule–reporter system, and the ease of incorporating bioorthogonal functional groups into the biomolecule. In the following, commonly used bioorthogonal reactions in chemical biology will be discussed.

3.1 The Aldehyde/Ketone–Hydrazide/Alkoxyamine Pair

Both aldehyde and ketone functionalities are adequate bioorthogonal chemical reactants [2]. Aldehyde and ketone functionalities are easily accessible by synthesis, for instance via periodate cleavage of 1,2-diols. The small size enables incorporation into biomolecules via the biosynthetic machinery. Both groups are moderately inert under physiological conditions. However, especially aldehydes tend to react with nucleophiles and both functional groups are target to oxidation and reduction. A pH optimum between ~5–6 is required for an amine group to attack the electrophilic carbonyl carbon atom. In case of unreactive amines such as aliphatic amino groups, the equilibrium of this reaction favors the free reactants and not the reversibly formed Schiff base adduct. When reactive amines, especially hydrazides and alkoxyamines, are used, stable hydrazone and oxime adducts, respectively, were formed due to the α-effect of nearby lone pair electrons [3]. Limiting factor for intracellular application of the carbonyl/reactive amine pair is the cross-reaction of reactive amines with other highly abundant electrophiles in cells such as pyruvate or oxaloacetate. Furthermore, the need for an acidic environment restricts this bioorthogonal labeling reaction to the extracellular space

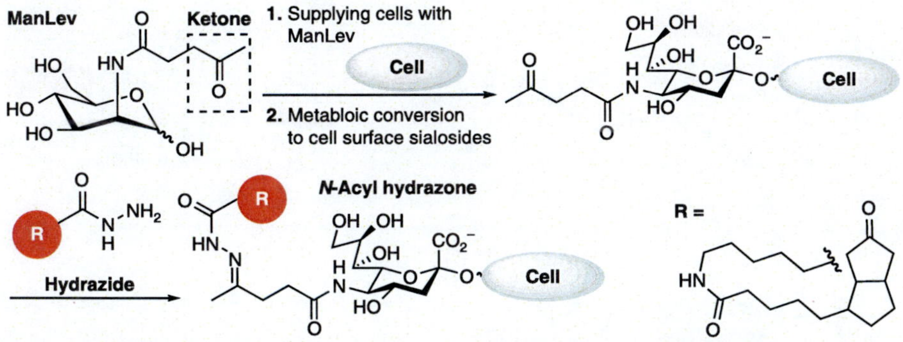

Fig. 1 Incorporation and transformation of ketone-bearing *N*-levulinoyl mannosamine (ManLev) into cell surface-exposed sialic acid via metabolic pathways followed by specific reaction with a hydrazine-modified biotinyl derivative yielding in a covalent *N*-acyl hydrazone adduct. R (red) can be either biotin or any other probe functionalized with a reactive amine resulting in a covalently labeled cell surface glycoconjugate

or lysosomal compartments. However, the potential of this chemistry was demonstrated on cell surface oligosaccharide derivatives bearing a ketone group [4]. Jurkat cells were treated with *N*-levulinoyl mannosamine resulting in the installation of cell surface ketones via the native biosynthetic machinery (Fig. 1). Covalent labeling with biotin hydrazide led to the display of a specified molecular target onto the cell surface. In the next step, a ricin A chain–avidin conjugate was used to selectively target biotin-modified cells. Due to the modular principle of bioorthogonal reactions, probes other than biotin were modified with hydrazide or alkoxyamine moieties and reacted with ketone- or aldehyde-bearing biomolecules. The established hydrazone coupling as well as the oxime ligation is based on this bioorthogonal carbonyl/amine reaction pair. Both couplings can be used to covalently ligate protein and peptide fragments or to couple nonpeptidic building blocks to proteins and peptides [2].

3.2 *The Azide–Phosphine Pair*

The detection of biomolecules that are not directly encoded in the genome, such as glycans, lipids, and other metabolites, can be achieved using the azide as a bioorthogonal chemical reporter (see also Sect. 3.3). One of the first highly selective and bioorthogonal reactions applied to living organisms is the Staudinger reaction that converts azides into primary amines upon treatment with phosphine reagents and subsequent hydrolysis [2]. Advancements in the design of the phosphine reagent by Bertozzi and coworkers permit the trapping of the aza-ylide intermediate by a nearby electrophilic carbonyl group yielding a stable covalent adduct [5]. This modification, also referred to as Staudinger-ligation, transforms

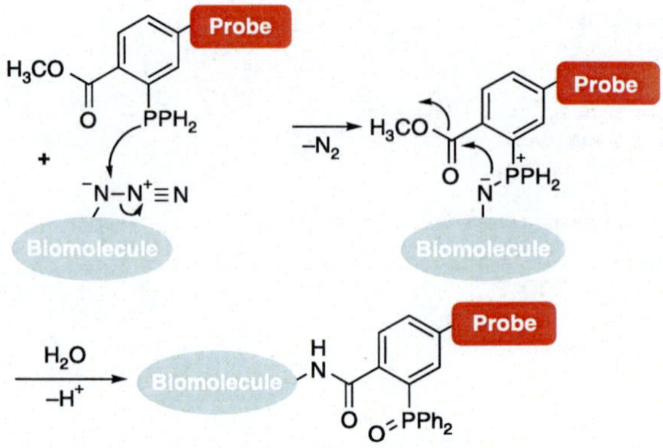

Fig. 2 Bioorthogonal labeling via Staudinger ligation of azides and triarylphosphines resulting in a covalent linkage between the azide-bearing biomolecule (gray; either amino acid or any other biomolecule) and the phosphine-functionalized probe (red)

azides into secondary amine adducts upon treatment with phosphine reagents (Fig. 2). The method is highly bioorthogonal, forms a stable amide bond and works in living cells. Drawbacks are the bulkiness of the phosphine reagent and its sensitivity to oxidation by air or metabolic enzymes which makes the use of large excesses of phosphine reagents necessary. Azides can be reduced by glutathione, which is highly abundant inside and outside cells. Moreover, phosphines are known to interact with proteinogenic disulfide bonds. Thus, the Staudinger ligation is best suited for applications on cell surfaces. In a seminal work, the metabolic pathway for the biosynthesis of cell surface glycans was harnessed to install azide-bearing mannosamine building blocks [5]. Human Jurkat cells were grown in peracetylated *N*-azidoacetylmannosamine ($Ac_4ManNAz$)-containing medium yielding azide-functionalized cell surfaces. In the next step, the metabolically generated azido sialic acid (SiaNAz) moieties were covalently linked to biotin-containing phosphine reagents via the Staudinger reaction both in vitro and in vivo. The same technology for incorporating bioorthogonal azides into cell surface glycans was performed in living animals using FLAG-modified phosphines [6]. The FLAG-tag is an octapeptide (*H*–DYKDDDDK–*OH*) which is used in many different assays that require recognition by an antibody. Glycoconjugate labeling in vivo was observed under the fluorescence microscope by injecting the fluorescein isothiocyanate (FITC)-labeled anti-FLAG antibody (FITC-anti-FLAG) into the organism.

3.3 The Azide–Alkyne Pair

To date, the reactions with the probably largest influence on click chemistry and bioorthogonal labeling are based on the 1,3-dipolar [3+2] cycloaddition described by

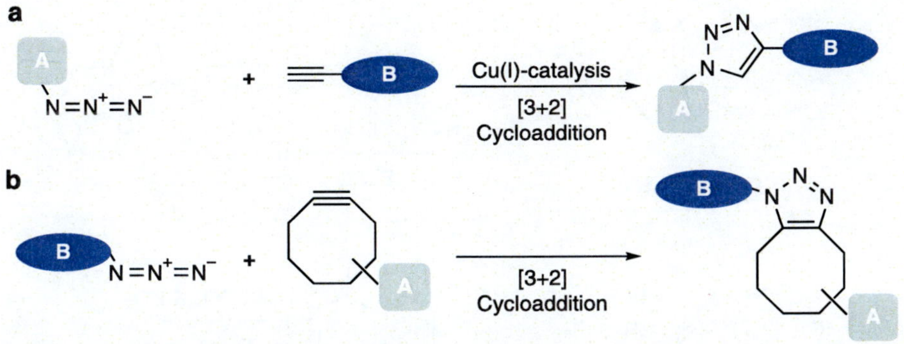

Fig. 3 Covalent labeling via modified Huisgen 1,3-diploar [3+2] cycloaddition between azides and alkynes suitable for biomolecular labeling: (**a**) copper (Cu)-catalyzed reaction; (**b**) strain-promoted and Cu-free reaction; both **A** (gray) and **B** (blue) can be any biomolecule, probe, or label

Huisgen almost five decades ago [7]. Azides react with alkynes upon heating to yield triazoles. A dramatic rate acceleration was achieved when copper(I) (Cu(I)) ions were used as catalyst as was uncovered by the groups of Sharpless and Meldal (Fig. 3a) [8, 9]. Both the azide and alkyne moiety are very small functional groups and are highly bioorthogonal. Additionally, the ligation product – the five-membered ring of the triazole – is small and the azide–alkyne pair is very universal with respect to potential applications to biopolymers, small biomolecules, or in material science. The Cu(I)-catalyzed azide–alkyne 1,3-dipolar cycloaddition reaction proceeds extremely efficient under physiological conditions; in fact, water has a rate-accelerating effect due to its favorable Cu(I) ligand properties compared to organic solvents such as acetonitrile [2]. Commonly, this Cu(I)-catalyzed Huisgen cycloaddition is widely referred to as click chemistry, although the original definition of the click chemistry concept includes diverse reaction types (see Sect. 2) [1]. This type of modified Huisgen cycloaddition has been employed in many biological studies due to its matchless functional group tolerance and its fast reaction kinetics. Amongst those, biological processes and applications involving lipid distribution in living and in fixed cells as well as in animals [10, 11], virus surface remodeling [12], site-selective protein modification in vitro and in vivo [13, 14], labeling of nucleotides for imaging DNA, RNA [15] as well as peptide nucleic acid (PNA) [16], and activity-based protein profiling [17] have been investigated using this powerful covalent labeling reaction. The major drawback of Cu(I) catalyzed Huisgen cycloaddition reactions is the cytotoxicity of copper ions which precludes long-term in vivo experiments (but see [107]). This limitation was overcome by the development of metal-free azide–alkyne 1,3-dipolar [3+2] cycloaddition reactions employing highly strained cyclic alkyne derivatives, such as cyclooctynyl groups (Fig. 3b) [18]. The Cu-free azide–alkyne coupling proceeds at room temperature under physiological conditions and comes very close to a perfect bioorthogonal ligation reaction. The new driving force is the release of the ring strain of the spring-loaded alkyne reactant.

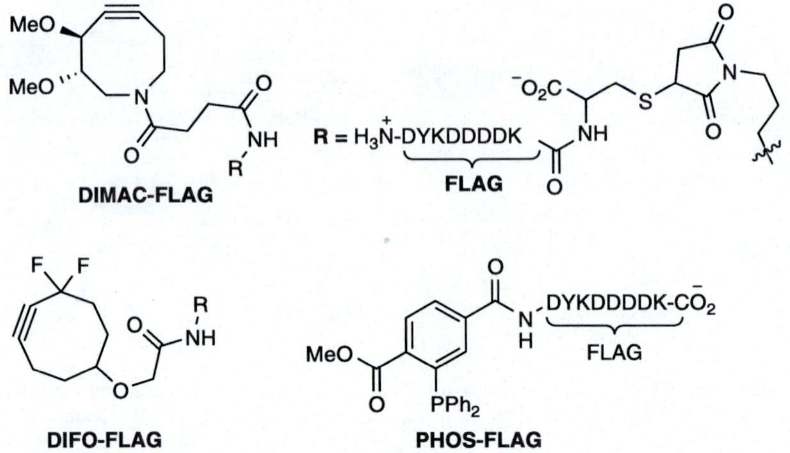

Fig. 4 FLAG conjugates suitable for bioorthogonal labeling of azides in living animals via Cu-free 1,3-dipolar [3+2] cycloaddition or Staudinger ligation, respectively; DIMAC dimethoxy azacyclooctyne, DIFO difluorinated cyclooctyne, PHOS phosphine

The Bertozzi lab used mouse models to demonstrate the applicability of copper-free click chemistry to living animals [19]. Similar to the exploitation of the scope of the Staudinger ligation for in vivo applications (see Sect. 3.2), mice were administered with peracetylated $Ac_4ManNAz$ to metabolically label cell surface sialic acids with azides. Feeding mice with an $Ac_4ManNAz$-containing diet (300 mg/kg in 70% DMSO) resulted in the incorporation of SiaNAz into cell surface glycoconjugates harnessing the native metabolic pathway. A panel of cyclooctyne–FLAG peptide conjugates was examined regarding water solubility, bioavailability, and labeling efficiency (Fig. 4). The highest intrinsic reactivity was measured for a difluorinated cyclooctyne (DIFO) in Jurkat T cells. The cyclooctyne–FLAG probes were injected into mice and glycoconjugate labeling was observed by adding the FITC-anti-FLAG antibody. Surprisingly and in contrast to phosphine's in vitro reaction efficiency, the best two reagents for Cu-free click chemistry, dimethoxy azacyclooctyne (DIMAC)–FLAG and DIFO–FLAG, produced less reaction products in vivo than the same dose of PHOS–FLAG linked to SiaNAz via Staudinger ligation [19]. Nevertheless, Cu-free click chemistry has significant advantages regarding the bioorthogonality of reactants over the Staudinger ligation, particularly for intracellular applications. A comparative study of bioorthogonal reactions with azides was presented by the Bertozzi lab [20]. Staudinger ligation, Cu(I)-catalyzed azide–alkyne cycloaddition, and the strain-promoted [3+2] cycloaddition were investigated in the context of various biological applications. For example, the labeling of biomolecules in cell lysates and on live cell surfaces using the already mentioned SiaNAz as target was performed. The Cu(I)-catalyzed cycloaddition has proven to be the most efficient reaction for detecting azides in proteins in in vitro applications. Due to the cytotoxicity of Cu, only the Staudinger ligation and the strain-promoted [3+2] cycloaddition were applied to label azides on live cell surfaces. Both reactions proceed comparatively efficient and fast with only minor differences. It was shown that the best most suitable reaction type depended on the specific structure of the azide, the nature of the azide-modified biomolecule, and the coupling partner [19, 20].

Besides designing probes for rational entry into metabolic pathways, minimally modified biomolecules can also be employed for direct analysis of the activities of small molecules within cells. Our group described recently the first successful labeling of lipids by azide–alkyne 1,3-dipolar cycloaddition both in fixed and in living cells (Fig. 5) [10]. For this purpose, bioactivatable derivatives of phosphatidic acid (PA) were synthesized bearing a terminal triple bond or cyclooctyne ether, respectively. PA is an important signaling and regulatory phospholipid. The investigation of PA distribution demonstrated significant differences between the use of prelabeled lipids and those labeled in vivo via 1,3-dipolar cycloaddition.

Probably very soon, the 1,3-dipolar cycloaddition will become even more valuable since it is possible to genetically incorporate unnatural amino acids (UAAs) of different origin (Phe, Lys) into proteins bearing either azide or alkyne functional groups (see Sect. 4).

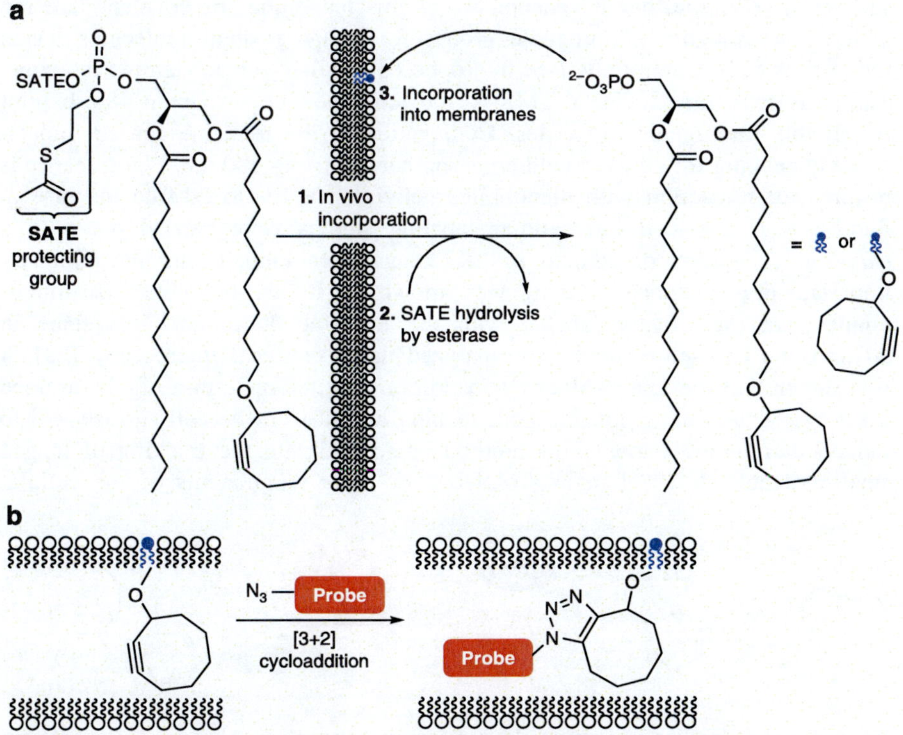

Fig. 5 In vivo labeling of lipids: (**a**) incorporation of minimally modified phosphatidic acid (PA) derivatives into cell membranes; (**b**) selective bioorthogonal labeling of the PA derivative with a azide-functionalized probe (red) via strain-promoted 1,3-diploar [3+2] cycloaddition

3.4 The Tetrazine–Alkene Pair

Very recently, the potential of strain-promoted inverse electron-demand hetero-Diels–Alder cycloaddition reactions of 1,2,4,5-tetrazines for bioconjugation and bioorthogonal labeling of biomolecules under physiological conditions was explored by the groups of Fox and Weissleder [21, 22]. Either modified norbornene or substituted *trans*-cyclooctene was used as dienophil for chemo- and regioselective [4+2] cycloaddition with asymmetric tetrazines functioning as diene coupling partner (Fig. 6). For example, Her2/neu receptors on live human breast cancer cells were targeted with a monoclonal antibody modified with a norbornene and conjugated selectively to a near-infrared fluorophore-bearing tetrazine [22]. Due to the slow reaction rate of norbornene in serum at 20°C, all following studies utilized ring-strained *trans*-cyclooctene derivatives [24].

In proof-of-principle cell experiments, the commercially available antibody for the epidermal growth factor receptor (EGFR) cetuximab was labeled with *trans*-cyclooctene succinimidyl carbonate. EGFR overexpressing cancer cells were first exposed to the cetuximab–cyclooctene conjugate. In the second step, pretargeted cells were selectively labeled with a benzylaminotetrazine linked to a fluorophore such as VT680. Further advancements of this technique for covalent labeling include bioorthogonal fluorogenic probes for imaging small molecules inside living cells [23]. Commonly used fluorophores coupled to benzylamino tetrazine, especially tetrazine–BODIPY TMR-X and tetrazine–Oregon Green 488, exhibit remarkable fluorogenic properties. Conjugation to the tetrazine caused fluorescence quenching for all dyes with emission between 400–600 nm. Fluorescence is restored after reaction with dienophiles delivering a 15- to 20-fold increase in fluorescence. Live-cell intracellular labeling studies were carried out with a *trans*-cyclooctene-modified taxol in PtK2 kangaroo rat kidney cells. The clinically important drug paclitaxel (taxol), well-known for its ability to stabilize microtubules, was selectively labeled with the tetrazine–fluorophore conjugate in 20 min at room temperature and analyzed using confocal microscopy [23]. It was shown that the Diels–Alder cycloaddition of a tetrazine to a highly strained *trans*-cyclooctene is applicable both to biomolecules on the cell surface and to intracellular biomolecules. This method could facilitate the tracking of tagged small-molecule drugs, signaling proteins, or other components of the cellular

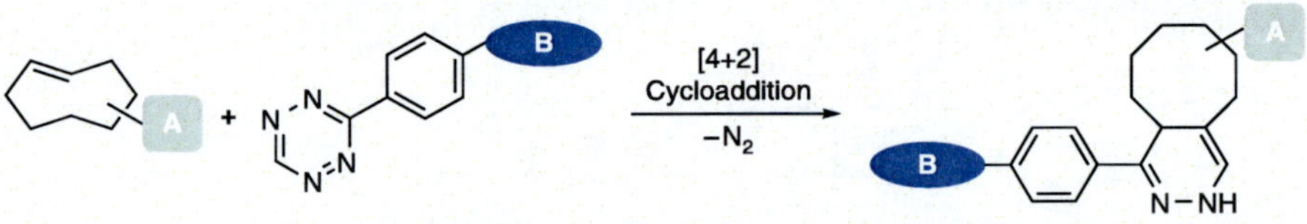

Fig. 6 Covalent labeling via inverse electron-demand hetero-Diels–Alder [4+2] cycloaddition between a tetrazine and a strained alkene derivative; both **A** (gray) and **B** (blue) can be any biomolecule, probe, or label. Some fluorophores exhibit fluorogenic properties when fused to tetrazine and subsequent reaction [23]

machinery within live cells. Moreover, it might be feasible to site-specifically incorporate either tetrazine- or *trans*-cyclooctene-substituted UAAs into proteins and then to express them in living cells (see Sect. 4). Given its speed, stability, efficiency, and sensitivity in whole serum, this covalent labeling reaction should be suitable for many diverse in vivo imaging applications comparable to the scope of 1,3-dipolar [3+2] cycloadditions.

3.5 The Tetrazole–Alkene Pair

Apart from Cu-catalyzed and Cu-free 1,3-diploar [3+2] cycloadditions, there is another type of cycloadditions belonging to the same family and referred to as photoclick chemistry [25]. Based on early investigations of Huisgen and coworkers, a photoactivated 1,3-diploar cycloaddition between a diaryltetrazole and an alkene derivative could be adjusted recently to physiological conditions possessing excellent bioorthogonality. Mechanistically speaking, the tetrazole is photoactivated by ultraviolet (UV) light (302 nm) producing a nitrile imine as highly reactive intermediate. Subsequent to photoinduction, nitrile imines react regioselectively with alkenes yielding a pyrazoline cycloadduct in good tolerance to other functional groups, excellent solvent compatibility including water, and high yields (Fig. 7) [2]. The utility of this photoinducible cycloaddition for in vivo labeling of proteins was demonstrated in proof-of-principle experiments in *E. coli* [26]. Bacteria cells overexpressing an alkene containing Z-domain protein (alkene-Z) were incubated with tetrazoles, photoirradiated and finally analyzed by fluorescence microscopy. Only cells expressing the alkene-Z protein exhibited strong fluorescence. It is not necessary to attach a fluorophore to the alkene because the pyrazoline cycloadduct emits fluorescent light in the range of cyan fluorescent protein (CFP) after excitation with the corresponding laser. In the near future, reaction kinetics might be improved by optimizing the reactivity of the reaction pair to allow faster and more efficient labeling. Thereby, photoclick chemistry might be a suitable chemical tool for the time- and space-resolved study of biological processes. Already now the scope of the bioorthogonal tetrazole–alkene pair is remarkable. When combined with the incorporation of UAAs into proteins (see Sect. 4), the tetrazole–alkene pair might

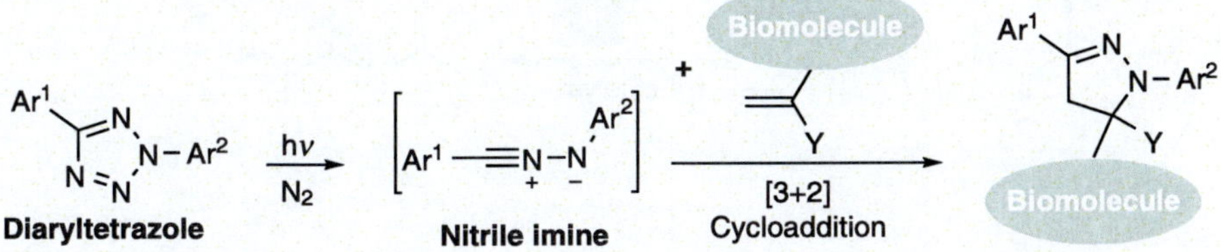

Fig. 7 Photoinduced 1,3-dipolar [3+2] cycloaddition reaction between an alkene and a 2,5-diaryltetrazole. Photoclick chemistry can be used to label diverse biomolecules (gray) with an autofluorescent pyrazoline moiety or with probe-bearing tetrazoles

become a powerful tool for covalent labeling of proteins. Apart from the intrinsic fluorescence of the pyrazoline, other probes fulfilling the requirements of bioorthogonality can be attached to the alkene in order to add new functionalities to the POI.

3.6 Transition Metal-Catalyzed Reactions

Transition metal-catalyzed reactions are amongst the most powerful and efficient organic transformations, particularly when it comes to the formation of new C–C bonds. Very recently, long established reactions of organic chemistry were adapted to benign solvents and physiological conditions and became a valuable tool for protein modification and labeling of biomolecules in general.

The basic requirement for olefin metathesis is an alkene–alkene pair. Usually, ruthenium (Ru) catalysts, such as the Hoveyda–Grubbs second-generation catalyst, are employed to covalently link two alkenes with each other yielding a single C–C double bond (Fig. 8a) [2]. For example, cross-metathesis was used to modify proteins containing an allyl sulfide group (*S*-allylcysteine, Sac) in aqueous buffer with allyl alcohol [27]. Various site-specific modifications of the serine protease subtilisin *Bacillus lentus* (SBL), such as mini-PEGylation and glycosylation, were successfully demonstrated with this metathesis reaction.

Another type of transition metal-catalyzed transformation is the palladium (Pd)-catalyzed cross-coupling reaction suitable for protein functionalization. The arene–alkene/alkyne/arene pair permits chemoselective covalent labeling of biomolecules under aqueous cross-coupling conditions, with high yield and with excellent functional group tolerance. Cross-coupling reactions known from organic chemistry, such as the Mizoruki–Heck reaction, the Sonogashira reaction, and the

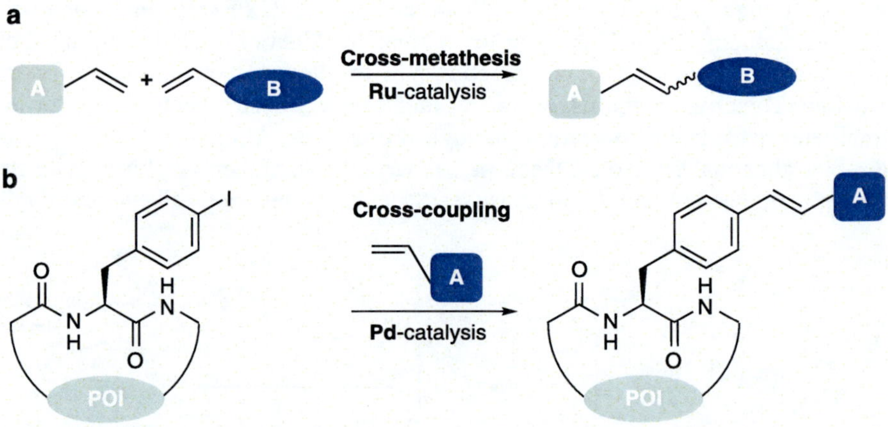

Fig. 8 Transition metal-catalyzed reactions: (**a**) ruthenium (Ru)-catalyzed olefin cross-metathesis; both **A** (gray) and **B** (blue) can be any biomolecule, probe, or label. (**b**) Functionalization of the protein of interest (POI; gray) via palladium (Pd)-catalyzed cross-coupling reactions; **A** (blue) can be any biomolecule, probe, or label

Suzuki cross-coupling reaction, were successfully applied to label proteins in vitro with biotin derivatives or other small molecule probes (Fig. 8b) [2].

In further attempts to adapt the tools of organic chemistry to synthetic biology, the scope of the Suzuki–Miyaura cross-coupling reaction on peptides and proteins was explored [28]. Outstanding features of this Pd-catalyzed reaction are its good yields on amino acid and peptide models under physiological conditions (37°C in buffered water, no organic solvent needed), short reaction time (complete conversion after 30 min), its need for a comparably small excess of catalyst (50 equiv) and cross-coupling partner (500 equiv), and its tolerance to oxygen. The catalytic complex formed by the sodium salt of the ligand 2-amino-4,6-dihydroxypyrimidine and $Pd(OAc)_2$ was used in initial experiments to covalently link halogenated phenylalanine derivatives, such as the genetically encodable UAAs *p*-bromo-L-phenylalanine and 3-iodo-L-tyrosine (see Sect. 4.2), to phenylboronic acid. Next, the bioorthogonality of this labeling technique in competition to endogenous functional groups was successfully tested. Therefore, an SBL mutant was chemically modified containing an aryliodide side chain suitable for subsequent cross-coupling reaction. The reaction proceeded to completion and vindicated extensive efforts to genetically incorporate aromatic halides as cross-coupling partners into proteins (see Sect. 4).

3.7 *Miscellaneous*

Protein modification is successfully achieved by strain-promoted alkyne–nitrone cycloaddition (SPANC) [29]. It was shown that 1,3-dipolar cycloadditions of cyclooctynes with nitrones containing ester or amide α substituents proceed with much faster kinetics than comparable reactions with azides. A one-pot three-step protocol was implemented for the site-specific installment of carbonyl groups in proteins and peptides (e.g., via serine oxidation) followed by reaction with reactive cyclooctyne derivatives such as 4-dibenzocyclooctynol and the more polar azacyclooctyne. SPANC is a metal-free click reaction and anticipated to be an attractive alternative to the well-established oxime ligation due to the ease of synthesis of nitrones, the stability of reactants, and its potential for chemical biology applications.

One important type of covalent modification of proteins is doubtlessly the ligation of two protein or peptide fragments. Since the impact of semisynthetic proteins on basic research, drug discovery, pharmaceutical industry, and biotechnical applications is continuously increasing, robust methods to isolate and modify recombinant proteins have been developed and further progress is supposable [30]. A powerful tool for chemical protein synthesis is the native chemical ligation (NCL). The most widespread method besides normal amino acid coupling in peptide synthesis that leads to the formation of a native amide bond between two peptide segments was introduced by Kent and colleagues in 1994 [31]. Mechanistically speaking, NCL involves a rapid equilibration of thioesters (transthioesterification) that is interrupted by an irreversible intramolecular reaction with the N-terminal amine of the fused protein (S–N acyl shift), ultimately forming a native amide bond [32]. Therefore, a C-terminal α-thioester and an N-terminal cysteine residue are required. Both functionalities can be either produced synthetically or recombinantly. Advantages of the NCL are the smooth reaction (aqueous buffer), widely orthogonal conditions, and its ease of application. NCL has been used to study protein structure and function by producing semisynthetic protein analogs with tailor-made properties [33]. Semisynthetic proteins of diverse protein families were synthesized via NCL, such as histone modifications [34, 35], semisynthetic ubiquitylated proteins [36], and potassium channel membrane proteins with noncanonical amino acids [37]. Furthermore, the NCL method was employed for the intein-mediated synthesis of lipidated Ras proteins [38] and to fluorescently label prenylated GTPases [39]. In general, the scope of the NCL is restricted to in vitro use and limited to proteins and peptides. However, NCL works also in living cells as shown for the in vivo generation of a circular protein domain (Src homology 3, SH3) in *E. coli* [40]. Besides the ligation of two peptidic molecules, NCL can also be used for cyclization of peptides. For example, linear precursors of cyclic plant-defense proteins – named cyclotides – synthesized via solid-phase peptide synthesis (SPPS) can be cyclized after cleavage from the solid support in basic buffer at room temperature [41].

4 Unnatural Amino Acids as Reactants

Traditionally, it is difficult to covalently label the protein under investigation with small molecule probes inside living cells. The multiplicity of functional groups present in cells and the lack of unique chemical reactivities among the 20 canonical amino acids made protein-, residue-, or site-specific in vivo labeling of endogenously expressed proteins close to impossible [37]. Diverse fusion protein and peptide tags of different size have been developed that allow the covalent attachment of chemical probes with distinct properties (see Sect. 5). Drawbacks of the fusion tag strategy are the potential disturbing influence of the tag on protein structure or function and the size of the tag itself – one of the smallest reported tags for covalent in vivo labeling is the tetracysteine (TC) tag with a minimal size of six amino acid residues. This technique is extensively reviewed in this volume. Here, the alternative to genetically encoded fusion proteins suitable for covalent labeling with small molecules is described. In comparison to fusion tags, the expanded genetic code concept for incorporating UAAs requires the modification of only a single amino acid of the POI. Encoding rationally designed and chemically synthesized amino acids with bioorthogonal functional groups opens up new possibilities for covalent labeling using bioorthogonal reactions out of the toolbox of click chemistry (see Sects. 2 and 3). In addition to bioorthogonal labeling, the incorporation of fluorescent UAAs or UAAs with eligible properties for different detection methods allows the direct labeling of a protein without further modifications or disturbing posttranslational labeling procedures.

4.1 Genetic Code Expansion

The genetic code specifies the information transfer from the linear sequence of the four letter alphabet of the DNA into the 20 amino acid language of proteins. With four possible nucleotides at each position, the total number of permutations of triplets is 64 ($4 \times 4 \times 4$), but only 61 encode the 20 amino acids leaving three stop codons. Proteins carry out an almost endless diversity of functions and duties in living organisms. Protein function and structure is determined by their primary structure and therewith, by the limited number of functional groups of proteinogenic amino acids augmented partially by cofactors and posttranslational modifications (PTMs). For decades, one of the dogmas of the genetic code has been its limitation to the same canonical 20 amino acid building blocks for all known organisms. Two rare exceptions that can be additionally encoded are the noncanonical amino acids selenocysteine found in some species of archaea, bacteria, as well as in eukaryotes and pyrrolysine used by only some representatives of archaea and bacteria [42]. This dogma (universality, uniqueness) and other important features of the genetic code, especially its degeneracy, triplet codon usage, and its redundancy of codons, are of course still valid for genetically unmodified genomes and organisms – but remarkable achievements in cloning and understanding of the biosynthetic process of proteins have led to some momentous advancements establishing the concept of the expanded genetic code [43–45].

Fundamental biological processes such as DNA replication, transcription, loading amino acids to specific tRNAs by aminoacyl-tRNA synthetases (aaRSs), and translation of mRNA into proteins can be harnessed for the incorporation of UAAs with tailor-made properties. The fidelity of tRNA aminoacylation is determined by both high substrate specificity and subsequent proofreading of the amino acid–tRNA adduct. The ability to paste UAAs into proteins bearing optical or physical probes, bioorthogonal functional groups, photocleavable cages of active site residues, or structure-relevant side chain analogs [45] will allow the design and synthesis of proteins with improved or completely novel properties. Finally, an expanded genetic code might provide unprecedented powerful tools in the evolution of new function on the molecular and cellular level [46].

About three decades back, first biosynthetic approaches for the in vitro synthesis of proteins containing UAAs were developed using truncated tRNAs [47, 48]. Chemically aminoacetylated nucleotides were enzymatically ligated to tRNA and used in cell-free expression systems for protein biosynthesis. The position of the UAA was specified by either blank (nonsense or frameshift codons) or coding codons (codon bias, depletion conditions). Despite the technically demanding procedure, this in vitro method has proven its value for studying protein function and structure, amongst others through the introduction of optical probes or non-proteinogenic functional groups [45]. In the next step toward the in vivo biosynthesis of UAA-containing proteins, UAAs grafted onto tRNAs were injected into cells making use of the biased nature of the genetic code and noncoding codons again. For example, a chemically aminoacetylated truncated amber (UAG) suppressor tRNA^{Gln} (CUA) derived from *Tetrahymena thermophilica* was shown to incorporate amino acids at the amber stop codon position with high fidelity and efficiency when injected into *Xenopus* oocytes [49]. Self-evident drawbacks of these methods, particularly the need for stoichiometrical and continuous supply of aminoacyl-tRNAs, the limited synthetical or biological accessibility, as well as the potentially troublesome injection and transfection techniques, encouraged several specialized laboratories to dive deeper into the complex process of protein biosynthesis.

Some years later, Tirrell and coworkers introduced a novel, robust approach for the residue-specific incorporation of UAAs applicable to living cells [43, 50, 51] based on earlier findings of Cohen and coworkers from the 1950s [52, 53]. This technique employs wild-type aaRSs that tolerate close structural analogs of their native amino acid substrates. In order to reliably insert an UAA into proteins, a strain auxotrophic for the natural amino acid corresponding to the harnessed tRNA/aaRS pair has to be used [43]. A drawback of this wholescale replacement of a canonical amino acid is that host cell cultures have to be starved and cannot maintain exponential growth. Noteworthy, either bacteria or eukaryotic cells can be used without limitations a priori [54] and the gene encoding the protein under investigation does not necessarily have to be mutated. Nevertheless, nondividing cells can survive and still overexpress the UAA-modified protein in good yields. In these recombinant proteins, the UAA residue specifically replaces the canonical amino acid – as well as in all simultaneously expressed proteins using the same codon (Fig. 9a) [43, 45]. The total replacement enables chemical modification at multiple sites and substantial changes in the physical properties of the POI. The technique became even more precious after broadening its scope with the help of elaborate screening strategies for aaRS active-site engineering [55, 56], aaRS overexpression [50], and editing domain mutations [57]. Applications include the incorporation of heavy atoms to solve the phase problem of X-ray crystallography (e.g., methionine substituted with selenomethionine) or the synthesis of protein analogs bearing spectroscopic probes suitable for fluorescence microscopy [43]. In summary, this method allows residue-specific global replacement of canonical amino acids with UAAs. However, so far only those UAAs can be incorporated that have a close structural similarity to a canonical amino acid.

Despite the wide scope of the residue-specific replacement technique, in other cases it is desirable to substitute only a single amino acid with an UAA. To date, the most sophisticated approach for site-specifically adding new chemistries to the genetic code is the generation of organisms that can genetically encode 21 or more amino acids. This epoch-making technology with its promising potential was mainly discovered and introduced by P. G. Schultz [44–46, 58] and his numerous coworkers. UAAs, rationally designed and synthesized at the chemistry bench, are encoded and site-specifically incorporated into proteins ideally with the same fidelity, efficiency and genetic manipulability of natural amino acids by using tRNA/aaRS pairs of donor organisms that are orthogonal to their new host organisms (Fig. 9b) [45].

This site-specific concept makes use of exactly the same biological principles (e.g., triplet code, mRNA as template and messenger, uniqueness, and tRNA as

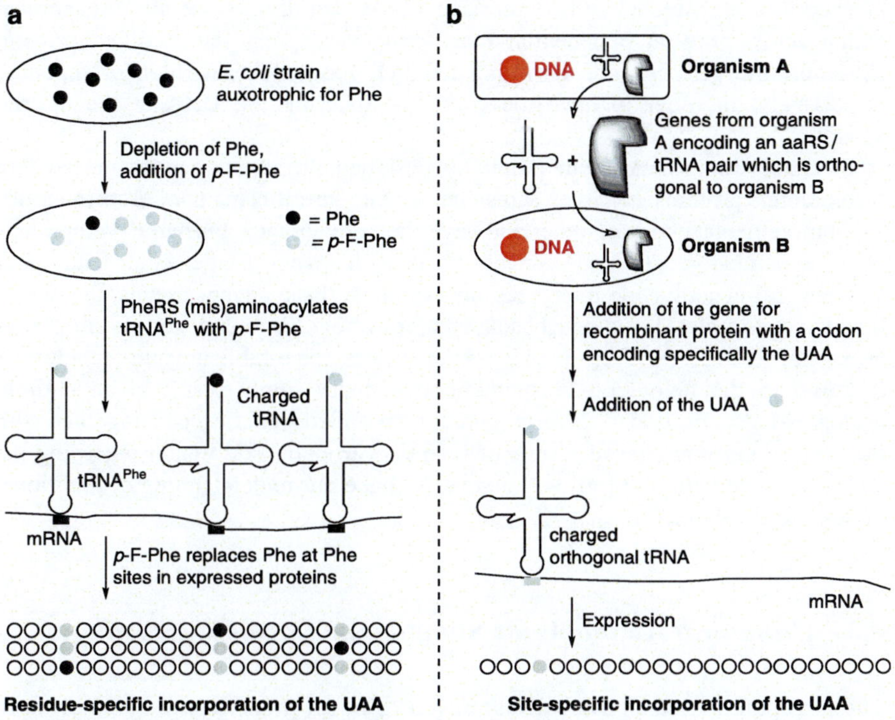

Fig. 9 Schematic representation of methods for incorporating unnatural amino acids (UAAs) into recombinant proteins. (**a**) A general method for residue-specific incorporation of UAAs; Phe phenylalanine, *p*-F-Phe *p*-fluorophenylalanine, PheRS phenylalanyl-tRNA synthetase; (**b**) a general method for site-specific incorporation of UAAs

adapter) and biosynthetic machineries (e.g., polymerases, aaRSs, and ribosomes) essential for the natural protein biosynthesis. Every amino acid is encoded by a specific set of codons (only one codon each for Met and Trp), but the codon usage differs from organism to organism. For example, the amber nonsense codon (UAG) is the least used stop codon of *E. coli* and *S. cerevisiae* and was harnessed to encode UAAs [45] using orthogonal tRNA/aaRS that encode an amino acid in response to UAG. All components involved in the in vivo translation with new amino acids, such as UAAs, tRNAs, tRNA/aaRSs, elongation factors (EFs), and the ribosome, must meet several criteria: first, the UAA has to be cell-permeable as well as metabolically stable and possess good cellular bioavailability. Second, the UAA should not be a substrate for endogenous tRNA/aaRS pairs; its unique codon must be recognized exclusively by the new tRNA/aaRS pair and must be accepted by EF-Tu and the ribosome. Third, the new tRNA/aaRS pair must be functional in the host organism, highly specific for the UAA, and orthogonal to endogenous tRNA/aaRS pairs in the host cell.

The concept of Schultz in comparison to previously described methods has the advantage that only the UAA has to be added to the growth medium.

Furthermore, transfected cells expressing UAAs can divide, use all 20 canonical amino acids, grow at exponential rates, and overexpress the modified protein. However, the genetic code encoding the POI has to be rewritten or mutated, respectively, in order to have the codon deciphering the UAA at the desired position.

In conclusion, UAAs with rationally designed properties can be pasted into recombinant proteins in either a residue- or site-specific manner. Both methods are complementary to each other and have their advantages. Further developments of the expanded genetic code concept break with several fundamental features of the classical genetic code interpretation. Amongst these advancements, the evolution of a quadruplet-decoding ribosome that is orthogonal to endogenous ribosomes is probably the most striking one [59]. In the future, the establishment of quadruplet codons and the belonging biosynthetic machinery might result in an entirely orthogonal genetic code. An orthogonally expanded genetic code together with the very recently published results of scientists around J. C. Venter regarding the creation of synthetic life [60] will definitely shake our understanding of traditional biology to the core.

4.2 Unnatural Amino Acids Suitable for Covalent Labeling

The incorporation of UAAs with improved or novel properties in comparison to natural amino acids can contribute to obtain new insights into both protein structure and function. Investigations may be performed in vitro as well as in vivo and probes or functional groups suitable for fluorescence microscopy, nuclear magnetic resonance (NMR), electron paramagnetic resonance (EPR), or chemical biology applications can be introduced. In the following, UAAs bearing bioorthogonal functional groups suitable for in vivo labeling and subsequent imaging will be discussed.

As mentioned in Sect. 3.3, azides and alkynes are the most widely used bioorthogonal functional groups when it comes to efficient labeling of both extracellular as well as intracellular biomolecules. Other functionalities, such as alkenes and ketones, are particularly valuable for in vitro protein modification and cell surface labeling. The first UAAs introduced site-specifically were analogs of the aromatic amino acid phenylalanine. For example, *p*-azido-L-phenylalanine, *p*-acetyl-L-phenylalanine, *O*-(2-propynyl)-L-tyrosine, *p*-iodo-L-phenylalanine, and *O*-allyl-L-tyrosine (Fig. 10) have been successfully added to the genetic code of *E. coli*, yeast, and mammalian cells [44, 61]. Even if bioorthogonal taggable Phe derivatives could be used for many studies [44], their aromatic nature limited the scope because not every canonical amino acid can be replaced by an aromatic residue without causing misfolding or loss of function. Recent publications reported the utilization of robust tRNA/aaRS pairs that allow the incorporation of aliphatic lysine, nonaromatic pyrrolysine, and other amino acid derivatives with distinct side chains [62, 63]. Amongst those, the *Methanosarcina barkeri* MS pyrrolysyl tRNA synthetases/$tRNA_{CUA}$ (MbPylRS/$MbtRNA_{CUA}$) pair is a new orthogonal pair in *E. coli* and in mammalian cell lines [64]. As the name indicates, the MbPylRS/MbtRNA$_{CUA}$ pair incorporates UAAs in response to the amber stop codon. Aliphatic azide, alkene and alkyne lysine derivatives (Fig. 10) synthesized in two steps from Boc-L-Lys-OH and the corresponding chloroformates permit bioorthogonal labeling using 1,3-diploar [3+2] cycloaddition chemistry.

In a proof-of-principle study, myglobin–his6 bearing the alkyne-modified lysine was covalently attached to a fluorophore azide (By3 azide) or a biotin azide via Cu (I)-catalyzed Huisgen [3+2] cycloaddition (Fig. 11) [64].

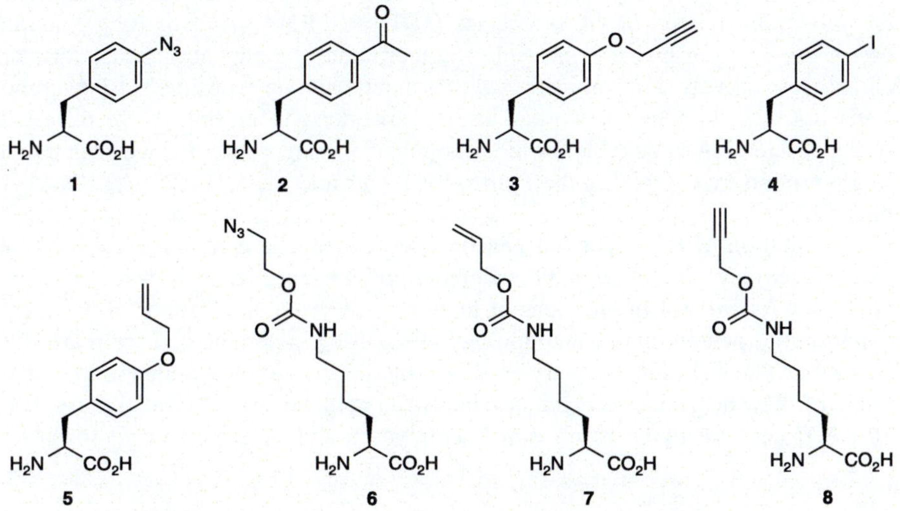

Fig. 10 A selection of unnatural amino acids suitable for click chemistry that have been introduced into proteins. *p*-azido-L-phenylalanine (**1**), *p*-acetyl-L-phenylalanine (**2**), *O*-(2-propynyl)-L-tyrosine (**3**), *p*-iodo-L-phenylalanine (**4**), *O*-allyl-L-tyrosine (**5**), azidoethyloxycarbonyllysine (**6**), allyloxycarbonyllysine (**7**), and propynyloxycarbonyllysine (**8**)

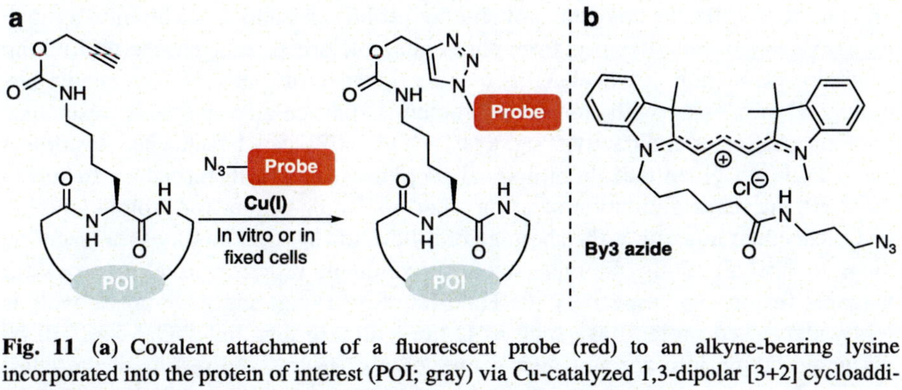

Fig. 11 (**a**) Covalent attachment of a fluorescent probe (red) to an alkyne-bearing lysine incorporated into the protein of interest (POI; gray) via Cu-catalyzed 1,3-dipolar [3+2] cycloaddition; (**b**) chemical structure of the azide-modified fluorophore By3 azide

Therefore, *E. coli* was transformed with pBKPylS (a plasmid which encodes *Mb*PylRS) and pMyo4TAGPylT–his6 (a plasmid which encodes *Mb*tRNA$_{CUA}$ and a C-terminally hexahistidine (his6) tagged myoglobin gene with an amber codon at position four) [64]. Protein expression and labeling reactions were analyzed using SDS-PAGE and ESI-MS and resulted in site-specific incorporations into proteins in good yields (3 mg UAA/L culture medium) and a labeling efficiency of 90–100%, respectively. Directed evolution of the *Mb*PylRS/*Mb*tRNA$_{CUA}$ pair enabled the efficient incorporation of additional UAAs into genetically determined sites [45, 63, 65]. In addition, the *Mb*PylRS/*Mb*tRNA$_{CUA}$ pair is mutually orthogonal in its aminoacylation specificity to the aromatic UAAs incorporated by the *Methanococcus jannaschii* tyrosyl-tRNA synthetase (*Mj*TyrRS/tRNA$_{CUA}$) pair [66]. Making use of this relationship, it is possible to incorporate the aliphatic lysine azide or alkyne, respectively, in combination with aromatic alkyne- or azide-bearing amino acids [64, 66]. In a recent work, Liu and coworkers incorporated two different lysine-derived UAAs suitable for bioorthogonal labeling at two defined sites into a single protein by combining the PylRS–PylT pair and the *Mj*TyrRS–*Mj*tRNATyr pair [67].

A limitation of the expanded genetic code concept is predominantly its very recent discovery. Evolution and development of enhanced or novel ways to incorporate UAAs are still in the focus of interest. Advanced applications of the great opportunities penetrating far into biology or resulting in commercial profit are still emerging. The situation will very likely change in the near future due to the large scope of new academic research and industrial applications, as in total more than 70 UAAs can already be encoded in *E. coli*, yeast, and mammalian cells [45].

5 Reactive Protein Fusions

In an alternative to genetically encoded AFPs, small molecule probes suitable for specific covalent attachment to particular protein tags, especially physical probes or organic dyes, offer the advantage of smaller probe sizes and application to a much higher variety of functionality, from photoinducible processes such as cross-linking of proteins, imaging at wavelengths outside the spectrum of AFPs, to imaging of biological processes with nonoptical methods, particularly magnetic resonance imaging (MRI) or radioactivity assays. While AFPs can be attached to almost every POI and given that its biological properties are not disturbed by fusion to the AFP, the major limitation of using small molecule probes for protein labeling and subsequent imaging is the shortage of robust and specific methods for targeting them to specific POIs. In this chapter, prominent representatives of covalent taggable fusion tags, especially the Halo-tag, SNAP-tag, and ACP-tag as well as less widely used protein tags such as Q-tag, cutinase tag, and β-Gal-tag, will be discussed taking into account their scope, ease of usage, and results of the latest research.

5.1 Self-Labeling by Alkylguanine Transferases

Shortly after the introduction of the fluorescein arsenical hairpin (FlAsH)-tag technology for in vivo and in vitro labeling of fusion proteins by Tsien and coworkers [68], the group of Kai Johnsson established a novel protein tag suitable for covalent labeling with small synthetic molecules based on the human DNA repair protein O^6-alkylguanine-DNA alkyltransferase (human AGT, hAGT) [69].

The native function of the 21 kDa hAGT is to recognize O^6-alkylated guanine in the DNA and to transfer the alkyl group to an active site cysteine residue, yielding alkylated enzyme and – most importantly – repaired DNA. hAGT is sufficiently unspecific to react with the monomeric and cell-permeable nucleobase O^6-benzylguanine (BG) and its derivatives substituted at position four of the benzyl ring. A broad variety of substrates turned out to be useful for labeling hAGT when fused to the protein under investigation (Fig. 12) [69]. Mechanistically, an active site cysteine thiol group reacts with a probe-functionalized BG derivative via nucleophilic substitution to yield a covalently modified hAGT and guanine as leaving group.

Many of the optical, physical, and chemical probes are commercially available as activated esters and can simply be coupled to the free amine group of BG derivative in a one-step reaction [70]. Former limitations of the hAGT/BG labeling technology, such as the inability to distinguish between the BG probe and alkylated DNA for both the protein tag hAGT as well as the endogenous AGT, were overcome using mutants of the protein tag [71]. One of these mutants – called SNAP-tag – is slightly smaller (19 kDa) and reacts 50-fold faster with its BG derivative than hAGT. Background labeling of endogenous AGT is not an issue when the fast hAGT-based mutant SNAP is used and can be completely eliminated by using a specific inhibitor for native AGT [71]. The scope of the SNAP-tag exceeds most of the other tags described in the following subchapters: proteins in any cellular compartment can be labeled without any restriction a priori. SNAP fusion proteins were successfully and efficiently labeled on the cell surface, in the cytoplasm, in the endoplasmic reticulum (ER), and in the nucleus [72, 73].

In initial but representative experiments, the nuclear protein human estrogen receptor α (hERα) was fused to hAGT. The cellular localization of hERα was determined in hAGT-deficient HeLa cells by labeling the hAGT–hERα construct

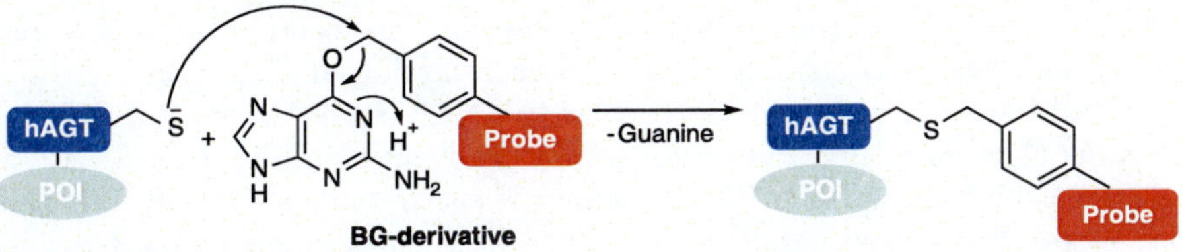

Fig. 12 The human O^6-alkylguanine-DNA alkyltransferase (hAGT; *blue*)-tag fused to the protein of interest (POI, *gray*) reacts with probe (*red*)-bearing benzylguanine (BG) derivative, resulting in the covalent attachment of the label (*red*) to the active-site cysteine of hAGT

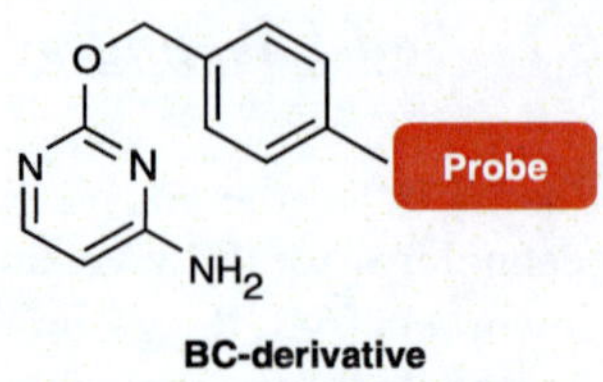

Fig. 13 O^2-Benzylcytosine (BC) derivative bearing a probe (red) suitable for covalent labeling of CLIP-tag fusion proteins and subsequent imaging

with fluorescein by using BG–fluorescein [70]. In addition, the SNAP-technology was successfully applied to measure calcium Ca(II) concentrations in defined locations in living cells [74]. A BG derivative was synthetically linked to the Ca(II)-sensitive dye Indo-1 which in turn was covalently attached to nucleus-localized SNAP-tag fusion proteins inside living cells. This approach combines the spatial specificity of genetically encoded SNAP-tag-based Ca(II) indicators with the beneficial spectroscopic properties of Indo-1 upon Ca(II) binding (change in fluorescence).

The former limitation of the SNAP-tag to allow labeling of only a single intracellular protein was overcome by a novel AGT-based tag generated via directed evolution [75]. This probe, called CLIP-tag, reacts specifically with O^2-benzylcytosine (BC) derivatives (Fig. 13) yielding a covalent adduct and is orthogonal to the SNAP-tag (1,000-fold less efficient reactivity with BC than with BG derivatives [76]) regarding its substrate selectivity [75]. Thus, CLIP and SNAP fusion proteins can be used for simultaneous and orthogonal protein tagging in two colors in vivo with different small molecule probes fused to BC or BG, respectively [76]. Compared to SNAP, the CLIP has a similar size, similar selectivity as well as speed of labeling, and its substrates can be easily synthesized from commercially available precursors. Furthermore, BC derivatives are not substrates for original AGTs – an important fact for the sometimes problematic background staining observed for other tags, in particular in the case of the TC/FlAsH system.

Specific double labeling with more than 99% selectivity could be achieved in cell extracts as well as in living cells [76]. HEK293T cells were used transiently coexpressing either FK506 binding protein (FKBP) fused to the C-terminus of CLIP-tag (CLIP-FBBP) plus the FKBP-rapamycin binding domain (FRB) fused at the C-terminus of SNAP-tag (SNAP-FRB) or vice versa. Cells were incubated with a mixture of BC- and BG-derivatives and no significant cross-labeling or background staining was observed [76]. Furthermore, the self-labeling SNAP- and CLIP-tag technologies can be used for cross-linking of interacting proteins using bifunctional probes that contain the substrates of both tags connected via a fluorophore [77]. Very recently, photoactivatable and photoconvertible fluorescent probes for covalent protein labeling have been developed [78]. These new and advanced tools can be selectively coupled to SNAP-tag fusion proteins in living cells via click chemistry. Photosensitive BG derivatives designed for SNAP-tag labeling exhibit a change in their fluorescence spectrum upon labeling. To make use of the fluorogenic nature of the probe, a photocleavable linker based on 4,5-dimethoxy-2-nitrobenzyl derivatives is cleaved by irradiation with UV-light (365 nm). Thereby, donor and acceptor fluorophore are separated and exhibit distinct spectra compared to unreacted intact or attached intact probe. Photosensitive protein labels are powerful tools and expand the scope of protein tags to study cellular processes at low molecular concentrations.

5.2 *The Halo-Tag*

The commercially marketed Halo-tag [79] is based on a modified ~33 kDa haloalkane dehalogenase domain. In general, dehalogenases degrade compounds with haloalkane functionality in a two-step process: First, the alkyl chain is transferred by a nucleophilic displacement mechanism to the side chain of an active site Asp which is deeply buried within a hydrophobic pocket [80]. The formation of an ester bond is coupled with the loss of the halide ion. Second, hydrolysis of the alkyl–enzyme linkage and regeneration of the Asp side chain is catalyzed by an active site His residue [81]. Wood and coworkers have developed a mutant which lacks the capability of hydrolysis and therewith, preserves the alkyl–enzyme adduct permanently (Fig. 14) [82].

The enzymatic reaction is not depending on PTMs or cofactors which broadens the scope of this technology. Trapping the covalent intermediate offers the possibility to attach bioorthogonal chemical functionalities or small molecule probes, for instance fluorophores, irreversibly and permanently with high specificity and

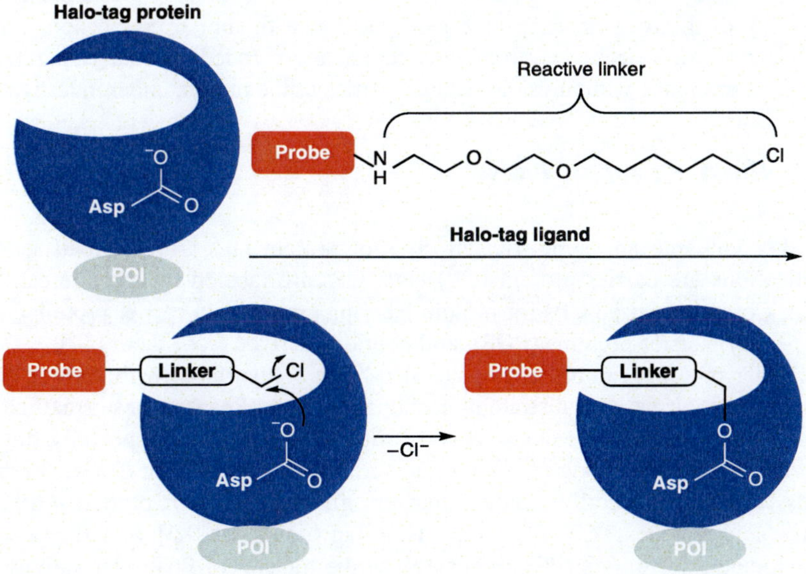

Fig. 14 Schematic representation of the covalent labeling approach based on a haloalkane dehalogenase domain (Halo-tag; blue) fused to the protein of interest (POI; gray). The Halo-tag ligand bearing any kind of probe (red) forms a covalent adduct with an active-site aspartate

efficiency. The newly generated ester bond turned out to be stable under both physiological and denaturing conditions such as boiling or SDS treatment [83]. Cell-permeable haloalkanes bearing different fluorophores or biotin are either commercially available or can be synthesized easily. As a result, the Halo-tag technology permits specific fluorescent labeling of fusion proteins with efficient interchangeability of tag properties both in living and in fixed cells as well as in in vitro assays and by immobilizing the fusion protein onto solid supports [83]. The reporting dehalogenase can be used in a manner similar to other genetically encoded fusion tags such as AFPs and AGT-tags.

The utility of this modular protein tagging system was demonstrated for protein immobilization and cellular imaging by means of analyzing multiple molecular processes associated with the nuclear factor (NF)-κB physiology [83]. A single genetic mutation was used in multiple analytical methods including the labeling and subsequent imaging of temporally different expressed target proteins. A set of alternative chloralkane ligands was employed to analyze subcellular localization, translocation, and degradation of NF-κB proteins. Assays in HeLa cells showed convincingly that protein labeling was completed within 15 min at room temperature and that no unspecific labeling occurred. It is worth mentioning that the length of the linker between the functional reporter and the reactive terminal halogen atom affects the efficiency of the halide release and has proven to be optimal for carboxyfluorescein (FAM) with 14 atoms in length [83]. Since there are no enzymes homologous to bacterial dehalogenases in eukaryotic cells, background staining in cells is insignificant.

Limitations of the Halo-tag technology might be cytotoxicity of the haloalkane, the solubility of the reagent, unspecific side reaction with similar endogenous nucleophiles, and drawbacks known from other genetically encoded fusion tags, particularly size or interference with physiological properties of the protein under investigation.

5.3 *Carrier Protein Tags*

Another valuable and sophisticated class of protein tags with a broad scope of applications are carrier protein (CP) tags. CPs are labeled by enzyme-catalyzed PTM. Commonly used CPs for protein labeling are the acyl carrier protein (ACP), the peptityl carrier proteins (PCP), and peptides derived from them [84].

Carrier proteins can be labeled covalently on cell surfaces with various probes attached to their native substrate phosphopantetheine (Ppant). These small adapter proteins play important roles in diverse metabolic pathways, especially the fatty acid and polyketide biosynthesis. CPs, as for instance the ACP of *E. coli* (77 amino acids [85]), are generally smaller than the green fluorescent protein (GFP; 238 amino acid residues, ~27 kDa) or the Halo-tag (~33 kDa) [86]. Specific phosphopantetheine transferases (PPTases) catalyze the transfer of Ppant derivatives from coenzyme A (CoA) onto a conserved Ser residue of CPs in order to form a phosphate ester bond (Fig. 15) [86]. During biosynthesis, metabolic intermediates are bound to Ppant via a thioester bond. Since PPTases generally have proven to

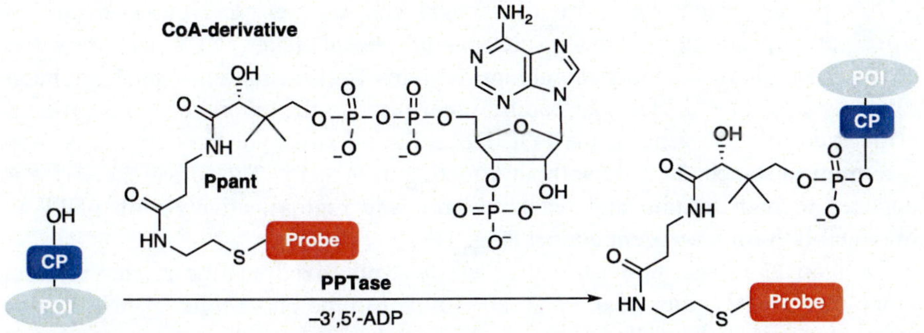

Fig. 15 Covalent protein labeling of the protein of interest (POI; gray) using the carrier protein (CP; blue)-tag techniques. Phosphopantetheine transferase (PPTases) transfers probe (red)-bearing phosphopantetheine (Ppant) from coenzyme A (CoA) derivatives to a conserved serine on the CP fusion protein by releasing 3′,5′-adenosine bisphosphate (3′,5′-ADP)

accept thiol-modified derivatives of CoA, fusion proteins of a CP and the cell surface POI have been designed and studied toward labeling and imaging with diverse set of probes such as fluorophores, quantum dots, or other chemical probes with tailored properties [85, 87, 88]. Either labeling of one fusion CP with several probes for studying time-dependent protein expression or the selective labeling of different CPs in the same approach is feasible and has been demonstrated [84].

Notably, a study of the Johnsson laboratory showed the use of CP fusion tags for sequential in vivo labeling followed by live-cell imaging in budding yeast [84]. Newly synthesized cell wall protein α-agglutinin (Sag1p) of *Saccharomyces cerevisiae* was tagged at different time points. Sag1p is known to be transported to sites of active cell growth and to be linked covalently to a specific cell wall glucan. Multicolor imaging of CP fusion proteins with a clear separation of differently colored tags occurred and demonstrated the power of this technology.

The protein tag can be adapted to differing experimental conditions and requirements without modifying the genetic construct due to the interchangeable design of the CoA-linked Ppant derivatives. The major limitation of this technology is, as mentioned already, that all CP-tags reported so far can only be used to label cell surface proteins due to the intracellular competition with endogenous CPs. Bioorthogonal mutants of CPs and their substrates might overcome this limitation in the future.

5.4 *Emerging Fusion Tags for Protein Labeling with Small Molecule Probes*

5.4.1 Suicide Inhibitors for Protein Labeling

Unlike the SNAP- or CLIP-tag, other protein–ligand pairs suitable for the covalent attachment of imaging probes rely on the targeted use of irreversible active-site inhibitors often called suicide substrates.

A reporter system based on the fungal cell wall enzyme cutinase and its corresponding suicide substrate *p*-nitrophenyl phosphonate (pNPP) was shown to be very useful for studying membrane proteins [89]. Cutinase, a small globular serine esterase of ~22 kDa size, can be either fused to the termini of target proteins or inserted into the gene of the POI due to the proximity of the cutinase N- and C-termini (distance: 28.2 Å) without affecting its activity. The active-site serine is opposed to both termini and reacts chemo- and regioselectively with pNPP to irreversibly form a covalent adduct (Fig. 16).

Derivatives of the inhibitor can be easily synthesized bearing a reactive thiol group (pNPP–SH) connected via an alkyl linker to the phosphorus of pNPP using organic chemistry. The alkyl chain of pNPP sticks out from the catalytic pocket of cutinase and allows the attachment of probes suitable for imaging [90]. Making use of the nucleophilicity of the thiol group, many commercially available labels in the form of their activated esters, for instance fluorophores and quantum dots, can be fused to pNPP–SH. pNPP adducts are stable under physiological conditions and therefore allow to study protein dynamics.

In a study by Bonasio et al. pNPP derivatives were used to covalently attach a variety of labels to the integrin lymphocyte function-associated antigen-1 (LFA-1) on the surface of living cells followed by live cell imaging. The cutinase was inserted into a domain which was known to localize to the outer leaflet of the plasma membrane after stimulation facilitating the reaction with the probe-bearing pNPP derivative [89].

Since cutinase is orthogonal in mammalian cells, background staining is negligible. Finally, the cutinase/pNPP technology turned out to be a flexible strategy to covalently label specific proteins noninvasively in vivo as well as in vitro.

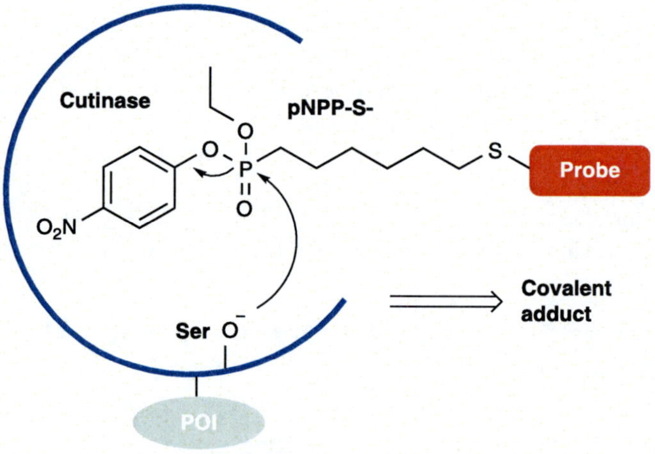

Fig. 16 Schematic representation of the covalent labeling method based on cutinase and its suicide substrate *p*-nitrophenyl phosphonate (pNPP). Cutinase is fused to the protein of interest (POI). An active-site serine attacks the pNPP–probe conjugate to yield a covalently probe-labeled POI–cutinase fusion protein

However, the scope of the cutinase/pNPP system has not yet been exploited intensively [91].

5.4.2 Enzyme–Substrate Labeling Technologies

Apart from the covalent labeling tags described above, several other modified enzyme/substrate, enzyme/inhibitor (see Sect. 5.4.1) or enzyme/cosubstrate systems have been developed but often are not yet applied extensively. Among these tags are, for instance, technologies based on the *E. coli* β-galactosidase (β-gal) [92, 93], *E. coli* biotin ligase (BirA) [94], guinea-pig liver transglutaminase (TGase) [95].

Nagano and coworkers [92] have chosen β-gal for labeling fusion proteins in living cells because of its high substrate specificity for the glycan unit, its high turnover number for β-galactopyranosyl-protected phenols, and its wide tolerance for the aglycan unit. β-Galactosidase of *E. coli* is encoded by the LacZ gene and is frequently used for genetics and molecular biology for decades. The designed probe coumarin–mandelate–fluorescein–β-galactopyranosyl (CMFβ-gal) consists of two fluorophores forming a FRET pair – 7-hydroxycoumarin as donor and fluorescein as acceptor – which is cleaved upon labeling resulting in an increase in donor fluorescence (Fig. 17).

The dramatic spectral change allows real-time imaging of the formation of labeled product. In the study presented, CMFβ-gal was used for ratiomeric imaging in LacZ-positive and LacZ-negative HEK293 cells [92].

Another approach to tag surface proteins is provided by the two-step labeling protocol based on BirA [94]. First, BirA ligates sequence-specifically biotin or its ketone isoster to an acceptor peptide (AP) of 15 amino acids length (first labeling step) [96]. In the next step, a hydroxylamide- or hydrazide-functionalized small molecule probe bearing for instance a fluorophore is attached to the ketone functional group in the presence of ATP (Fig. 18). The AP is very small compared to other fusion tags and thus might be advantageous for investigations in which the size of the tag is an issue. Furthermore, bacteria have only one natural substrate for BirA and endogenous mammalian proteins are not modified by BirA.

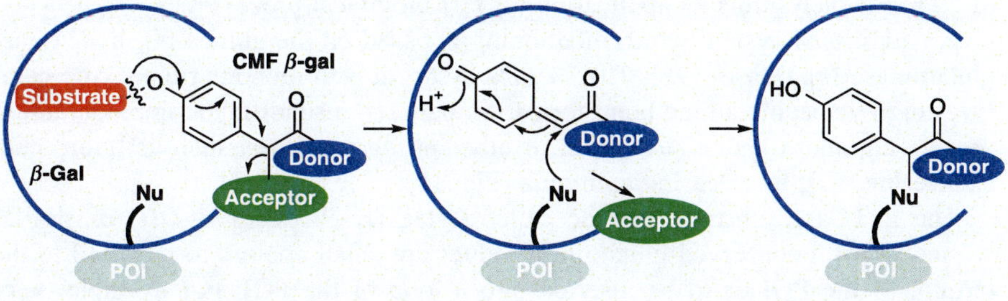

Fig. 17 Covalent protein labeling using the β-galactosidase (β-gal)-tag fused to the protein of interest (POI; gray). The probe coumarin (blue)–mandelate–fluorescein (green)–β-galactopyranosyl (red) (CMFβ-gal) forms a FRET pair that is cleaved by an active-site nucleophile (Nu) upon labeling

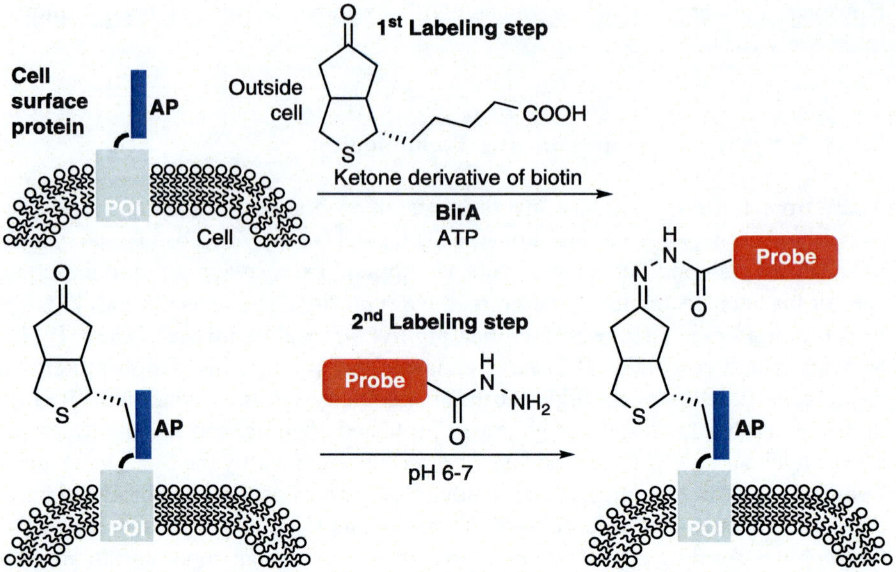

Fig. 18 General approach for labeling an acceptor peptide (AP; blue)-tagged protein of interest (POI; gray) on the cell surface with diverse probes (red). The first labeling step is catalyzed by biotin ligase (BirA) ligating the ketone derivative of biotin to the AP. In the second step, subsequent bioorthogonal ligation between the ketone and a hydrazide enables the introduction of a probe (red)

The laboratory of Ting applied the BirA/reactive biotin analog/hydrazine technology to in vitro and cell lysate labeling of AP-fusion proteins and an AP fused to the N-terminus of human epidermal growth factor receptor (hEGFR) in HeLa cells [94]. To date the BirA labeling approach is restricted to cell surfaces, although BirA can be expressed in mammalian cells. However, some endogenous small molecules contain ketone and aldehyde functionalities and might prevent specific labeling. Likewise, reduction of the ketone group in the cell environment might take place. Another drawback is the two-step labeling procedure with a minimal labeling time of 20 min which limits its application for fast biological processes.

An additional system worth mentioning is based on the guinea-pig liver transglutaminase (gpTGase) [95]. The 77 kDa cytosolic and monomeric enzyme catalyzes highly specific amine bond formation between a small glutamine-containing peptide tag and a lysine side chain or other primary amines, such as fluorescein cadaverine [97], by releasing ammonia (Fig. 19).

The gpTGase substrate peptide called Q-tag (*H*–PKPQQFM–*OH*) or similar sequences with conserved glutamine residues are small enough to be fused to the termini of the POI or to be inserted into a loop of the POI. For example, very efficient and highly specific gpTGase-catalyzed labeling of Q-tagged cell surface proteins was accomplished for the EGFR [95]. For probes, biotin (followed by streptavidin–Alexa 568 conjugate for imaging) or Alexa 568 cadaverine itself were

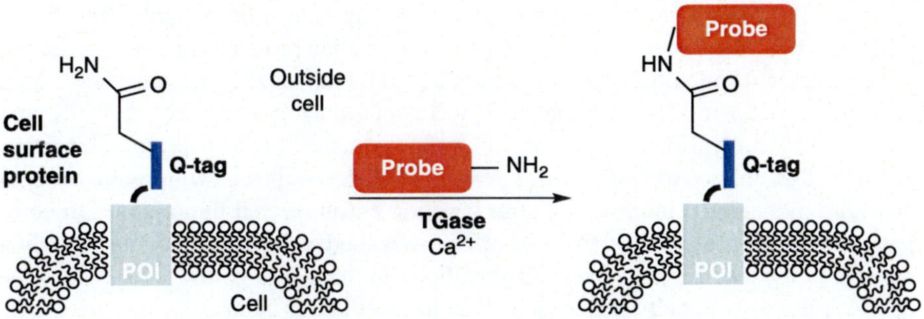

Fig. 19 A general method for labeling specific peptide (Q)-tagged cell surface proteins via a transglutaminase (TGase) catalyzing the amide bond formation between a glutamate side chain and a reactive amine. Probes of different kind bearing for instance a lysine can be used as substrates resulting in a probe (red)-labeled Q-tag (blue)–protein of interest (POI; gray) fusion

used and neither tagging nor labeling interfered with the biological function of the EGFR. Possible in vitro applications include the cross-linking of proteins or attaching fluorophores for visualization on SDS-PAGE which also proves the covalent nature of the protein tag [95]. However, it is unlikely that the TGase labeling technology can be applied to label intracellular proteins due to competing endogenous TGases substrates.

Amongst the peptide tags, a very versatile method for site-specific covalent protein labeling referred to as sortagging (sortase-mediated transpeptidation) convinces with its small size (<2 kDa) and interchangeability of attached probes [98]. The natural function of bacterial sortases, such as the *Staphylococcus aureus* sortase A ($SrtA_{staph}$), is to covalently attach proteins to the bacterial cell wall. Engineered variants of these enzymes can be used to label site-specifically proteins at both the C- [98] and the N-terminus [99]. In initial experiments, the C-terminal recognition site (*H*–LPXTG–*OH*) was fused to the POI and was sequence-specifically cleaved by $SrtA_{staph}$ between the Thr and Gly prior to the linkage with the N-terminus of a pentaglycine nucleophile which has to be provided and can be modified with a probe or peptide fragment. Sortases suitable for catalyzing the site-specific transpeptidation transformation can be expressed and purified easily in *E. coli* and used for the modification of proteins. The sortagging technique was successfully applied to label proteins in vitro on cell surfaces of living cells [98] as well as to attach different chemical labels at the two ends of the same POI by using two orthogonal sortases [99].

Very recently, a method for fluorescent labeling of peptide-fused recombinant proteins on cell surfaces [100] as well as in living cells [101] was introduced by Ting and coworkers. This so-called PRIME (PRobe Incorporation Mediated by Enzymes) approach for covalent labeling is based on the enzyme lipoic acid ligase (LplA) from *E. coli*. LplA is absent from the mammalian proteome and catalyzes in its native form the ligation of lipoic acid to three *E. coli* proteins of the oxidative metabolism. The mutant generated by mutagenesis recognizes 7-hydroxycoumarin as substrate and conjugates this fluorophore to a 13-amino acid peptide referred to as LAP (LplA

Acceptor Peptide). It was shown that fluorophore ligation is both highly specific for LAP fusions to the POI over all endogenous mammalian proteins and highly efficient (labeling completed within 10 min at 37°C) [101]. Humanized genes for both LplA and the most kinetically efficient LAP were reproducibly expressed in HEK and HeLa cells. Finally, PRIME was used for imaging nuclear actin. Therefore, LAP was fused to β-actin, coexpressed with LplA in COS-7, incubated with coumarin, and analyzed by confocal microscopy after washing out unreacted fluorophore. In comparison to other protein tags, PRIME offers several advantages, especially regarding the size of the tag, the efficiency and specificity of the labeling, low toxicity, and the applicability to both cell surface as well as intracellular proteins.

Most of the above-mentioned labeling methods do not exhibit fluorogenic properties. The treated cells need to be washed prior to microscopic investigations to eliminate background fluorescence resulting from unreacted probe. Recently, a new technique for covalent labeling of proteins was published that satisfies the dual criteria of specificity and fluorogenicity [102]. Bacterial β-lactamase that usually hydrolyzes antibiotics containing a β-structure was evolved to sustain the acyl–enzyme intermediate which is normally cleaved straight away after acylation (Fig. 20). A FRET probe referred to as CCD for the class A β-lactamase mutant ^{E166N}TEM (29 kDa) was developed consisting of the FRET donor 7-hydroxycoumarin, the enzyme substrate cephalosporin, and 4-(4′-dimethylaminophenylazo) benzoic acid (DABCYL) as FRET acceptor. Experiments with ^{E166N}TEM fused to maltose binding protein (MBP; 42 kDa) in HEK293T cell lysate as well as in vivo studies with ^{E166N}TEM fused to the N-terminus of EGFR in HEK293T cells proved high specificity and negligible background staining through unspecific labeling. As expected, upon labeling and after the loss of the DABCYL group, the cyan fluorescence of coumarin was restored due to a loss of FRET. Fast, covalent and complete reaction in less than 45 min was observed by a 30-fold increase of the fluorescence signal compared to unreacted CCD.

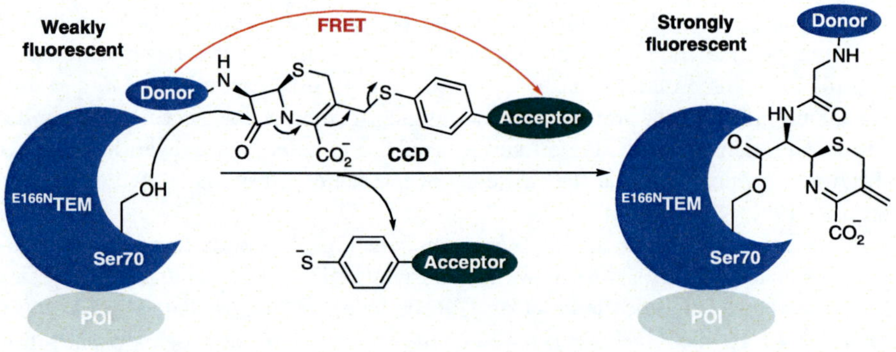

Fig. 20 Labeling mechanism of the β-lactamase (^{E166N}TEM; blue)–substrate technology. The protein of interest (POI; gray) is fused to ^{E166N}TEM. An active-site serine of ^{E166N}TEM cleaves the β-lactam structure and thus, destroying the FRET of the CCD (7-hydroxyCoumarin (blue)–Cephalosporin–4-(4′-Dimethylaminophenylazo)benzoic acid (green)) probe

An interesting approach to evaluate protein activity employs activity-based probes. This kind of probes were designed to react covalently with their specific enzyme and thus allow labeling of the adduct via a bioorthogonal reaction (see Sects. 2 and 3). In a recent study, functional subclasses of phospholipases (PLAs) belonging to the serine hydrolase (SH) superfamily of enzymes were selectively targeted by activity-based probes [103]. Fluorophosphonate (FP) activity-based protein-profiling (ABPP) probes bearing a terminal alkyne suitable for click chemistry were linked at different positions (*sn*-1 or *sn*-2, respectively) to phosphatidylcholine (PC) and shown to selectively react with either PLA_1 or PLA_2 activity.

New protein fusion tags for covalent labeling of the target protein with small molecule probes will very likely be introduced in the near future and will expand the toolbox of genetically encodable fusion tags of this type. For instance, protein farnesyltransferases (name of the tag: farnesylation motif) [104] or formyl glycine generating enzymes (name of the tag: formyl glycine tag) [105] have been successfully harnessed to label proteins in vitro. However, their adaptability to cell surfaces or intracellular imaging has not yet been demonstrated.

5.4.3 Ligand-Directed Protein Labeling

One of the major limitations of all methods for protein labeling mentioned earlier in this chapter is that they all rely on either metabolic or genetic manipulations that might disturb native protein function.

Recently, Hamachi and coworkers reported a novel ligand-directed tosyl (LDT)-based covalent labeling technique [106]. The LDT-approach permits the site-selective, covalent and traceless attachment of diverse synthetic small molecule probes to endogenous proteins in living cells, tissues and living animals without genetic or metabolic alterations. In this affinity-based approach, a modified protein ligand bearing a probe binds to the target protein at a specific ligand-binding site. Very importantly, the probes have to be attached to the native ligand via an electrophilic phenylsulfonate ester group. Thereby, a S_N2-type chemical reaction of the modified ligand with a nucleophilic amino acid side chain on the protein surface in close proximity is enabled. The outcome is a probe-labeled protein and a ligand bearing only the toluene sulfonic acid that can leave the ligand-binding site by simple diffusion (Fig. 21). The advantage of this approach is the noncovalent nature of the ligand–protein adduct (in contrast to the covalent attachment of the probe to the protein surface) which allows the protein to regain its native function after the probe transfer. The method combines excellent target selectivity and site specificity with the possibility to attach different kinds of probes, provided that a suitable ligand for the POI is available and that a nucleophilic amino acid side chain is able to react with the probe. The LDT-labeling technology allows tagging of proteins inside cells as well as on the cell surface.

The scope of the so-called tosyl chemistry technique was initially tested on carbonic anhydrase II (CAII) as the POI and benzenesulfonamide-containing synthetic probes for three different detection modalities (fluorophores, biotin tag,

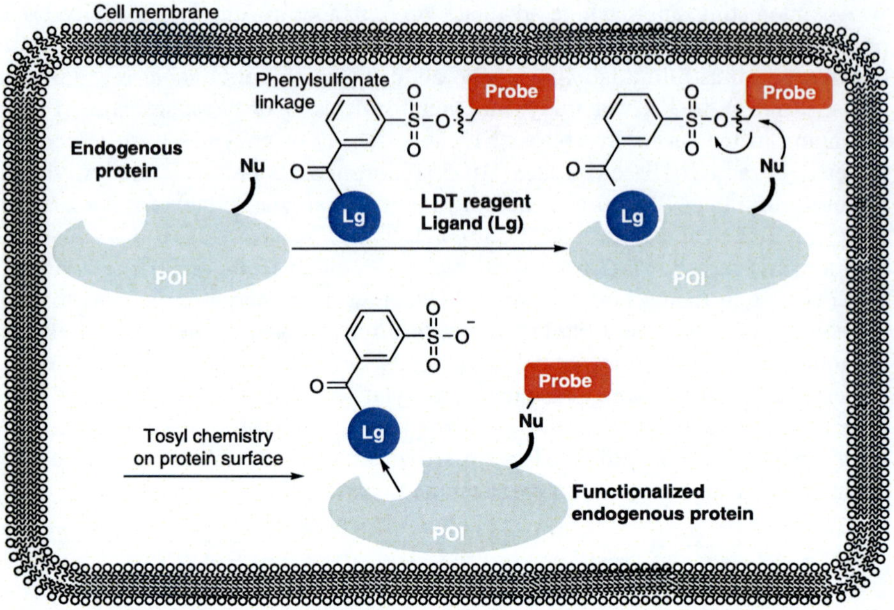

Fig. 21 Ligand-directed tosyl (LDT) chemistry for labeling endogenous proteins in living cells. A ligand (Lg; blue) specific for the protein of interest (POI; gray) binds noncovalently to the active site of the POI. A nucleophile (Nu) on the POI surface reacts with the LDT reagent bearing a probe (red) forming a covalent adduct

^{19}F-NMR probe) [106]. Experiments in test tubes (in vitro), human red blood cells, as well as in live mice showed very specific labeling and no disturbing effect of the probe on the catalytic activity of CAII after the removal of the small molecule reagents. Since the ligand was specific for CAII, no background signal was observed after washing out the unreacted ligand.

In the near future, it might be possible to label several distinct proteins with different probes in the cell using LDT chemistry due to the ligand-determined target selectivity. LDT chemistry might also be a powerful tool when it comes to studying and engineering of biologically complex processes. In summary, the LDT approach is part of the ongoing progress of expanding the power of organic chemistry in biological systems.

6 Conclusion and Outlook

The number of methods for labeling proteins in living cells rapidly expands. Especially, there are now several new bioorthogonal reactions available that are applicable for this purpose. This development is essential as experiments performed in cells become more and more important for many areas in the life sciences. What is particularly needed in the future are methods that permit tagging of endogenous proteins because this will permit quantifying protein levels more accurately and will reduce artifacts from protein overexpression. In this respect, the expansion of the genetic code will hopefully penetrate further into the scientific community. Another aspect that will develop in the near future is the use of several orthogonal labeling techniques within a single experiment. This will largely increase the information content and will allow following the relative timing of cellular events simultaneously.

References

1. Kolb HC, Finn MG, Sharpless KB (2001) Click chemistry: diverse chemical function from a few good reactions. Angew Chem Int Ed 40:2004–2021
2. Lim RKV, Lin Q (2010) Bioorthogonal chemistry: recent progress and future directions. Chem Commun 46:1589–1600
3. Sander EG, Jencks WP (1968) Equilibria for additions to the carbonyl group. J Am Chem Soc 90:6154–6162
4. Mahal LK, Yarema KJ, Bertozzi CR (1997) Engineering chemical reactivity on cell surfaces through oligosaccharide biosynthesis. Science 276:1125–1128
5. Saxon E, Bertozzi CR (2000) Cell surface engineering by a modified Staudinger reaction. Science 287:2007–2010
6. Prescher JA, Dube DH, Bertozzi CR (2004) Chemical remodelling of cell surfaces in living animals. Nature 430:873–877
7. Huisgen R (1963) 1, 3-Dipolar cycloadditions. Past and future. Angew Chem Int Ed Engl 2:565–598
8. Rostovtsev VV, Green LG, Fokin VV, Sharpless KB (2002) A stepwise Huisgen cycloaddition process: copper(I)-catalyzed regioselective "ligation" of azides and terminal alkynes. Angew Chem Int Ed 41:2596–2599
9. Tornoe CW, Christensen C, Meldal M (2002) Peptidotriazoles on solid phase: [1–3]-triazoles by regiospecific copper(I)-catalyzed 1, 3-dipolar cycloadditions of terminal alkynes to azides. J Org Chem 67:3057–3064
10. Neef AB, Schultz C (2009) Selective fluorescence labeling of lipids in living cells. Angew Chem Int Ed 48:1498–1500
11. Jao CY, Roth M, Welti R, Salic A (2009) Metabolic labeling and direct imaging of choline phospholipids in vivo. Proc Natl Acad Sci USA 106:15332–15337
12. Bruckman MA, Kaur G, Lee LA, Xie F, Sepulvecla J, Breitenkamp R, Zhang X, Joralemon M, Russell TP, Emrick T, Wang Q (2008) Surface modification of tobacco mosaic virus with "click" chemistry. Chembiochem 9:519–523
13. Deiters A, Cropp TA, Mukherji M, Chin JW, Anderson C, Schultz PG (2003) Adding amino acids with novel reactivity to the genetic code of *Saccharomyces cerevisiae*. J Am Chem Soc 125:11782–11783
14. Link AJ, Tirrell DA (2003) Cell surface labeling of Escherichia coli via copper(I)-catalyzed [3+2] cycloaddition. J Am Chem Soc 125:11164–11165
15. Weisbrod SH, Marx A (2008) Novel strategies for the site-specific covalent labelling of nucleic acids. Chem Commun 5675–5685
16. Hüsken N, Gasser G, Köster SD, Metzler-Nolte N (2009) "Four-potential" ferrocene labeling of PNA oligomers via click chemistry. Bioconjug Chem 20:1578–1586
17. Speers AE, Adam GC, Cravatt BF (2003) Activity-based protein profiling in vivo using a copper(I)-catalyzed azide-alkyne [3 + 2] cycloaddition. J Am Chem Soc 125:4686–4687
18. Becer CR, Hoogenboom R, Schubert US (2009) Click chemistry beyond metal-catalyzed cycloaddition. Angew Chem Int Ed 48:4900–4908

19. Chang PV, Prescher JA, Sletten EM, Baskin JM, Miller IA, Agard NJ, Lo A, Bertozzi CR (2010) Copper-free click chemistry in living animals. Proc Natl Acad Sci USA 107:1821–1826
20. Agard NJ, Baskin JM, Prescher JA, Lo A, Bertozzi CR (2006) A comparative study of bioorthogonal reactions with azides. ACS Chem Biol 1:644–648
21. Blackman ML, Royzen M, Fox JM (2008) Tetrazine ligation: fast bioconjugation based on inverse-electron-demand Diels-Alder reactivity. J Am Chem Soc 130:13518–13519
22. Devaraj NK, Weissleder R, Hilderbrand SA (2008) Tetrazine-based cycloadditions: application to pretargeted live cell imaging. Bioconjug Chem 19:2297–2299
23. Devaraj NK, Hilderbrand S, Upadhyay R, Mazitschek R, Weissleder R (2010) Bioorthogonal turn-on probes for imaging small molecules inside living cells. Angew Chem Int Ed 49: 2869–2872
24. Devaraj NK, Upadhyay R, Haun JB, Hilderbrand SA, Weissleder R (2009) Fast and sensitive pretargeted labeling of cancer cells through a tetrazine/trans-cyclooctene cycloaddition. Angew Chem Int Ed 48:7013–7016
25. Wang Y, Rivera Vera CI, Lin Q (2007) Convenient synthesis of highly functionalized pyrazolines via mild, photoactivated 1, 3-dipolar cycloaddition. Org Lett 9:4155–4158
26. Song W, Wang Y, Qu J, Lin Q (2008) Selective functionalization of a genetically encoded alkene-containing protein via "photoclick chemistry" in bacterial cells. J Am Chem Soc 130:9654–9655
27. Lin YA, Chalker JM, Floyd N, Bernardes GJL, Davis BG (2008) Allyl sulfides are privileged substrates in aqueous cross-metathesis: application to site-selective protein modification. J Am Chem Soc 130:9642–9643
28. Lin YA, Chalker JM, Floyd N, Bernardes GJL, Davis BG (2009) A Convenient catalyst for aqueous and protein Suzuki-Miyaura cross-coupling. J Am Chem Soc 131:16346–16347
29. Ning X, Temming RP, Dommerholt J, Guo J, Ania DB, Debets MF, Wolfert MA, Boons GJ, van Delft FL (2010) Protein modification by strain-promoted alkyne-nitrone cycloaddition. Angew Chem Int Ed 49:3065–3068
30. Muir TW (2009) Chemical biology in a time of transition. ACS Chem Biol 4:241–243
31. Dawson PE, Muir TW, Clarklewis I, Kent SBH (1994) Synthesis of proteins by native chemical ligation. Science 266:776–779
32. Sletten EM, Bertozzi CR (2009) Bioorthogonal chemistry: fishing for selectivity in a sea of functionality. Angew Chem Int Ed 48:6974–6998
33. Muir TW (2008) Studying protein structure and function using semisynthesis. Biopolymers 9:743–750
34. Chatterjee C, Muir TW (2010) Chemical approaches for studying histone modifications. J Biol Chem 285:11045–11050
35. Karukurichi KR, Wang L, Uzasci L, Manlandro CM, Wang Q, Cole PA (2010) Analysis of p300/CBP histone acetyltransferase regulation using circular permutation and semisynthesis. J Am Chem Soc 132:1222–1223
36. McGinty RK, Chatterjee C, Muir TW (2009) Semisynthesis of ubiquitylated proteins. Methods Enzymol 462:225–243
37. Bayley H, Cheley S, Harrington L, Syeda R (2009) Wrestling with native chemical ligation. ACS Chem Biol 4:983–985
38. Gottlieb D, Grunwald C, Nowak C, Kuhlmann J, Waldmann H (2006) Intein-mediated in vitro synthesis of lipidated Ras proteins. Chem Commun 260–262
39. Durek T, Alexandrov K, Goody RS, Hildebrand A, Heinemann I, Waldmann H (2004) Synthesis of fluorescently labeled mono- and diprenylated Rab7 GTPase. J Am Chem Soc 126:16368–16378
40. Camarero JA, Fushman D, Cowburn D, Muir TW (2001) Peptide chemical ligation inside living cells: in vivo generation of a circular protein domain. Bioorg Med Chem 9:2479–2484
41. Daly NL, Love S, Alewood PF, Craik DJ (1999) Chemical synthesis and folding pathways of large cyclic polypeptides: studies of the cystine knot polypeptide kalata B1. Biochemistry 38:10606–10614

42. Ambrogelly A, Palioura S, Söll D (2007) Natural expansion of the genetic code. Nat Chem Biol 3:29–35
43. Link AJ, Mock ML, Tirrell DA (2003) Non-canonical amino acids in protein engineering. Curr Opin Biotechnol 14:603–609
44. Wang L, Schultz PG (2005) Expanding the genetic code. Angew Chem Int Ed 44:34–66
45. Liu CC, Schultz PG (2010) Adding new chemistries to the genetic code. Annu Rev Biochem 79:413–444
46. Wu X, Schultz PG (2009) Synthesis at the interface of chemistry and biology. J Am Chem Soc 131:12497–12515
47. Hecht SM, Alford BL, Kuroda Y, Kitano S (1978) "Chemical aminoacylation" of tRNA's. J Biol Chem 253:4517–4520
48. Noren CJ, Anthony-Cahill SJ, Griffith MC, Schultz PG (1989) A general method for site-specific incorporation of unnatural amino acids into proteins. Science 244:182–188
49. Saks ME, Sampson JR, Nowak MW, Kearney PC, Du F, Abelson JN, Lester HA, Dougherty DA (1996) An engineered Tetrahymena $tRNA^{Gln}$ for in vivo incorporation of unnatural amino acids into proteins by nonsense suppression. J Biol Chem 271:23169–23175
50. Kiick KL, van Hest JCM, Tirrell DA (2000) Expanding the scope of protein biosynthesis by altering the methionyl-tRNA synthetase activity of a bacterial expression host. Angew Chem Int Ed 39:2148–2152
51. Hendrickson TL, de Crécy-Lagard V, Schimmel P (2004) Incorporation of nonnatural amino acids into proteins. Annu Rev Biochem 73:147–176
52. Cohen GN, Munier R (1956) Incorporation of structural analogues of amino acids in bacterial proteins. Biochim Biophys Acta 21:592–593
53. Cowie DB, Cohen GN (1957) Biosynthesis by Escherichia coli of active altered proteins containing selenium instead of sulphur. Biochim Biophys Acta 26:252–261
54. Beatty KE, Liu JC, Xie F, Dieterich DC, Schuman EM, Wang Q, Tirrell DA (2006) Fluorescence visualization of newly synthesized proteins in mammalian cells. Angew Chem Int Ed 45:7364–7367
55. Datta D, Wang P, Carrico IS, Mayo SL, Tirrell DA (2002) A designed phenylalanyl-tRNA synthetase variant allows efficient in vivo incorporation of aryl ketone functionality into proteins. J Am Chem Soc 124:5652–5653
56. Tanrikulu IC, Schmitt E, Mechulam Y, Goddard WA 3rd, Tirrell DA (2009) Discovery of Escherichia coli methionyl-tRNA synthetase mutants for efficient labeling of proteins with azidonorleucine in vivo. Proc Natl Acad Sci USA 106:15285–15290
57. Döring V, Mootz HD, Nangle LA, Hendrickson TL, de Crécy-Lagard V, Schimmel P, Marlière P (2001) Enlarging the amino acid set of Escherichia coli by infiltration of the valine coding pathway. Science 292:501–504
58. Anthony-Cahill SJ, Griffith MC, Noren CJ, Suich DJ, Schultz PG (1989) Site-specific mutagenesis with unnatural amino acids. Trends Biochem Sci 14:400–403
59. Neumann H, Wang K, Davis L, Garcia-Alai M, Chin JW (2010) Encoding multiple unnatural amino acids via evolution of a quadruplet-decoding ribosome. Nature 464:441–444
60. Gibson DG, Glass JI, Lartigue C, Noskov VN, Chuang RY, Algire MA, Benders GA, Montague MG, Ma L, Moodie MM, Merryman C, Vashee S, Krishnakumar R, Assad-Garcia N, Andrews-Pfannkoch C, Denisova EA, Young L, Qi ZQ, Segall-Shapiro TH, Calvey CH, Parmar PP, Hutchison CA 3rd, Smith HO, Venter JC (2010) Creation of a bacterial cell controlled by a chemically synthesized genome. Science. doi: 10.1126/science.1190719
61. Liu W, Brock A, Chen S, Chen S, Schultz PG (2007) Genetic incorporation of unnatural amino acids into proteins in mammalian cells. Nat Methods 4:239–244
62. Neumann H, Peak-Chew SY, Chin JW (2008) Genetically encoding N^{ε}-acetyllysine in recombinant proteins. Nat Chem Biol 4:232–234
63. Yanagisawa T, Ishii R, Fukunaga R, Kobayashi T, Sakamoto K, Yokoyama S (2008) Multistep engineering of pyrrolysyl-tRNA synthetases to genetically encode N^{ε}-(o-azido-benzyloxycarbonyl) lysine for site-specific protein modification. Chem Biol 15:1187–1197

64. Nguyen DP, Lusic H, Neumann H, Kapadnis PB, Deiters A, Chin JW (2009) Genetic encoding and labeling of aliphatic azides and alkynes in recombinant proteins via a pyrrolysyl-tRNA synthetase/tRNACUA pair and click chemistry. J Am Chem Soc 131:8720–8721
65. Chen PR, Groff D, Guo J, Ou W, Cellitti S, Geierstanger BH, Schultz PG (2009) A facile system for encoding unnatural amino acids in mammalian cells. Angew Chem Int Ed 48:4052–4055
66. Neumann H, Slusarczyk AL, Chin JW (2010) De novo generation of mutually orthogonal aminoacyl-tRNA synthetase/tRNA pairs. J Am Chem Soc 132:2142–2144
67. Wan W, Huang Y, Wang Z, Russell WK, Pai PJ, Russell DH, Liu WR (2010) A facile system for genetic incorporation of two different noncanonical amino acids into one protein in Escherichia coli. Angew Chem Int Ed 49:3211–3214
68. Griffin BA, Adams SR, Tsien RY (1998) Specific covalent labeling of recombinant protein molecules inside live cells. Science 281:269–272
69. Keppler A, Gendreizig S, Gronemeyer T, Pick H, Vogel H, Johnsson K (2003) A general method for the covalent labeling of fusion proteins with small molecules in vivo. Nat Biotechnol 21:86–89
70. Keppler A, Kindermann M, Gendreizig S, Pick H, Vogel H, Johnsson K (2004) Labeling of fusion proteins of O^6-alkylguanine-DNA alkyltransferase with small molecules in vivo and in vitro. Methods 32:437–444
71. Juillerat A, Heinis C, Sielaff I, Barnikow J, Jaccard H, Kunz B, Terskikh A, Johnsson K (2005) Engineering substrate specificity of O^6-alkylguanine-DNA alkyltransferase for specific protein labeling in living cells. Chembiochem 6:1263–1269
72. Gronemeyer T, Chidley C, Juillerat A, Heinis C, Johnsson K (2006) Directed evolution of O^6-alkylguanine-DNA alkyltransferase for applications in protein labeling. Protein Eng Des Select 19:309–316
73. Keppler A, Pick H, Arrivoli C, Vogel H, Johnsson K (2004) Labeling of fusion proteins with synthetic fluorophores in live cells. Proc Natl Acad Sci USA 101:9955–9959
74. Bannwarth M, Corrêa IR, Jr SM, Pouvreau S, Fellay C, Aebischer A, Royer L, Ríos E, Johnsson K (2009) Indo-1 derivatives for local calcium sensing. ACS Chem Biol 4:179–190
75. Heinis C, Schmitt S, Kindermann M, Godin G, Johnsson K (2006) Evolving the substrate specificity of O^6-alkylguanine-DNA alkyltransferase through loop insertion for applications in molecular imaging. ACS Chem Biol 1:575–589
76. Gautier A, Juillerat A, Heinis C, Corrêa IR, Jr KM, Beaufils F, Johnsson K (2008) An engineered protein tag for multiprotein labeling in living cells. Chem Biol 15:128–136
77. Gautier A, Nakata E, Lukinavičius G, Tan KT, Johnsson K (2009) Selective cross-linking of interacting proteins using self-labeling tags. J Am Chem Soc 131:17954–17962
78. Maurel D, Banala S, Laroche T, Johnsson K (2010) Photoactivatable and photoconvertible fluorescent probes for protein labeling. ACS Chem Biol 5:507–516
79. Los GV, Darzins A, Karassina N, Zimprich C, Learish R, McDougallM G, Encell LP, Friedman-Ohana R, Wood M, Vidurgiris G, Zimmerman K, Otto P, Klaubert DH, Wood KV (2005) Halotag interchangeable labeling technology for cell imaging and protein capture. Promega Cell Notes 11:2–6
80. Janssen DB (2004) Evolving haloalkane dehalogenase. Curr Opin Chem Biol 8:150–159
81. Pries F, Kingma J, Krooshof G, Jeronimus-Stratingh C, Bruins A, Janssen DB (1995) Histidine 289 is essential for hydrolysis of the alkyl-enzyme intermediate of haloalkane dehalogenase. J Biol Chem 270:10405–10411
82. Los G, Wood K (2007) The HaloTag: a novel technology for cell imaging and protein analysis. Methods Mol Biol 356:195–208
83. Los GV, Encell LP, McDougall MG, Hartzell DD, Karassina N, Zimprich C, Wood MG, Learish R, Friedman-Ohana R, Urh M, Simpson D, Mendez J, Zimmerman K, Otto P, Vidugiris G, Zhu J, Darzins A, Klaubert DH, Bulleit RF, Wood KV (2008) HaloTag: a novel protein labeling technology for cell imaging and protein analysis. Chem Biol 3:373–383
84. Vivero-Pol L, George N, Krumm H, Johnsson K, Johnsson N (2005) J Am Chem Soc 127:12770–12771

85. George N, Pick H, Vogel H, Johnsson N, Johnsson K (2004) Specific labeling of cell surface proteins with chemically diverse compounds. J Am Chem Soc 126:8896–8897
86. O'Hare HM, Johnsson K, Gautier A (2007) Chemical probes shed light on protein function. Curr Opin Struct Biol 17:488–494
87. Johnsson N, George N, Johnsson K (2005) Protein chemistry on the surface of living cells. Chembiochem 6:47–52
88. Yin J, Liu F, Walsh CT (2004) Labeling proteins with small molecules by site-specific posttranslational modification. J Am Chem Soc 126:7754–7755
89. Bonasio R, Carman CV, Kim E, Sage PT, Love KR, Mempel TR, Springer TA, von Andrian UH (2007) Specific and covalent labeling of a membrane protein with organic fluorochromes and quantum dots. Proc Natl Acad Sci USA 104:14753–14758
90. Longhi S, Nicolas A, Creveld L, Egmond M, Verrips CT, de Vlieg J, Martinez C, Cambillau C (1996) Dynamics of fusarium solani cutinase investigated through structural comparison among different crystal forms of its variants. Proteins 26:442–458
91. Lin MZ, Wang L (2008) Selective labeling of proteins with chemical probes in living cells. Physiology 23:131–141
92. Komatsu T, Kikuchi K, Takakusa H, Hanaoka K, Ueno T, Kamiya M, Urano Y, Nagano T (2006) Design and synthesis of an enzyme activity-based labeling molecule with fluorescence spectral change. J Am Chem Soc 128:15946–15947
93. Urano Y, Kamiya M, Kanda K, Ueno T, Hirose K, Nagano T (2005) Evolution of fluorescein as a platform for finely tunable fluorescence probes. J Am Chem Soc 127:4888–4894
94. Chen I, Howarth M, Lin W, Ting AY (2005) Site-specific labeling of cell surface proteins with biophysical probes using biotin ligase. Nat Methods 2:99–104
95. Lin CW, Ting AY (2006) Transglutaminase-catalyzed site-specific conjugation of small-molecule probes to proteins in vitro and on the surface of living cells. J Am Chem Soc 128:4542–4543
96. Beckett D, Kovaleva E, Schatz PJ (1999) A minimal peptide substrate in biotin holoenzyme synthetase-catalyzed biotinylation. Protein Sci 8:921–929
97. Wolff C, Lai CS (1988) Evidence that the two amino termini of plasma fibronectin are in close proximity: a fluorescence energy transfer study. Biochemistry 27:3483–3487
98. Antos JM, Chew GL, Guimaraes CP, Yoder NC, Grotenbreg GM, Popp MW, Ploegh HL (2009) Site-specific N- and C-terminal labeling of a single polypeptide using sortases of different specificity. J Am Chem Soc 131:10800–10801
99. Popp MW, Antos JM, Grotenbreg GM, Spooner E, Ploegh HL (2007) Sortagging: a versatile method for protein labeling. Nat Chem Biotechnol 3:707–708
100. Fernández-Suárez M, Baruah H, Martínez-Hernández L, Xie KT, Baskin JM, Bertozzi CR, Ting AY (2007) Redirecting lipoic acid ligase for cell surface protein labeling with small-molecule probes. Nat Biotechnol 25:1483–1487
101. Uttamapinant C, White KA, Baruah H, Thompson S, Fernández-Suárez M, Puthenveetil S, Ting AY (2010) A fluorophore ligase for site-specific protein labeling inside living cells. Proc Natl Acad Sci USA 107:10914–10919
102. Mizukami S, Watanabe S, Hori Y, Kikuchi K (2009) Covalent protein labeling based on noncatalytic β-lactamase and a designed FRET substrate. J Am Chem Soc 131:5016–5017
103. Tully SE, Cravatt BF (2010) Activity-based probes that target functional subclasses of phospholipases in proteomes. J Am Chem Soc 132:3264–3265
104. Duckworth BP, Zhang Z, Hosokawa A, Distefano MD (2007) Selective labeling of proteins by using protein farnesyltransferase. Chembiochem 8:98–105
105. Carrico IS, Carlson BL, Bertozzi CR (2007) Introducing genetically encoded aldehydes into proteins. Nat Chem Biol 3:321–322
106. Tsukiji S, Miyagawa M, Takaoka Y, Tamura T, Hamachi I (2009) Ligand-directed tosyl chemistry for protein labeling in vivo. Nat Chem Biol 5:341–343
107. Hong V, Steinmetz NF, Manchester M, Finn MG (2010) Labeling live cells by copper-catalyzed alkyne-azide click chemistry. Bioconjugate Chem 21:1912–1916

Tetracysteine and Bipartite Tags for Biarsenical Organic Fluorophores

Carla Spagnuolo, María Joselevich, Federico Coluccio Leskow, and Elizabeth A. Jares-Erijman

Abstract Genetically encoded fluorescent tags enabling the direct determination of biomolecular functions, interactions and dynamics in living cells and organisms, have had a tremendous impact on cell biology. Key among them are the fluorescent proteins, which despite their great utility present a number of shortcomings. Thus, there has been a very active development of alternative approaches for chemically labeling proteins in live cells, with special interest in small probe molecules. This review depicts a comprehensive review of one of the most remarkable examples of such approaches for intracellular targeting, namely biarsenical ligands that selectively bind to tetracysteine motifs and bipartite dicysteine motifs incorporated into protein targets. The state-of-the-art with respect to small biarsenical molecules and peptide tags are presented, with consideration of their binding properties, labeling aspects, photophysical properties, and applications. The latter include purification of proteins, localization, trafficking and conformational changes of proteins, pulse-chase labeling, *chromophore* or *fluorophore-assisted light inactivation* (CALI or FALI), correlated fluorescence and electron microscopy (CLEM), and FRET-based investigations. The development and applications of *bimolecular* tetracysteine tags is a recent, promising extension of the method.

Keywords Biarsenical ligands · Bimolecular tetracysteine tags · CALI · CLEM · Tetracysteine motifs

Contents

C. Spagnuolo, M. Joselevich, F.C. Leskow, and E.A. Jares-Erijman (✉)
CIHIDECAR, CONICET, Universidad de Buenos Aires, Ciudad Universitaria, Pab. II, 3er piso, 1428 Buenos Aires, Argentina
e-mail: eli@qo.fcen.uba.ar

A.P. Demchenko (ed.), *Advanced Fluorescence Reporters in Chemistry and Biology III: Applications in Sensing and Imaging*, Springer Ser Fluoresc (2011) 10: 263–296,
DOI 10.1007/978-3-642-18035-4_8,

1 Introduction

The advent of genetically encoded fluorescent tags for the direct determination of biomolecular functions, interactions and dynamics in living cells and organisms has had an immense impact on cell biology. Expression probes are peptides, proteins or nucleic acids that the cell generates and locates either in its interior or on its surface. They display an intrinsic and specific fluorescent signal or constitute a target for a fluorescent probe supplied externally and/or taken up by the cell. Of the large family of expression probes (see Table 2 in [1]), the visible fluorescent proteins (VFPs) are in widespread use and constitute one of the most significant developments of the last years in live cell imaging, as exemplified by the award of the Nobel Prize for Chemistry in 2008. However, despite their extreme versatility, VFPs are unsuitable in many applications due to their *large size* (27 KDa). Smaller probes are desirable in that they offer reduced stereochemical interference, faster rates of labeling, and the ability to provide readouts in addition to or other than, fluorescence. Moreover, for the study of function over time, the *limited photostability* of VFPs poses a distinct disadvantage, particularly in imaging processes at the single-molecule level.

The family of biarsenical probes, introduced by Tsien and colleagues, constitute a labeling strategy capable of addressing several of the limitations described for VFPs. This strategy is based on a metal–chelation approach by which a genetically encoded –CCxxCC– tetracysteine (TC) tag binds with high affinity to

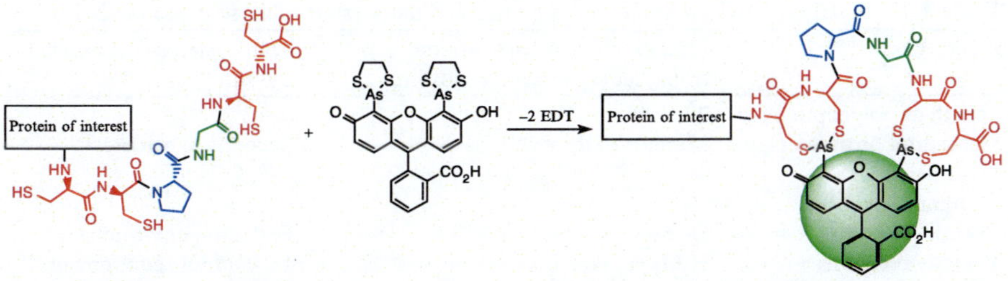

Fig. 1 Tetracysteine motif-biarsenical probe labeling strategy: complex formation between CCPGCC and FlAsH-EDT_2

membrane-permeant biarsenical fluorophores [2–4]. The TC tag is only a few amino acids in length (6–12), imposing fewer constraints than larger tags on the introduction by fusion with the protein sequence at either end, in loops or within domains. The fluorogenic nature of the biarsenical ligands is one of the unique features of the system. The biarsenical dye, initially in a nonfluorescent complex with EDT, becomes highly fluorescent upon binding to a target tag (Fig. 1). Griffin et al. proposed that the small size of EDT probably permits rotation of the aryl–arsenic bond allowing an excited state quenching by vibrational deactivation or photoinduced electron transfer. The peptide complex may evade such quenching because its more rigid conformation hinders conjugation of the lone electron pair of the arsenic atoms with the fluorescein orbitals [2].

A comparison between the properties and characteristics of fluorescent proteins and biarsenical ligands with tetracysteine motifs is presented in Table 1.

The great expectations generated by the original publication showing the excellent potential of the biarsenical ligands were somewhat tempered by the time and effort required to overcome certain difficulties of application, especially when trying to label species present at low concentration. The problems at the time included: background staining caused by nonspecific binding due to the presence of endogenous cellular proteins with thiols displaying a weak affinity for FlAsH, aggregation of the TC tags through the formation of intermolecular disulfide linkages, and other limitations such as the limited photostability of the probes or the difficulty of reproducing their synthesis. Furthermore, the application of the methodology was not possible in oxidizing environments since the reduced form of the TC motif is easily oxidized. By this time, many of these limitations have been overcome. Biarsenical compounds in combination with TC motifs are becoming increasingly applied for the specific labeling of proteins and peptides with small molecules. Key advances in the field include the introduction of high affinity and high quantum yield peptide sequences by Martin et al. [1]. The identification of two naturally occurring TC-containing sequences which have essentially the same binding affinity for FlAsH as current CCPGCC tags, but forms brighter complexes without intermolecular cross-linking was made by Wang's group [5]. Other important developments include photostable and pH insensitive compounds in the physiological pH range [6], and methods such as CALI and correlated fluorescence and electron microscopy (CLEM) (see below). The introduction of polar groups in the

Table 1 Comparison between fluorescent proteins and tetracysteine–biarsenical system [3]

Property	Fluorescent proteins derived from Aequorea and Discosoma	Tetracysteine–Biarsenical System
Length of polypeptide (aa)	238–900	6–12
Maturation or time delay to develop fluorescence	Minutes to days	Seconds to minutes
Chemical requirements	O_2 must be present	Reducing environment
Toxicity concerns	H_2O_2 generated	1,2-Dithiol requirement
Extinction coefficients ($mM^{-1}\ cm^{-1}$)	20–60	30–80
Fluorescence quantum efficiencies	0.2–0.8	0.1–0.6
Fluorescence detection limits	<1 μM	Several μM
Utility for fluorescence polarization assays	Limited because protein itself is large enough to be rotationally immobile, regardless of host	Very promising because tag is small and rigidly attached to its host
Fluorescence remaining after denaturing gel electrophoresis	None to slight	Good, if boiling with excess thiols avoided
Simultaneous multicolor labeling of different proteins	Yes	Yes
Sequential (pulse-chase) multicolor labeling of a single protein	Possible by photobleaching or without precise temporal control	Yes
Photoconvertible into electron-microscopic image	No	Yes
Utility for affinity chromatography	No	Yes

Adapted with permission from [3]. Copyright 2002 American Chemical Society

biarsenical structure led to a less hydrophobic molecule [7]. The lack of modular structures was tackled by the introduction of SplAsH and its derivatives [8]. SplAsH is a nonfluorescent fluorescein derivative that operates as a carrier for a second dye that can be virtually any molecule. Perhaps the main drawback of this approach is the general lack of fluorogenicity.

2 Binding Properties

2.1 *Sequence Matters*

On the conformation of the complex. The original target peptide was designed to favor the binding of the biarsenical probe in the i, $i+1$, $i+4$, $i+5$ positions of an α-helix [2]. The name of the original biarsenical fluorophore, FlAsH-EDT$_2$

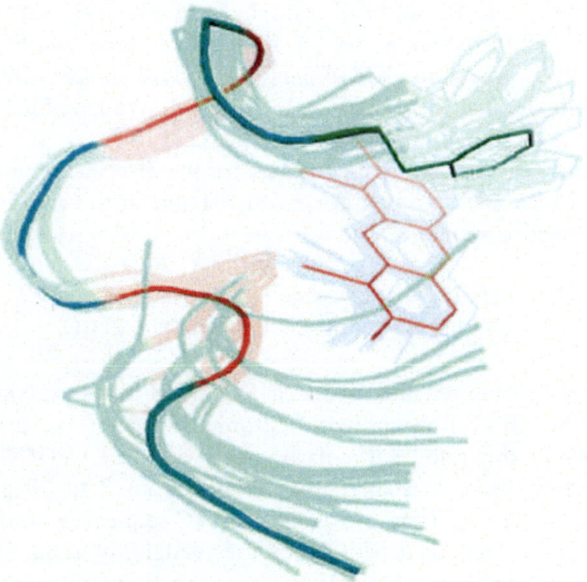

Fig. 2 Structure calculation of the peptide bound to ReAsH, based on the NOESY-derived distance constraints. The arsenic atoms are fixed to the cysteine thiol (*colored red*) at a distance of 2.25 Å. An ensemble of the 30 best structures, which appear as semitransparent, represents the peptide backbone and ReAsH, while the average peptide backbone structure from the ensemble, including the Phe1 side chain and the ReAsH moiety, is shown as a *solid line*. The overlay was made on the backbone atoms from residue 4 to 8 and also included ReAsH. Reprinted with permission from [9]. Copyright 2009 American Chemical Society

(Fluorescein Arsenical Helix binder, bis-ethanedithiol adduct) was based on the α-helical stereochemistry conceived for the binding site, the peptide tag [2]. This proposal has been since revised (see below). Sequence-affinity studies by the same group showed that placing proline and glycine as the central amino acids flanked by the two cysteine pairs led to complexes with the highest affinities [3]. However, these complexes displayed a preference for a hairpin rather than an α-helical conformation. Finally in 2009, Madani et al. determined by NMR the conformation of the complex formed by the 12-mer peptide FLNCCPGCCMEP and ReAsH [9]. A family of conformations (Fig. 2) shows a relatively tight hairpin interacting with the fluorophore and also closely with the N-terminal phenylalanine side chain. The backbone structure of the peptide is fairly well-defined, with a hairpin-like turn, similar to a type-II β-turn, formed by the central CPGC segment. Two clear NOESY cross-peaks between the Phe1 side chain and ReAsH confirmed the close positioning of the phenyl ring of Phe1 and ReAsH. Phe1 was found to have an edge-to-face geometry relative to ReAsH. The close interaction between Phe1 and ReAsH may be relevant for the fluorescence properties of the ReAsH complex.

As was mentioned before, FlAsH-EDT$_2$ is nonfluorescent but develops bright emission after addition of the TC peptide to displace EDT. The excitation and emission maxima are 508 and 528 nm, respectively, i.e., ~20-nm red-shifted compared to the corresponding spectra of free fluorescein. After binding to the CCPGCC motif, the quantum yield of the FlAsH–Cys4 complex is 0.49, more than 50,000 times brighter than unbound FlAsH [3]. The formation of the FlAsH–peptide is essentially irreversible in the absence of excess of EDT.

Adams et al. measured the rate constants for the dissociation and formation of the FlAsH–tetracysteine complex at neutral pH and in the absence of EDT [3], assuming the reaction sequence:

$$\begin{gathered}\text{FlAsH bis} - \text{arsenoxide} + \text{tetracysteinepeptide} \\ \rightarrow \text{FlAsH} - \text{tetracysteinepeptide} + 2\text{H}_2\text{O}\end{gathered}$$

It was found that the reaction of FlAsHO (FlAsH bis-arsenoxide) with excess peptide in 5 mM MES follows pseudo-first-order kinetics. The various complexes showed fast association rates, particularly those with the CCPGCC motif [3]. The equilibrium dissociation constant values ranged from 2 to 70 pM at different monothiol concentrations. Thus, biosynthesized TC sequences would become fluorescent in just a few seconds depending on the availability of the biarsenical dyes, which could be preloaded into the cell. By comparison, fluorescent proteins require minutes to hours to develop their emission. Since the reaction studied by Adams did not involve thiol exchange the reported binding rates might have overestimated the rates of reaction.

In 2005 Martin et al. developed a retrovirally transduced mammalian cell-based library approach for optimization of the residues flanking the TC motif, the goal being to maximize binding affinities and quantum yields [1]. They identified FLNCCPGCCMEP and HRWCCPGCCKTF as the best sequences, displaying the highest fluorescence quantum yields upon binding and a remarkable resistance of the TC–biarsenical complex to high concentrations of dithiol washes. The authors suggested that large aromatic residues could function as a shield to protect the complex from dissociation by competing thiols.

Wang et al. using a proteomic approach with cell lysates of the metal-reducing microorganism *Shewanella oneidensis*, identified in 2007 two naturally occurring tetracysteine containing sequences: S_peptide (GCCGGSGNDAGGCCGG) and E_peptide (GCCGGHGHDHGHEHGGEGCCGG) [5]. The two peptides leaded to complexes with improved quantum yields and most importantly, without formation of intermolecular cross-linking.

Moreover, also in 2007 Chen et al. applied a high-throughput peptide screening technique to identify alternate binding motifs, and found that the CCKACC (KA tag) has a brightness similar to that of the classical sequence (CCPGCC), but displays altered rates and affinities of association [10, 11]. The authors took advantage of this property to accomplish differential labeling of coexpressed proteins with the red probe ReAsH-EDT_2 and FLAsH-EDT_2. Thus, they followed the expression of PG- and KA-tagged subunits of RNA polymerase in *E. coli*. Specific labeling of two subunits of RNA polymerase in cellular lysates

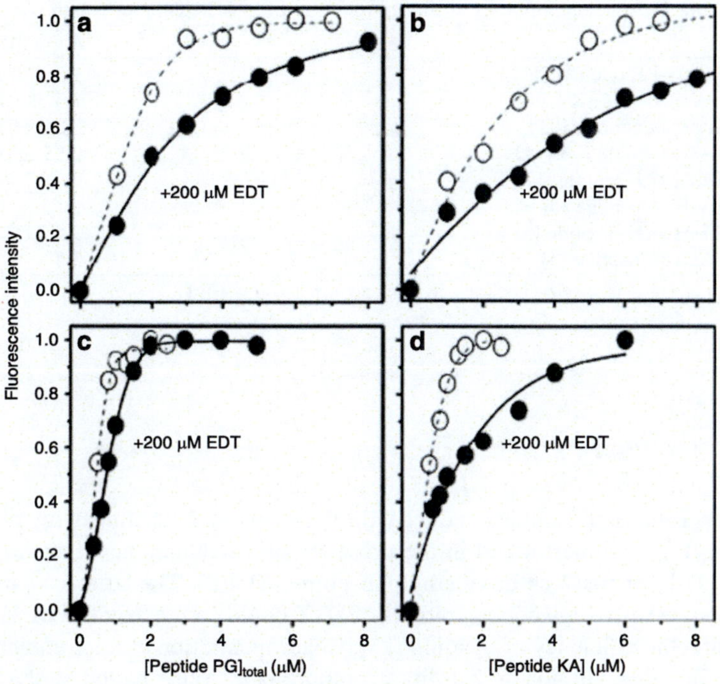

Fig. 3 Equilibrium binding of ReAsH-EDT$_2$ (**a**, **b**) or FlAsH-EDT$_2$ (**c**, **d**) to peptides AREACCPGCCK-CONH$_2$ (peptide PG) or AREACCKACCK-CONH$_2$ (peptide KA) in the absence (*open circle*) or presence (*closed circle*) of 200 μM EDT. Reprinted with permission from [10]. Copyright 2007 American Chemical Society

was achieved, whereby ReAsH-EDT$_2$ was shown to selectively label the PG-tag on RNA polymerase R-subunit prior to the labeling of the KA-tag sequence of the β-subunit of RNA polymerase with FlAsH-EDT$_2$. Their result showed that chelation of ReAsH or FlAsH in the presence of EDT reduces the rates of formation of the complex, yielding labeling times of approximately 10 min (Fig. 3). Furthermore, the apparent binding affinities are also reduced in the presence of EDT.

A very nice example of orthogonal labeling was recently developed by Zürn et al., who performed a sequential labeling method of two motifs: CCPGCC and FLNCCPGCCMEP allowing site-specific labeling with FlAsH and ReAsH, respectively [12]. The authors labeled the cell surface receptor for parathyroid hormone and its cytosolic binding protein, β-arrestin, and showed their selective visualization in intact cells. In addition, the orthogonal labeling allowed an analysis of the interaction by colocalization and fluorescence resonance energy transfer (FRET) [12].

A comparison of the affinities of the different sequences is shown in Table 2.

Table 2 Apparent affinities of FlAsH-EDT_2 and ReAsH-EDT_2 for peptide tags

	*K*app (μM)
ReAsH-Peptide PG [11]	1.3 ± 0.2
ReAsH-Peptide KA [11]	3.3 ± 1.2
FlAsH-Peptide PG [11]	0.24 ± 0.09
FlAsH-Peptide KA [11]	0.85 ± 0.23
FlAsH-PG_peptide [5]	2
FlAsH-S_peptide [5]	2
FlAsH-PG2_peptide [5]	3.5
FlAsH-E_peptide [5]	3.5

Apparent dissociation constant in presence of 200 μM EDT

2.2 *Labeling Aspects*

Complex formation requires reduced thiol groups and labeling with biarsenical probes is typically performed in the presence of a reducing agent such as dithiotreitol (DTT) or tris(2-carboxy)ethylphosphine (TCEP). The latter is considered a better choice since high concentrations of DTT could compete with FlAsH, thereby decreasing the efficiency of labeling [13]. Coadministration of micromolar concentrations of ethanodithiol or 2,3-dimercaptopropanol (also called BAL) prevents unspecific binding of the biarsenical probes, but this procedure reduces the affinity to the nanomolar range. Millimolar concentrations of these compounds can competitively remove the biarsenical from the TC motif, allowing the labeling to be reversed. On the other hand, the presence of monothiols like 2-mercaptoethanol (ME) or 2-mercaptoethansulfonic acid (MES) as components of the buffer has shown to favor the labeling inducing the right conformation of the complex [3]. Within the cell, the same task falls to the endogenous glutathione. The apparent pK_a of the fluorescein chromophore in the FlAsH–peptide complex was initially measured as 5.4, and it was concluded that the fluorescence would not be sensitive to variations in cytosolic pH near 7. However, further research showed a strong dependence of the fluorescence upon pH and that the fluorescence signal associated with the FlAsH–peptide complex has a transition point pH of ~7 [7]. The fluorescein moiety of the FlAsH–peptide complex has multiple pK_a's, two of which were calculated to be 5.6 and 8.1 and assigned to the neutral-to-anion, and anion-to-dianion transitions, respectively.

A detailed protocol for the labeling of tetracysteine-tagged proteins in intact cells will be reported in *Nature Protocols* along with the publication of this book (Tsien Personal Communication). Hoffman et al. describe a general labeling methodology that can be applied to a wide range of proteins including those with low expression levels. The authors discuss the problems and considerations to take into account during the experiment in a very detailed fashion. The labeling procedure using FlAsH-EDT_2 as described is expected to take 2–3 h depending on the number of samples to be processed.

3 Biarsenical Probes: What Is Available or the State-of-the-Art

A recent review authored by Soh describes available probes for different metal–chelation methodologies for the chemical labeling of proteins [14]. FlAsH (**1**, 4′,5′-bis(1,3,2-dithioarsolan-2-yl)fluorescein), the first biarsenical introduced by the Tsien group in 1998 [2], was originally prepared in a single step by transmetallation of commercially available fluorescein mercuric acetate, followed by the addition of EDT to facilitate purification; the yields were poor. Improved general synthesis methods for FlAsH-EDT2, ReAsH-EDT2 and other derivatives were reported by the same group in 2002 [3] and in 2008 [15]. FlAsH-EDT2 and ReAsH-EDT2 are commercially available as Lumio™ Green and Lumio™ Red, respectively, as part of a labeling kit provided by Invitrogen Corporation (USA) (Fig. 4).

Several other biarsenical derivatives with complementary spectroscopic properties have been synthesized (Fig. 5).

3.1 ReAsH (2)

The phenoxazine analog of fluorescein was used to obtain a red biarsenical derivative. Adams et al. proposed that this fluorophore is a better choice than rhodamine for

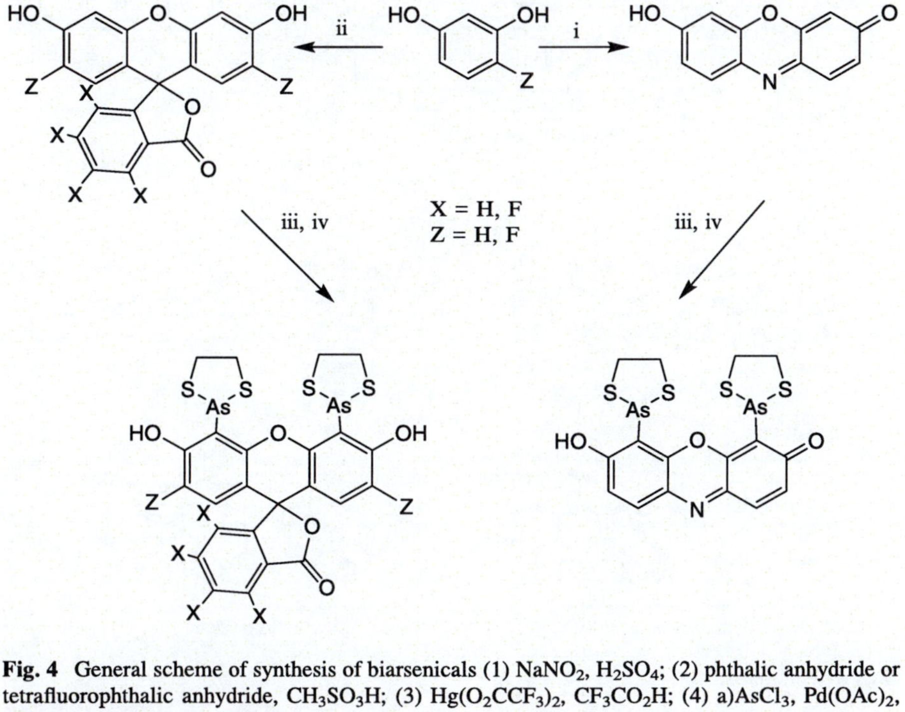

Fig. 4 General scheme of synthesis of biarsenicals (1) $NaNO_2$, H_2SO_4; (2) phthalic anhydride or tetrafluorophthalic anhydride, CH_3SO_3H; (3) $Hg(O_2CCF_3)_2$, CF_3CO_2H; (4) a)$AsCl_3$, $Pd(OAc)_2$, DIEA, NMP, b) H_2O, c) EDT

avoiding steric interactions in the complex between the peptide and 3′, 6′-alkylamino groups of rhodamine, which result in the loss of fluorescence [3]. As in the case of FlAsH-EDT$_2$, ReAsH-EDT$_2$ is nonfluorescent but rapidly forms a fluorescent complex with a TC-containing peptide (i.e., AcWEAAAREACCRECCARA-NH$_2$). This complex has excitation and emission maxima of 593 and 608 nm, respectively. Thus, it is slightly red-shifted to resorufin. As a particular feature, ReAsH generates reactive oxygen species upon illumination [16]. This characteristic may be problematic for extended time lapse imaging of live cells, but enables targeted knockout of protein function by chromophore–fluorophore-assisted light inactivation (CALI,

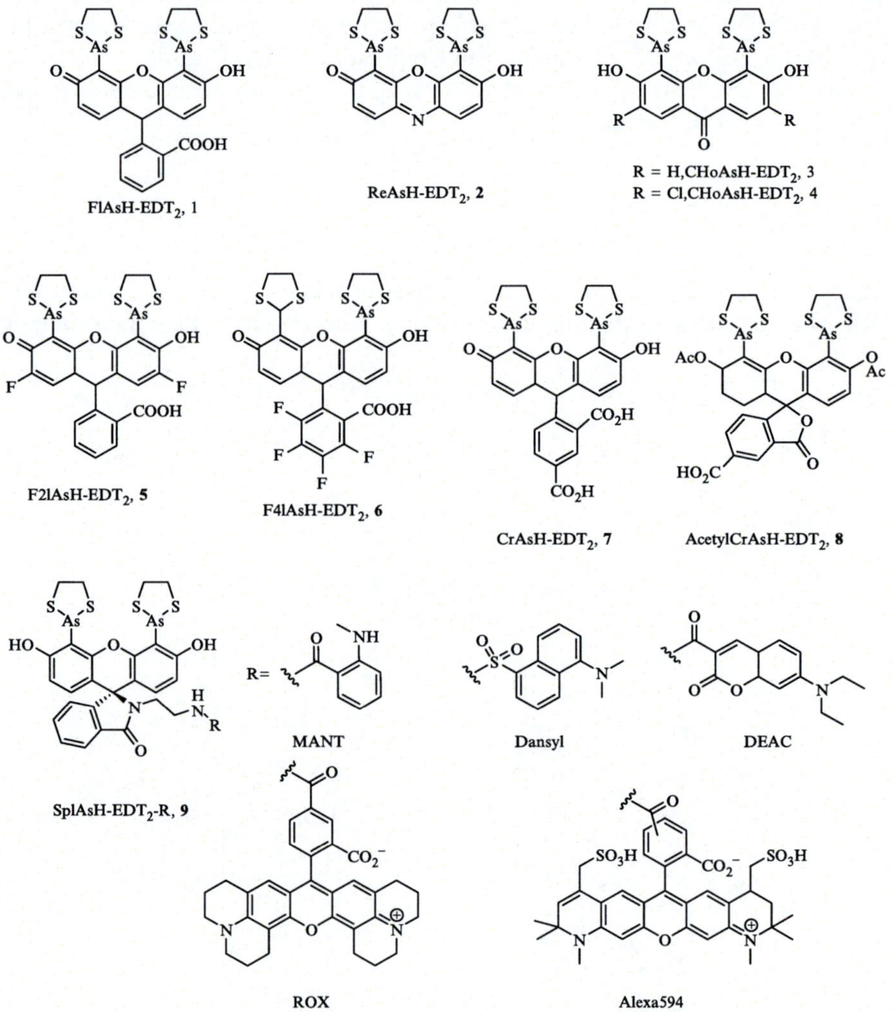

Fig. 5 (continued)

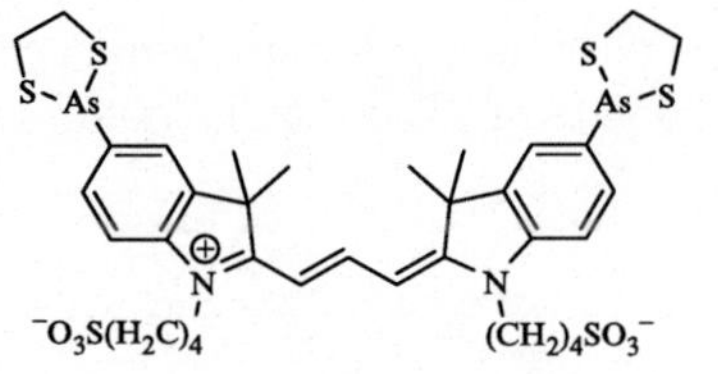

AsCy3, **10**

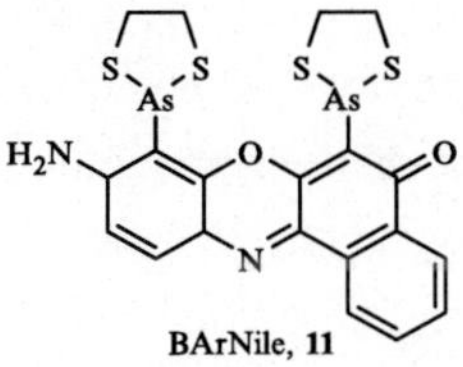

BArNile, **11**

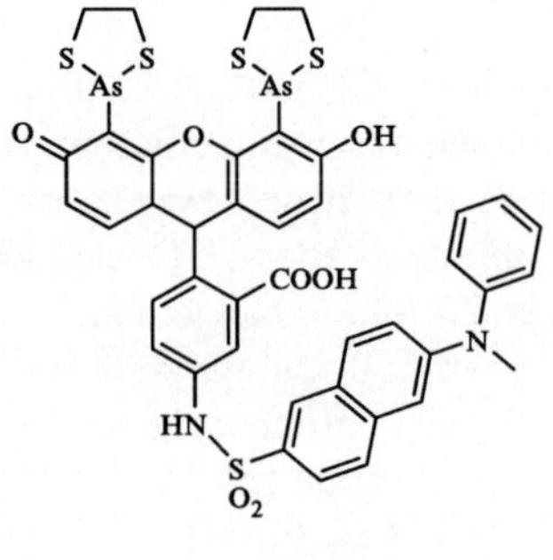

mansyl-FlAsH, **12**

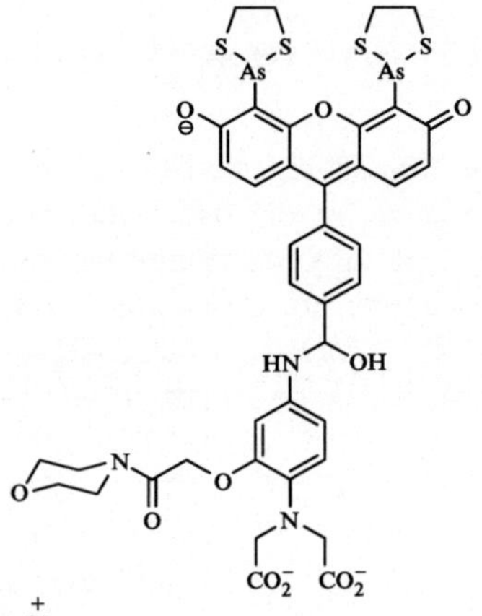

CaGF, **13**

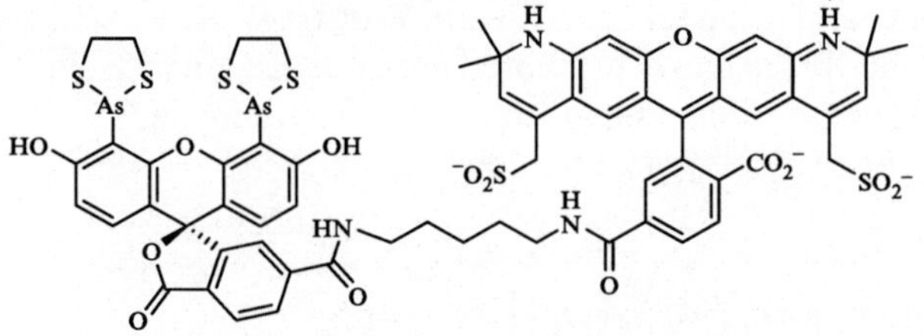

6-hydroxy FlAsH cadaveryl-Alexa Fluor 568 conjugate, **14**

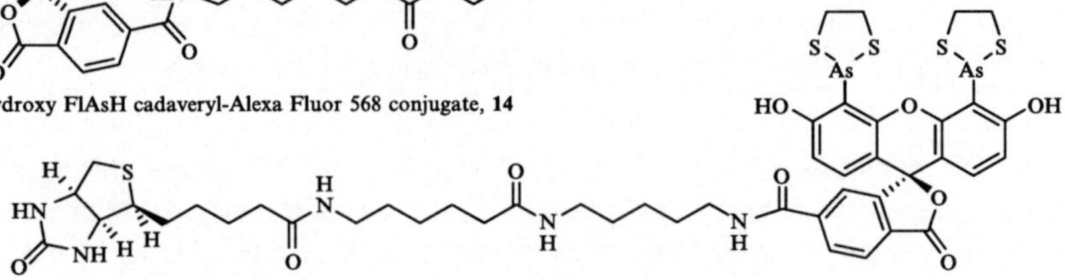

Bio FlAsH conjugate, **15**

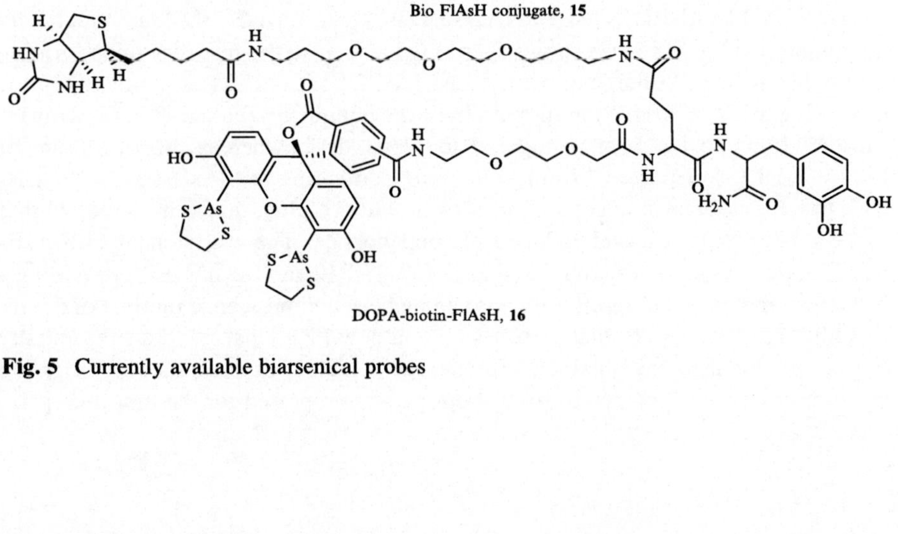

DOPA-biotin-FlAsH, **16**

Fig. 5 Currently available biarsenical probes

FALI) and specific staining for correlated electron microscopy (see below). The photostability and localization of ReAsH were studied at the single-molecule level, showing that it is possible to measure the step sizes of biomolecular motors using this probe [17].

3.2 HoXAsH-EDT$_2$ (3) and CHoXAsH-EDT$_2$ (4)

Blue-fluorescing biarsenicals (HoXAsH-EDT$_2$ and CHoXAsH-EDT$_2$) can be generated by the usual mercuration and subsequent transmetallation reactions upon 3,6-dihydroxyxanthone and its 2,7-dichloro derivative, respectively. These compounds form isolable complexes upon binding to a TC peptide (i.e., AcWEAAAREACCRECCARA-NH$_2$) with over 20-fold enhancement of fluorescence, and with excitation and emission maxima of 380 nm and 430 nm, respectively [3]. CHoXAsH-EDT$_2$ showed a brighter peptide complex (quantum yield of 0.35) and no pH sensitivity within the physiological range (pK_a < 5). HoXAsH-EDT$_2$ and CHoXAsH-EDT$_2$ are both less efficient than FlAsH or ReAsH with respect to contrast, brightness, and photostability of the complexes. CHoXAsH-EDT$_2$ might be useful as a FRET donor for GFP or YFP (see below) and for multicolor labeling with other biarsenicals or fluorescent proteins. Apart from the initial reports, there have been no further applications of these compounds.

3.3 F2FlAsH-EDT$_2$ (5) and F4FlAsH-EDT$_2$ (6)

The pH sensitivity and limited photostability of fluorescein derivatives in the physiological range constitute inherent limitations to the original biarsenical binding motif. In order to overcome these drawbacks F2FlAsH-EDT$_2$ (**5**) and F4FlAsH-EDT$_2$ (**6**), were synthesized and characterized as two new fluoro-substituted derivatives from FlAsH-EDT$_2$ [6]. F2FlAsH-EDT$_2$, and F4FlAsH-EDT$_2$ are practically nonfluorescent but F2FlAsH shows a striking increase in fluorescence upon forming a complex with a 12-mer sequence (FLNCCPGCCMEP, P12) as a model peptide target (Fig. 6). The absorption maximum of the complex F2FlAsH-P12 (522 nm) is shifted 11 nm to the blue, compared to FlAsH-P12, whereas the maximum of F4FlAsH-P12 is displaced 17 nm to the red. Comparing to FlAsH-P12, F2FlAsH-P12 has higher absorbance, larger Stokes shift, higher quantum yield, higher photostability (Fig. 6), and reduced pH dependence. The emission of F4FlAsH-P12 lies in a region intermediate to that of FlAsH-P12 and ReAsH-P12, providing a new color and excellent luminosity with an increased fluorescence lifetime of 5.2 ns. In addition, the two fluorinated probes form a new FRET pair with a substantially larger R_o value than any obtained with these dyes. Also in this article, lifetimes and anisotropies of the dyes bound to the peptide were reported for the first time [6].

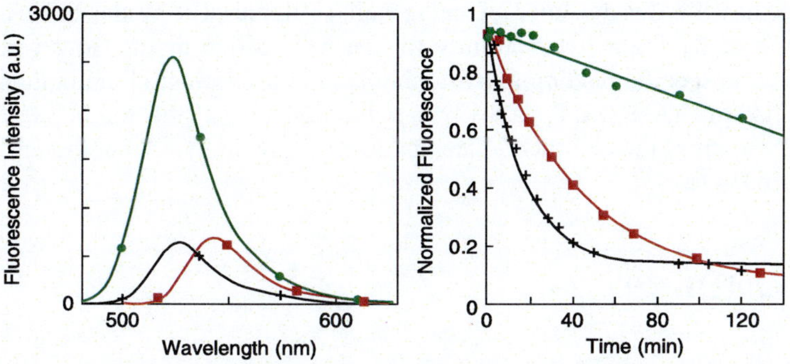

Fig. 6 Fluorescence emission spectra (*left panel*, 0.1 μM) and photobleaching (*right panel*) of F2FlAsH-P12 (*filled circle, green*), F4FlAsH-P12 (*filled square, red*) and FlAsH-P12 (*plus, black*). Complexes of the three dyes (10 μM) were irradiated with a mercury arc lamp through a 490–560 nm filter with an irradiance of 70 mW/cm. Reprinted with permission from [6]. Copyright 2006 American Chemical Society

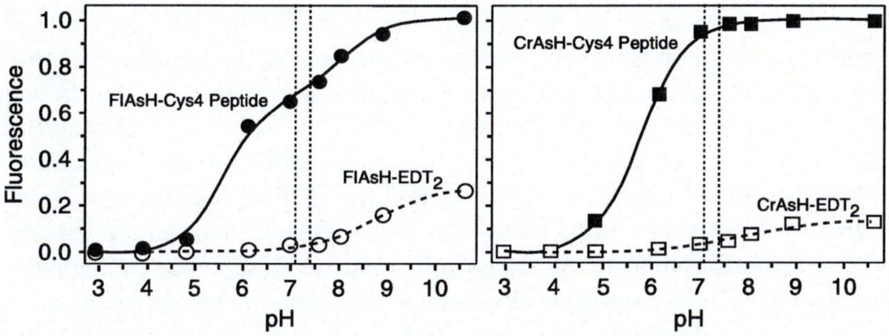

Fig. 7 Comparison of the fluorescence properties of free and peptide-bound FlAsH and CrAsH as a function of pH, using the following buffers: acetate (pH 3–5), phosphate (pH 6–8) and carbonate (pH 9–11) [7]. Reproduced by permission of The Royal Society of Chemistry

3.4 CrAsH-EDT$_2$ (7)

In 2006 Cao et al. proposed that a less hydrophobic fluorescein derivative should have reduced nonspecific binding [7]. They added a carboxyl group to fluorescein and utilized the 5-carboxy derivative of FlAsH-EDT$_2$ (CrAsH-EDT$_2$) originally synthesized by Adams et al. As in the case of FlAsH, CrAsH-EDT$_2$ exhibits a fluorogenic behavior and strong pH dependence upon binding to the TC motif.

However, the CrAsH–Cys4 complex reaches a maximum at about pH 7 and is stable at higher pH. In contrast, the fluorescence signal associated with the FlAsH–peptide complex is at a transition point near pH 7 (Fig. 7). CrAsH, therefore, should be a more sensitive probe for quantitative fluorescence measurements, especially in cases in which labeled proteins translocate between different cellular environments (with different pH) or when stress factors induce pH changes [7]. The CrAsH–peptide complex has the disadvantage of being half as fluorescent as the FlAsH–peptide complex and having a lower affinity for the TC peptide motif. However, CrAsH shows less unspecific binding to hydrophobic proteins such as albumin than FlAsH. Acetylation of CrAsH-EDT_2 as a way to reduce the polarity that might interfere with its ability to cross the cell membrane, led to the completely nonfluorescent, Acetyl-CrAsH-EDT_2 (**8**) [3].

3.5 SplAsH-EDT_2 (9)

Inspection of the structures of Fig. 5 shows that biarsenicals in general are not modular, i.e., the TC-binding moiety is coupled directly to the fluorophore. A type of biarsenical probes introduced by Bhunia et al., named SplAsH-EDT_2, was designed with the intention of extending the range of dyes that can be exploited with the TC motif [8]. The authors synthesized the precursor SplAsH-EDT_2-Boc starting from commercial fluorescein. To ensure that the fluorescein-based targeting moiety would not interfere with the fluorescence properties of the attached dye, they attached a primary amine group to the aromatic carboxylic group and thus favored the formation of a spirolactam moiety. SplAsH maintains the planarity of the xanthene ring, exhibits no inherent fluorescence, and provides a convenient handle for attaching fluorescent dyes. According to the authors, SplAsH-EDT_2-Boc was easily synthesized and readily purified. They also mentioned that the arsenic transmetallation reaction for this compound proceeded with high yield (95%). Removal of Boc and subsequent reaction with an active ester (or sulfonyl chloride) of the desired fluorophore leads to the respective SplAsH conjugate: *N*-methylanthranilate (MANT), dansyl, diethylamino-coumarin (DEAC), X-rhodamine (ROX), and Alexa-594 (Fig. 5). All of these dye conjugates were stable and could be purified. Each of them imparts the fluorescent properties of their respective payloads to the SplAsH conjugate and interestingly exhibits the same specificity for the TC tag (FLNCCPGCCMEP) as FlAsH under both native and denaturing conditions. In particular, the red rhodamine-based dyes SplAsH-ROX and SplAsH-Alexa594 exhibit great photostability typical for such type of fluorophores. The modularity of the SplAsH approach offers the flexibility for specifically tailoring other properties of the conjugate to suit a given application, such as solubility, cell-permeability, and photophysical properties.

Unfortunately, the SplAsH moiety itself lacks one of the most important characteristics of the FlAsH-type biarsenical derivatives: its fluorescence does not increase upon binding to a TC motif.

3.6 AF568-FlAsH (14) and Bio-FlAsH (15)

Novel modular biarsenical probes were introduced by Taguchi et al. in 2009 [18]. They synthesized two new FlAsH derivatives: AF568-FlAsH and Bio-FlAsH, consisting of conjugates to Alexa Fluor 568 dye and biotin, respectively [18]

(Fig. 5). These probes display unique advantages over FlAsH, such as improved fluorescence properties, versatility of detection, and ease of affinity purification. The authors exploited these probes in a new method for labeling extracellular and cell surface proteins, called *Instant with DTT, EDT, And Low temperature* (IDEAL-labeling). The high pH sensitivity and poor photostability of FlAsH precluded its use in IDEAL-labeling. The synthesis of both derivatives was accomplished by coupling the succynimidyl ester of CrAsH-EDT$_2$ with Alexa Fluor® 568 6-carboxyl cadaverine or a long chain derivative of biotin.

3.7 AsCy3-EDT$_2$ (10)

A biarsenical probe intended for simultaneous use with FlAsH-EDT$_2$ was designed and synthesized by Cao et al. [11]. AsCy3-EDT$_2$ is based on Cy3, a member of the cyanine dye family, whose brightness and photostability are well-recognized [11]. AsCys3-EDT$_2$ (Fig. 5) was designed to have a greater interatomic distance between the two arsenic atoms (~14.5 Å) than FlAsH-EDT$_2$ and its derivatives (~6 Å). As a consequence, AsCy3-EDT$_2$ and its complementary sequence do not interact with the components of the traditional TC-biarsenical systems. Therefore, both labeling systems can be used in the same cell in a biorthogonal manner. In addition, AsCy3 and FlAsH can act as a FRET couple.

The synthesis of AsCy3-EDT$_2$ involved five steps, starting from 2,3,3-trimethyl-indole and 1,4-butane sultone, with an overall yield of 38%. A complementary binding sequence was designed (CCKAEAACC-Cy3TAG) with an affinity (K_d = 80 nM) comparable to that of FlAsH with the CCPGCC motif [3]. The absorbance spectrum of AsCy3 is insensitive to the binding, whereas the fluorescence spectrum is red-shifted from 568 to 576 nm with a 6-fold increase in quantum yield (0.28). The fluorescence of AsCy3 is pH independent in the range 4–9. Compared to FlAsH and ReAsH, AsCy3 shows superior photostability and a minimal environmental sensitivity. The authors also claimed that binding of AsCy3 is rapid, and occurring in <15 s, whereas under similar conditions FlAsH and ReAsH require minutes for reaction.

3.8 BArNile-EDT$_2$ (11)

Environment-sensitive fluorophores have been extensively applied in vitro to assess conformational changes of proteins. Nakanishi et al. extended these studies to live cells using an ad hoc designed biarsenical derivative of Nile Red (9-diethylamino-5H-benzo[a]-phenoxazin-5-one) [19]. Nile Red is structurally similar to fluorescein; however, a bulky diethylamino group at the 9 position appears to hinder the entrance of ethanedithiol. Therefore, the equally environmentally sensitive 9-amino analog was preferable.

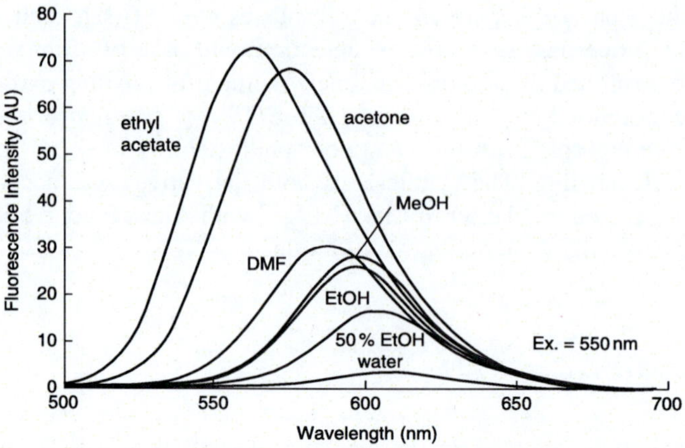

Fig. 8 Emission spectra of 9-amino-5-benzophenoxazone (the fluorophore of BArNile) in different solvents (excited at 550 nm). Reprinted with permission from [19]. Copyright 2001 American Chemical Society

BArNile-EDT_2 was synthesized in seven steps starting from 2-hydroxy-4-nitroaniline and 2-hydroxy-1,4-naphthoquinone with an overall yield of 0.25%. The fluorogenic nature of this compound is lower than that of FlAsH; that is, the fluorescence increase upon complex formation is not as pronounced. A possible rationale is that aminobenzophenoxazone, the fluorophore of BArNile, has an inherently lower quantum yield in polar environments such as water, and even a higher emission in nonpolar solvents (Fig. 8). In contrast, fluorescein (FlAsH) is highly fluorescent in water. FRET experiments were used to confirm that BArNile can bind to the TC motif inside live cells.

3.9 *Mansyl-FlAsH-EDT_2 (12)*

Following the intention of developing a protein conformation indicator with increased sensitivity than BArNile-EDT_2, Nakanishi et al. designed a fluorophore capable of undergoing photoinduced electron transfer (PET) [20], a property highly sensitive to the molecular environment.

The fluorescence quantum efficiency of fluorescein (the fluorophore of FlAsH) is known to be affected by variations of the energy level of its benzoic acid moiety via a PET-process [21]. Therefore, the authors attached a mansyl group to the benzoic acid moiety of FlAsH-EDT_2 so as to obtain mansyl-FlAsH-EDT_2 as a new fluorescent probe (Fig. 5). The extended benzoic acid moiety promotes a more flexible interaction of the fluorophore with protein surfaces. Thus, the fluorophore is more sensitive to the protein conformational change than BArNile, which is rigidly attached to the TC motif. 5-aminofluorescein was used as the starting material for the synthesis of mansyl-FlAsH-EDT_2 following the classic methodology of mercuration and transmetallation.

The fluorescence of both mansyl-FlAsH conjugates to calmodulin recombinants was far stronger than that of mansyl FlAsH-EDT_2. No emission peaks could be observed corresponding to the mansyl moiety (when excited at 330 nm, an emission peak was expected around 450 nm) suggesting that the excited energy of the mansyl moiety transferred to the fluorescein group within the same molecule. Nevertheless, the quantum efficiencies of mansyl FlAsH–tetracysteine complex were less than of FlAsH-TC, indicating a partial fluorescence quenching of mansyl FlAsH via the PET process as mentioned above. The maximal fluorescence changes of mansyl FlAsH for the conformational rearrangement of calmodulin recombinants were two times higher than the fluorescence change with BArNile. This finding confirms the prediction that the extended benzoic acid moiety enables flexible interactions between the fluorophore and the protein surface and shows that mansyl-FlAsH is highly dependent on protein conformational changes.

3.10 CaGF (13)

The genetic targetability of biarsenical dyes can be combined with the fast kinetics and small size of organically synthesized calcium indicators. In this framework, an interesting biarsenical compound that detects calcium was introduced recently by Tour et al. [22]. Calcium green FlAsH (CaGF-EDT_2, Fig. 5) was designed and synthesized as a triple hybrid combining a fluorescein with biarsenical substituents and a BAPTA-like chelator (BAPTA: 1,2-bis(*o*-aminophenoxy)ethane-*N*,*N*,*N*′,*N*′-tetraacetic acid). The BAPTA-like domain provides the Ca^{2+} binding site, the fluorescein structure is the fluorophore, and the two arsenic atoms that form four covalent bonds with the two pairs of cysteines of the tagged peptide provide genetic targetability [22].

As expected for biarsenical dyes, the CaGF fluorescence is highly quenched when the dye is not bound to a TC peptide. Peptide binding increases the fluorescence more than 4-fold, after which Ca^{2+} binding enhances the fluorescence a further tenfold with a $K_d = 100\ \mu M$ for Ca^{2+}. The rate constants for Ca^{2+} binding were too high for experimental resolution, yet using a stopped-flow apparatus the authors determined that the dissociation rate constant was $>2\ s^{-1}$. This value is consistent with the $>20{,}000\ s^{-1}$ rate constants of structurally similar calcium-binding molecules with comparable K_d's.

The presence of multiple carboxylates in CaGF imparts greater hydrophilicity compared to FlAsH or ReAsH. For this reason, CaGF was less membrane permeable than other biarsenicals and it was necessary to generate an ester derivative in order to test CaGF in living cells. CaGF/AM penetrated living cells, and after esterase cleavage successfully bound the target protein. CaGF-EDT_2 was added to HeLa cells expressing TC-tagged connexin 43. The gap junctions were

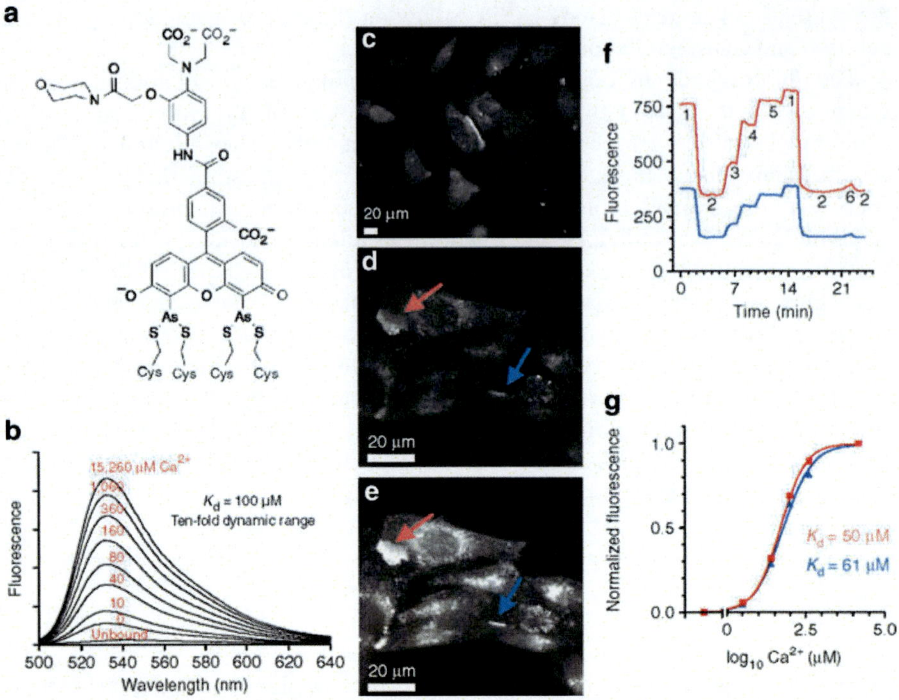

Fig. 9 In vitro and intracellular titration of CaGF fluorescence. (**a**) Structure of CaGF bound to a TC-containing peptide or protein. (**b**) In vitro calibration of CaGF; fluorescence increased ca. tenfold (concentration is indicated in *red* above the traces). (**c**) A large gap junction between two HeLa cells expressing recombinant connexin 43 tagged with a TC in its C-terminus and stained with CaGF. (**d–g**) Intracellular calibration of CaGF bound to cx43-GFPC4. A field of cells with an en-face and a perpendicular gap junction (*red and blue arrows*, respectively) bathed in zero calcium plus 1 mM EGTA (**d**); the same field in 15 mM calcium (**e**); raw data of CaGF titration (**f**). The calcium concentrations are indicated by *numbers*: 1 = 15 mM; 2 = zero calcium plus 1 mM EGTA; 3 = 25 μM; 4 = 100 μM; 5 = 400 μM and 6 = nominal zero estimated at 3 μM. (**g**) Dose–response curves constructed from the data shown in **f** (*red*, $K_d = 50$ μM; *blue*, $K_d = 61$ μM) [22]. Reprinted by permission from Macmillan Publishers Ltd: Nature Chemical Biology. Copyright 2007

successfully labeled and Ca^{2+} waves through gap junctions were monitored in real time (Fig. 9).

3.11 *DOPA–Biotin–FlAsH (16)*

Liu et al. developed a label transfer method that eliminates the need for covalent modification of the protein of interest [23]. The procedure involves tagging the "bait" protein with the TC-containing peptide (CCPGCC) and binding to a FlAsH derivative containing a tethered biotin and DOPA residues (Fig. 10). When DOPA is oxidized with sodium periodate, it is transformed into a reactive *ortho*-quinone

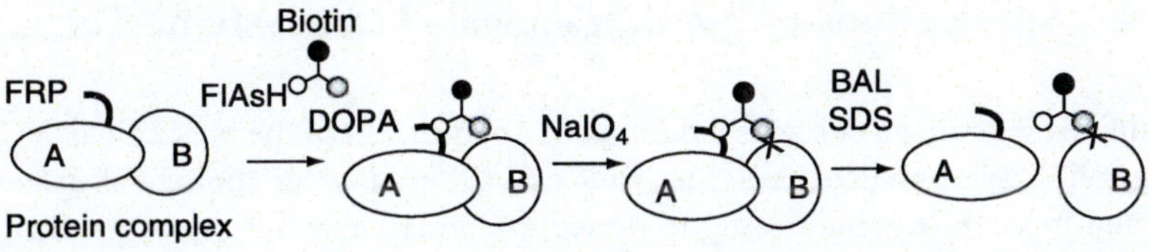

FRP = CCPGCC

Fig. 10 Label transfer system designed by Liu and col. Reprinted with permission from [23]. Copyright 2007 American Chemical Society

Table 3 Properties of biarsenical dyes and some of their conjugates

	λ_{ex} (nm)	λ_{em} (nm)	Fluorescence quantum yield	ϵ_{max} (M^{-1} cm^{-1})
Fluorescein	494	521	0.92	–
FlAsH-EDT$_2$	508	–	–	41,000
FlAsH-Cy4	508	528	0.49	–
FlAsH-P12	511	527		52,000
FlAsH- CaM4Cys	–	–	0.80	–
FlAsH-CAMpep	–	–	0.83	–
ReAsH-EDT$_2$	579	–	–	63,000
ReAsH-Cy4	593	608	0.2	–
ReAsH-AcWEAAAREACCRECCARA-NH$_2$	590	632	–	–
HoXAsH-Cy4	–	–	0.12	–
CHoXAsH-AcWEAAAREACCRECCARA-NH$_2$	388	433	0.35	–
BArNile- CaM4Cys	520	604	0.034	–
BArNile- CaMpep	–	–	0.037	–
Mansyl-FlAsH-CaM4Cys	480	540	0.14	–
Mansyl-FlAsH-CaMpep	–	–	0.053	–
F2FlAsH-P12	500	522	–	65,500
F4FlAsH-P12	528	544	–	35,100
AsCy3	560	568	0.28/6	–
AsCy3-Cy3TAG	560	576	0.28	180,000
SplAsH-MANT	332	415	–	–
SplAsH-Dansyl	351	497	–	–
SplAsH-DEAC	418	465	–	–
SplAsH-Alexa594	582	617	–	71,000
CrAsH-EDT$_2$	–	513	–	15,221
CrAsH-CCKACC	–	–	–	172,191
CrAsH-CCKACC	–	–	–	17,780

that can cross-link with nearby nucleophilic amino acids only when the reactive partners are held in close proximity. If this is the case, cross-linking takes place and the biarsenical complex can be dissociated by boiling in an excess of dithiol, resulting in transfer of the biotin label to nearby partner proteins [23].

The photophysical properties of the probes described above are summarized in Table 3.

4 Applications of the Biarsenical–Tetracysteine Technology

In this section we describe a number of applications of the biarsenical–TC complexes serving as analytical tools, as well as in the study of biological systems in vitro and in vivo. It is interesting to notice the remarkable variety of applications of this technology, showing its versatility and potential.

4.1 Purification of Proteins

The high specificity of the binding combined with the low occurrence of endogenous proteins containing TC motifs provide the basis for a convenient application of immobilized biarsenicals to purify proteins of interest. One example is provided by Adams et al. [3]. They immobilized the *N*-hydroxysuccinimide ester of 5-carboxy-FlAsH-EDT_2 to an amino modified agarose support. This affinity matrix was mixed with lysate from bacterial cells expressing a low level of calmodulin containing an inserted TC site. The desired protein was eluted with a high concentration of 2-dimercaptopropanesulfonate. SDS-PAGE of these fractions (Fig. 11) indicated a high and comparable purity to the same protein construct purified by a conventional polyhistidine: Ni^{2+}-NTA method. Unlike the latter, the biarsenical–TC method requires milder elution conditions. This result agrees those obtained earlier by Thorn & col. for the purification of kinesin [24].

In order to achieve effective high-throughput electrophoretic analysis, Feldman et al. investigated the binding of FlAsH to TC-tagged proteins fractionated by microplate array SDS-PAGE [25]. The proteins were treated with FlAsH prior to gel electrophoresis, becoming fluorescent and thus detectable upon exposure to UV light. Conveniently, the fluorescence signal was conserved even when the protein was denatured by the SDS, and the stoichiometry remained constant, i.e., at one fluorescent reporter molecule per protein molecule [25].

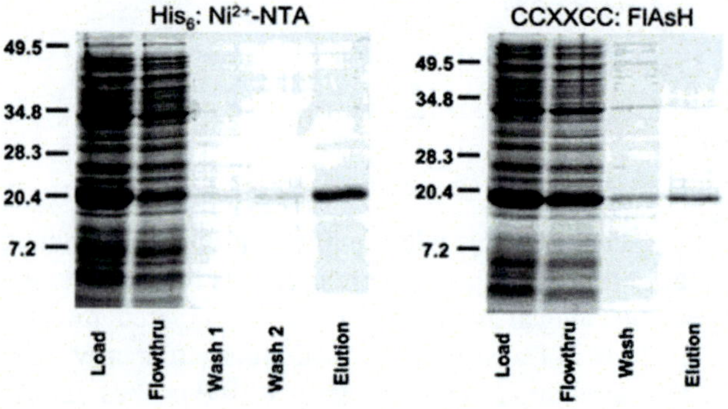

Fig. 11 Affinity purification by a tetracysteine tag yields protein of purity similar to that by conventional his6 tag. Reprinted with permission from [3]. Copyright 2002 American Chemical Society

4.2 Localization, Trafficking and Conformational Changes of Proteins

The PhoP/PhoQ two-component system regulates numerous virulence phenotypes in *Salmonella enterica*. Using the FlAsH labeling technique, Sciara et al. examined PhoP cytolocalization in response to extracellular Mg^{2+} in vitro and in vivo [26] (Fig. 12). They found that previous permeabilization of the cells envelope of gram-negative bacteria to the dye was unnecessary, as the same labeling efficiency was achieved without the treatment. They also determined that the 1 FlAsH : 10 EDT_2

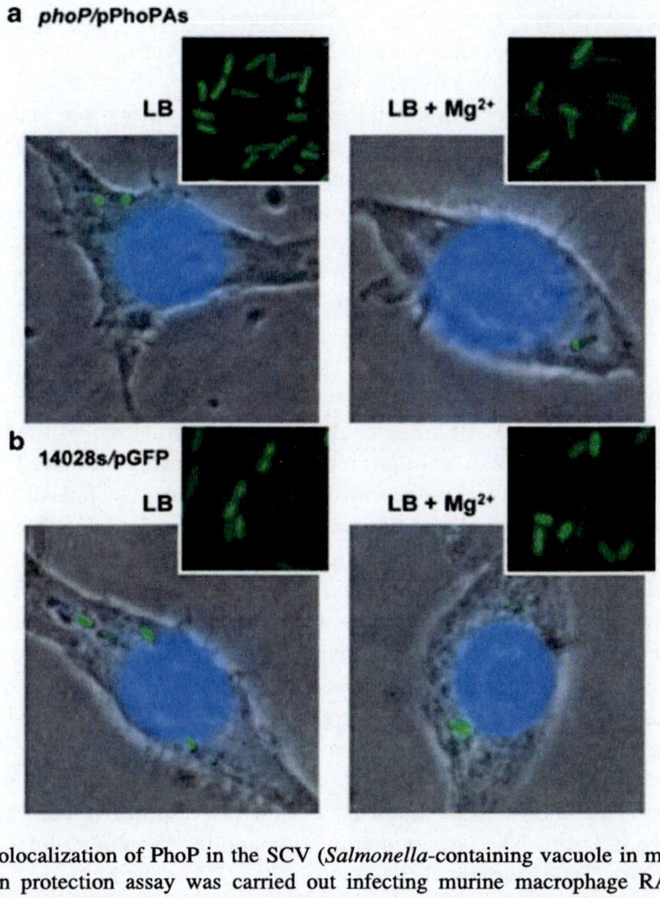

Fig. 12 Cytolocalization of PhoP in the SCV (*Salmonella*-containing vacuole in macrophages). A gentamicin protection assay was carried out infecting murine macrophage RAW264. The fluorescence pattern observed for the *phoP*/pPhoPAs (**a**) or the 14028s/pGFP (**b**) strains used for infection, previously grown in either LB without or with addition of Mg^{2+}, as indicated, is shown. DIC and fluorescence images of the macrophages after infection were acquired. An overlay of the *green* (GFP or FlAsH-stained PhoPAs) and *blue* (DAPI stain for DNA) fluorescence in macrophage-infected cells with bacteria grown in inducing (*left*) or repressing (*right*) Mg^{2+} conditions is illustrated. Reprinted by permission from [26]. Copyright 2008

ratio was the most effective in their labeling conditions, and demonstrated that a preferential localization of PhoP under inducing and repressing conditions, constituting a new insight into the mechanism of the PhoP/PhoQ system.

The advantages of TC tag labeling over VFPs were clearly demonstrated in another gram-negative model organism. Senf and coworkers studied the polar secretion of proteins via the Xcp type II secretions system in *Pseudomonas aeruginosa* [27]. Expression of chimeric XcpS and XcpR proteins fused to GFP did not restore the secretion of elastase, the mayor substrate of the Xcp system, in *xcpS* and *xcpR* mutants respectively. These results indicated that these GFP-fusion proteins were not functional. They then placed the TC tag in both the N-terminal and C-terminal of XcpS and XcpR. Surprisingly; only the C-terminally tagged proteins rescued the secretion system in the mutant strains. These functional recombinant proteins were used to determine the localization of the Xcp system in *P. aeruginosa* at the poles of the cells. Moreover, it was also shown by direct visualization of the Xcp-mediated protease secretion that this phenomenon takes place at the poles [27].

Another interesting example of the use of this labeling strategy in the localization of a protein of interest was reported by Panchal et al. [28]. The matrix protein VP40 plays a critical role in Ebola virus assembly and budding, a process that utilizes specialized membrane domains known as lipid rafts. The study involved the imaging of Ebola VP40 in live cells with FlAsH, revealing the localization of VP40 in plasma membrane rafts in human cells and the enrichment of VP40 oligomers in DRMs detergent-resistant membranes, virus-like particles (VLPs), and in the virus. The authors also identified C-terminal residues critical for membrane association and vesicular budding.

The work of Ignatova et al. has enabled the determination of the stability and aggregation of mammalian cellular retinoic acid-binding protein I (CRABP I) directly in bacterial cells [29]. The unfolding of labeled TC-CRABP I was accompanied by enhancement of FlAsH fluorescence, which made it possible to determine the free energy of unfolding of this protein by urea titration in cells and to follow in real time the formation of inclusion bodies by a slow-folding, aggregation prone mutant (FlAsH-labeled P39A tetra-Cys CRABP I). The aggregation in vivo displayed a concentration-dependent apparent lag time similar to observations of protein aggregation in purified in vitro model systems.

The conformational changes of proteins in live cells were monitored by Nakanishi et al. using Mansyl-FlAsH-EDT_2, the environment-sensitive biarsenical probe described above [19]. The high sensitivity of the fluorophore to its environment enabled monitoring of the conformational changes of the proteins in live cells as changes in fluorescence intensity. The present method was applied to calmodulin (CaM), which exposes hydrophobic domains depending on the Ca^{2+} concentration.

The use of TC tags has proven useful in real time studies of the relocalization of proteins within different cell compartments. In particular, Muñoz-Pinedo et al. showed by labeling different mitochondrial intermembrane-space proteins with FlAsH and ReAsH that their release during apoptosis is coordinately initiated but can vary in duration [30]. These results support the model by which the permeabilization of the mitochondrial membrane is due to the rapid formation of a pore that allows the release of many soluble proteins.

4.3 Pulse-Chase Labeling

Due to the high stability of the TC–biarsenical complex (complex stabilities of up to weeks [3]) a two-color pulse chase of protein turnover allowing one to distinguish between old and newly synthesized proteins, can be performed with the sequential addition of biarsenical probes. As opposed to the traditional method based on the incorporation of radioactive amino acids, this technique permits continuous imaging of the cell. The method was first introduced by Gaietta et al., to study the turnover of gap junctions in HeLa cells [16]. Gap junctions are channels formed by connexins subunits and they connect adjacent cells. The authors reported the modification of connexin43 with a C-terminal TC motif and the labeling of the old protein with FlAsH and the subsequent newly synthesized with ReAsH. Variations of the staining interval showed freshly synthesized protein (red) on the outside of the plaque and older protein removed (green) from its center, in addition to the protein trafficking through the cytoplasm.

Another protein of interest studied by means of the pulse-chase methodology was the AMPA receptor (AMPAR). Ju et al. investigated the trafficking and synthesis of the subunits GluR1 and GluR2 by labeling with FlAsH and ReAsH [31]. They used sequential ReAsH/FlAsH labeling to identify the AMPARs that had been recently synthesized in a known, finite time period. Their work provided a novel mechanism for synaptic modifications and established several advantages of using this labeling strategy in the study of trafficking and synthesis of proteins in a cellular compartments. The sequential staining provided high sensitivity while retaining high specificity.

4.4 Chromophore- or Fluorophore-Assisted Light Inactivation

CALI is a technique that allows a controlled inactivation of selected proteins with high spatial and temporal resolution in living cells. Normally, an antibody is used to direct a fluorophore to the protein of interest. A local generation of reactive oxygen species (ROS) is induced by illumination, which reacts with the protein and inactivates it. The use of TC tags to direct the fluorophore has several advantages over the use of antibodies.

Marek et al. first demonstrated this application in live organisms by introducing a TC tag into the cytoplasmatic tail of *Drosophila* synaptotagmin I (Syt I), a protein involved in transmitter release at synapses and neuromuscular junctions [32]. The TC tagged Syt I has been shown to be functional, inasmuch as it rescues the *syt I* null mutant fly to adult viability. The FlAsH labeled protein was correctly localized at the neuromuscular junction as seen by FlAsH staining in rescue larvae. This experiment demonstrated for the first time that FlAsH could be used to label proteins in live transgenic animals. Efficient labeling was achieved when incubating with FlAsH even for as little as 5 min of labeling followed by 5 min of rinsing. Control wild-type flies showed no fluorescence at the synapses or elsewhere when treated under identical labeling conditions. Recombinant tagged Sty I labeled with FlAsH was successfully inactivated after a few seconds of illumination with a mercury lamp in an epifluorescence microscope. This feature is of great advantage compared to probing with VFPs since no laser illumination is needed and higher levels of inactivation are achieved. Neurotransmitter release by Sty I causes changes in the excitatory postsynaptic potential (EPSP). After a minute of illumination, complete inactivation of Sty I was reached as the EPSP dropped to a plateau with similar levels of release as the null mutant. Importantly, continuous illumination for up to 10 min did not cause any further changes in EPSP. This powerful technique is a significant addition to the arsenal of tools available for studying protein function in vivo, and it was used in *Drosophila* to successfully inactivate other proteins such as clathrins light and heavy chains [33, 34]. Venken et al. developed an elegant procedure to generate TC protein fusions at either terminus in a genomic context using recombinant engineering [35]. This methodology allows an efficient and quick tagging of proteins under endogenous control of expression in flies. The authors successfully tagged and inactivated several genes. The recombinant proteins show normal expression, activity, and no toxicity when labeled with FlAsH, demonstrating the power of this technique.

4.5 Correlated Fluorescence and Electron Microscopy

Fluorescent staining of intracellular components with appropriate dyes may also be visualized at much higher resolution by electron microscopy (EM) using photoconversion of diaminobenzidine (DAB). Under intense illumination, some dyes catalyze the formation of singlet oxygen causing highly localized polymerization of DAB into an insoluble osmiophilic precipitate visible by EM.

Gaietta et al. illustrated how ReAsH (described earlier) can be used for correlating optical recording of live cells and ultrastructural analysis by electron microscopic images [16]. They used this technology to study the trafficking of connexin (the gap junction channel) to the cell surface monitoring by fluorescence and with electron microscopy snapshots of the photoconverted product, yielding detailed ultrastructural information (see Fig. 2 in [16]).

In subsequent work by the Ellisman group [36], the former methodology was applied to study the Golgi apparatus during cell division. Duplication of the Golgi is a complex and controversial process. During mitosis it undergoes vesiculation and fragmentation and its components are found scattered throughout the cytoplasm in the form of Golgi clusters and thousands of small vesicles. These vesicles, also known as the Golgi haze, are approximately 50 nm in size, i.e., below the resolution of light microscopy. Using the first 117 aminoacids of α-mannosydase II, a Golgi-resident enzyme fused to GFP with a C-terminal TC-tag, it was possible to monitor the changes occurring to the Golgi apparatus during mitosis. The fluorescent protein allowed direct live imaging without any labeling, and labeling with ReAsH was used both for live cell imaging and for FRET-based studies. In addition, the ReAsH labeling and further photoconversion of DAB led to high resolution analysis by electron microscopy. It was possible to label the recombinant Golgi-resident protein in the oxidizing environment of the Golgi lumen but it was necessary to treat the cells with membrane-permeant reducing agents such as tributylphosphine or trietylphosphine. These procedures allowed the authors to image Golgi fragmentation and reconstitution during mitosis both by light microscopy and EM, showing that the Golgi reassembly is preceded by the formation of four collinear clusters at telophase, two per daughter cell, that migrate and rejoin on the far side of each nuclei to asymmetrically reconstitute a single Golgi apparatus first in one daughter cell and then in the other.

5 FRET-Based Investigations

Imaging Fluorescence Resonance Energy Transfer (FRET) is a powerful tool for studying changes and molecular interactions of biomolecules in their physiological cellular environment [37, 38]. Biarsenicals constitute useful donor–acceptor pairs, in combination with other fluorophores, fluorescent proteins, or another biarsenical.

Hoffman et al. investigated the activation of the G protein-coupled receptor (GPCR) by FRET monitoring, comparing the use of CFP/FlAsH pair and CFP/ YFP [39]. The FRET signals gave an improved detection of the GPCR activation in single living cells without alteration of the receptor function.

Evans and Walker obtained significant results in a study of the ability of the endothelins ETA and ETB to form homo and heterodimers that could affect receptor trafficking and function. They measured conformational changes by means of FRET between CFP and FlAsH, instead of YFP which could interfere with the receptor function or orientation in the membrane [40].

In 2006 Liu et al. reported on a single-molecule FRET study of interactions between FlAsH labeled Calmodulin (CaM) and Texas Red labeled oligomer C28W, the effective CaM-binding sequence of the plasma membrane Ca-ATPase [41]. They also studied the single-molecule fluorescence polarization of unlabeled C28W peptide interacting with FlAsH labeled CaM. Only when CaM and C28W interact or bind to each other can the distance between FlAsH and Texas Red fall in the FRET-resolvable range (2–10 nm) (Fig. 13). The authors were able to resolve binding–unbinding motions of the N-terminal domain of the CaM in CaM/C28W complexes, providing strong evidence for a two-state binding interaction of CaM-mediated cell signaling.

Another example of FRET measurements based on a pair FlAsH-fluorophore was reported by Granier et al., who used fluorescence resonance energy transfer to examine ligand-induced structural changes in the distance between two

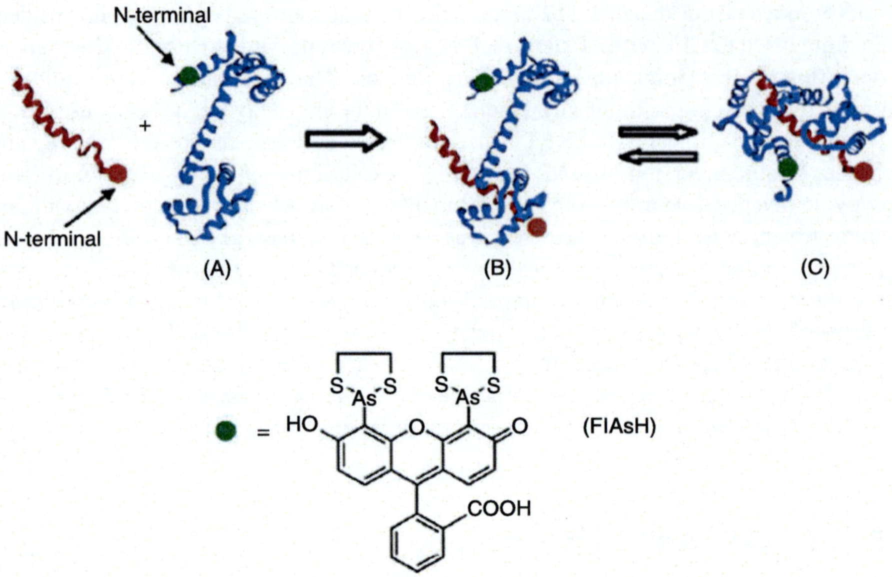

Fig. 13 CaM and C28W interaction process (*A*) CaM without C28W interaction; (*B*) the C-terminal domain of CaM binds to the N-terminal domain of C28W; (*C*) CaM tightly binds to C28W through both domains. Reprinted with permission from [41]. Copyright 2006 American Chemical Society

positions on the β2-adrenoreceptor (AR) C-terminus and cysteine 265 (Cys-265) at the cytoplasmatic end of the transmembrane domain [42]. The donor fluorophore FlAsH was attached at position 351–356 in the proximal C-terminus or at the distal C-terminus. As acceptor, Alexa Fluor 568 was attached to Cys-265. FRET analyses provided new insight into the structure of the C-terminus of the b2-AR as well as ligand-induced conformational changes, relevant to regulation and signaling.

5.1 *FRET Measurements Between Members of the Biarsenical Family*

Robia et al. performed the site-directed labeling with FlAsH and ReAsH of several positions of phospholamban (PLB), a protein associated with dilated cardiomyopathy and heart failure in humans [43]. The calculated transfer radii were comparable to distances predicted by a computer molecular model of the phospholamban pentamer constructed from NMR solution structures.

In 2007 Roberti et al. generated a recombinant α-synuclein (α-synuclein-C4) bearing a TC target for fluorogenic biarsenical compounds [44]. α-Synuclein is a major component of intraneuronal protein aggregates constituting a distinctive feature of Parkinson disease. Direct imaging of α-synuclein aggregation was possible inside living cells via the monitoring of FRET between FlAsH and ReAsH in microaggregates, confirming the close association of fibrillized α-synuclein-TC molecules (Fig. 14). The results indicate that reliable FRET between FlAsH and ReAsH, and presumably between improved biarsenical FRET pairs like F2FlAsH–F4FlAsH [6], constitutes a valuable reporter of dynamic protein– protein interaction in associated and aggregated forms of α-synuclein and other systems.

Recently, the Hoffman group developed a sequential labeling procedure of intracellular proteins in intact cells with minimal disturbance of their function using

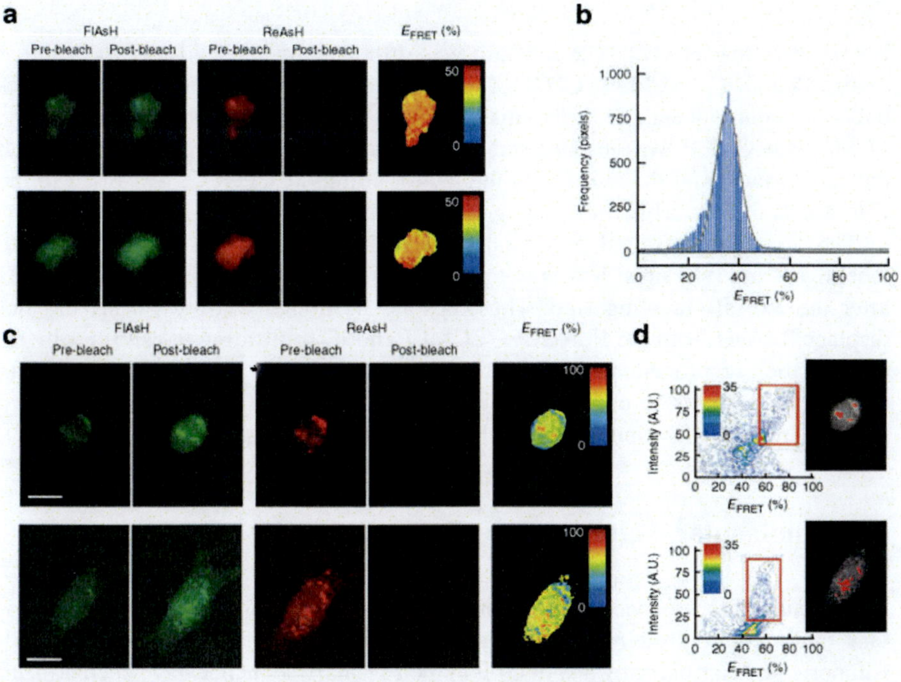

Fig. 14 Characterization of aggregated α-synuclein-C4 in vitro and in cells by apbFRET microscopy. **(a)** Determination of E_{FRET} in aggregates prepared from a 1:4 monomer mixture of α-synuclein-C4–FlAsH and α-synuclein-C4–ReAsH in vitro. FlAsH emission before and after ReAsH photobleaching (*left*), ReAsH emission channel before and after photobleaching (*middle*) and E_{FRET} map (*right*). **(b)** Histogram of the E_{FRET} values obtained from the set of aggregates (number of samples, $n = 5$), calculated on a pixel-to-pixel basis. The distribution was fitted to a Gaussian function, mean $E_{FRET} = 36\%$. **(c)** Living SH-SY5Y cells transiently expressing α-synuclein-C4, labeled with a 1:4 mixture of FlAsH and ReAsH. FlAsH emission channel before and after ReAsH photobleaching (*left*), ReAsH emission channel before and after photobleaching (*middle*) and E_{FRET} (*right*). Scale bars, 10 mm (*top*) and 20 mm (*bottom*). **(d)** Two-dimensional histogram of the correlation between FlAsH fluorescence intensity and E_{FRET}. For each sample, the values of the histogram within the *red box* were mapped back onto the FlAsH fluorescence image (*right*), to highlight the regions (*red*) that jointly displayed high FlAsH fluorescence and high E_{FRET}. The *color scale bars* represent frequency (number of pixels). Reprinted by permission from [44]. Copyright 2007

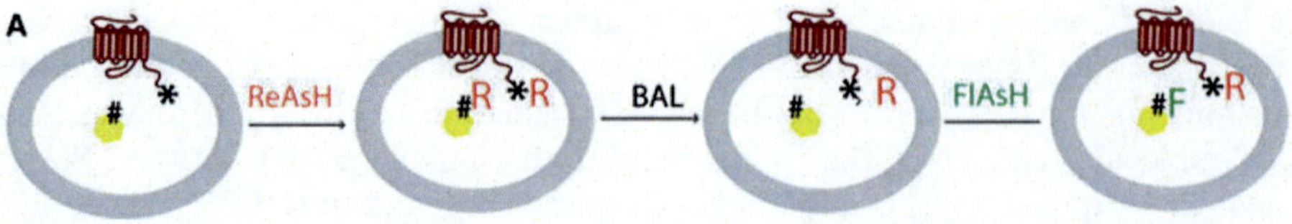

Fig. 15 Schematic representation of the labeling protocol. Initially, the FLNCCPGCCMEP motif in the PTH receptor (*asterisk*) as well as the CCPGCC motif in β-arrestin2 (*hash*) were both covalently labeled with ReAsH (R). In a washing step with BAL, ReAsH was selectively removed from the CCPGCC motif in β-arrestin2. Finally, the latter motif was again labeled with FlAsH (F). Reprinted with permission from [12]. Copyright 2010 American Chemical Society

FlAsH and ReAsH [12]. The labeling was orthogonally directed to two different motifs CCPGCC and FLNCCPGCCMEP, which bind FlAsH and ReAsH, respectively, as a consequence of differential affinities. The motif with the highest affinity, FLNCCPGCCMEP, was inserted into the C-terminus of a PTH receptor located on the cell surface, and the motif with the lowest affinity, CCPGCC, was fused to the C-terminus of β-arrestin2, its binding protein in the cytosol (Fig. 15).

The procedure resulted in virtually no background, but essentially complete labeling of the two sites. It was not possible to label the ReAsH-site with FlAsH after the ReAsH labeling step. Thus, despite its higher affinity FlAsH did not displace ReAsH from the FLNCCPGCCMEP motif, confirming the site specificity of the labeling procedure. This orthogonal and specific nature of the labeling allowed the authors to demonstrate the agonist-dependent interaction of the two proteins by colocalization and by FRET between the labels.

6 Bimolecular Tetracysteine Tags

Bimolecular fluorescence complementation has been widely used to monitor protein folding and protein–protein interactions as an alternative to FRET [45]. Chimeric proteins carrying a split GFP in their primary sequence have been utilized as biosensors by taking advantage of the reconstitution of GFP to monitor conformational changes [46]. Similarly, protein interactions have been approached by means of reconstituting the fluorescence of two chimeric interacting proteins, each carrying complementary halves of VFPs. Although this method has been successfully used [45], it presents several constrains such as the time it takes to restore the fluorescence and the size of the proteins.

In contrast, TC tags have proved to be ideal for this purpose. The strategy is to split the TC motif by placing two cysteine pairs on interacting proteins or distal in the primary sequence of a single protein such as they come into close proximity when the proteins interact or when the protein is folded in a particular conformation. Thus, the bulky VFPs commonly used in cellular fluorescence assays and FRET-based biosensors are replaced by just two Cys–Cys pairs that when assembled into close proximity (about 7 Å), bind to biarsenical dyes and form a stable fluorescent complex.

The changes in fluorescence upon binding of a biarsenical to the reconstituted motif are higher than those due to FRET, as was first demonstrated by Luedtke et al. by reconstituting intra- and intermolecular bipartite TC displays on four structurally characterized polypeptides and protein domains [47]. Reconstitution of intramolecular bipartite TCs was achieved using avian pancreatic polypeptide (aPP) and Zip4, two well-folded polypeptides forming α-helix and β-sheet structures respectively, which were modified to contain one-half of TC motif at each terminus [47]. Intermolecular bipartite TC display reconstitution was also accomplished by using the protein–protein dimerization domains from the basic region leucine zipper (bZIP) proteins GCN4 and Jun. These were modified to contain a single dicysteine motif at the C- and N-terminus, respectively. After dimerization the dicysteine-peptides formed stable complexes with FlAsH with apparent dissociation constants (K_{app}) comparable to those of optimized linear tetracysteine sequence. The brightness, quantum yield, and extinction coefficient of the FlAsH complexed to Jun, aPP, and GCN4 at saturation are comparable to that of the optimized linear tetracysteine complex and show a 1:1 protein FlAsH stoichiometries [47].

Using the same approach it is possible to differentiate folded and misfolded polypeptides. By introducing destabilizing point mutations in aPP (F24P, Y31P) and Zip4 (W9A, W16A), without affecting their secondary structure, their emission intensities after binding biarsenical dyes diminish two- to tenfold as compared to the wild-type complexes. Analogously, bipartite TC motives serve to detect intermolecular protein assemblies. Weakening the formation of GCN4 and Jun dimers by introducing point mutations at the dimerization interface (L20P), drastically diminish the affinity to FlAsH and ReAsH and lower fluorescence brightness of the complexes [47]. Thus, biarsenical affinity and brightness are sensitive to the precise tetracysteine arrangement.

These experiments proved the principle that well-folded polypeptides and protein–protein binding domains, when properly adapted for bipartite TC motives, bind biarsenicals with high affinity and generate fluorescence signals that rival the optimized linear TC display. Moreover, bipartite TC display can be used to sense discrete conformational states and to trace stable protein–protein interactions.

It is interesting to consider related experiments by Krishnan and Gierasch describing intramolecular bipartite display between adjacent β-strands in the cellular retinoic acid-binding protein I (CRABP I) [48]. CRABP I is a 136 aminoacid protein containing ten antiparallel β-strands in a close β-barrel. The authors constructed several bipartite CRABP 1 mutants by introducing two pair of cysteines in alternating positions (Cys–Xaa–Cys) on adjacent β-strands. The four thiols are situated so as to form a FlAsH binding site on the folded protein. Although the authors showed successful coordination of FlAsH to these variants under permissive conditions, a lower affinity and quantum yield of the bound FlAsH was observed, as compared to the C-terminus liner TC tag. This finding indicates that biarsenical binding is sensitive to the inherent flexibility of the TC configuration, suggesting that these might best be placed in flexible regions.

The Schepartz group went one step forward in examining the requirement of a flexible environment by evaluating the viability of reconstituting intramolecular bipartite moieties between adjacent loops of the oncogenic protein p53, and between adjacent β-strands of the robust β-barrel folded EmGFP [49]. Under stringent conditions, only on those proteins whose thiols are located within flexible regions bound biarsenicals with high affinity and generated fluorescence complexes. These regions were always located within protein loops; neither binding nor fluorescence was detected when thiols were located within rigid β-strands. It can be concluded that the utility of bipartite TC display will depend on how easily and predictably the split Cys–Cys motif can be introduced into proteins of complex structure. In the case of intramolecular bipartite display, the chances of success are clearly increased when the Cys–Cys pairs are located in loops rather than stiff β-strands.

7 Conclusion: Pros and Cons

The specific targeting with biarsenical compounds presents *unique features*. In first place, the biarsenical derivatives show a dramatic *increase in fluorescence* upon binding to their target, minimizing the deterioration in contrast from background contributions in labeling experiments.

Secondly, the TC motif is sufficiently *small* so that it can be fused not only to the C-terminus of a protein, but it can also be incorporated into loops or on the outer surface of the protein of interest, thereby decreasing the chance of interference with the target protein function. In addition, the probes are generally membrane permeable and do not require complicated procedures such as microinjection.

Additionally, the stability of the biarsenical–TC motif interaction and the availability of different colors enable the consecutive labeling of fusion proteins in pulse-chase experiments.

A further advantage of the biarsenical targeting is that it allows the fluorescence detection to be confirmed by *electron microscopy*. This requires labeling with ReAsH-EDT_2, which photoconverts diaminobinzidine into an electron-rich precipitate.

Finally, a very important quality of biarsenical labels is that they can be used as mediators of specific techniques such as (CALI).

Currently, several biarsenicals, such as the Lumio™ dyes, are commercially available, facilitating the advances of different specific targeting techniques.

Some *limitations* of this methodology have precluded, especially in former times, of its extension to a number of applications. Today, many of these limitations have been overcome. One general restriction is the need of a reducing environment. As the cysteines of the receptor tag must be in their reduced form, labeling of proteins in cellular oxidizing environments becomes problematical. Another limitation is that some of the biarsenical derivatives can lead to a reduced signal at low pH (for example in endosomal compartments or in lysosomes). This restriction has been overcome by the synthesis of new biarsenical probes with a low pK_a value, such as CHoAsH-EDT_2 and the fluorinated derivatives F2FlAsH-EDT_2 and F4FlAsH-EDT_2.

The pioneering work by the Tsien group has inspired many researchers to develop new selective chemical labeling of proteins with small fluorescent molecules. This achievement has required an interdisciplinary effort of biologists and chemists to improve the tag sequence and the molecular probes in order to overcome the limitations of the method. The variety of applications described here demonstrate the remarkable contribution that this methodology is making to the elucidation of protein biochemistry and function in living cells.

References

1. Martin BR, Giepmans BN, Adams SR, Tsien RY (2005) Mammalian cell-based optimization of the biarsenical-binding tetracysteine motif for improved fluorescence and affinity. Nat Biotechnol 23:1308–1314
2. Griffin BA, Adams SR, Tsien RY (1998) Specific covalent labeling of recombinant protein molecules inside live cells. Science 281:269–272
3. Adams SR, Campbell RE, Gross LA, Martin BR, Walkup GK, Yao Y, Llopis J, Tsien RY (2002) New biarsenical ligands and tetracysteine motifs for protein labeling in vitro and in vivo: synthesis and biological applications. J Am Chem Soc 124:6063–6076
4. Adams SR (2007) The biarsenical-tetracysteine protein tag: chemistry and biological applications. In: Schreiber SL, Kapoor TM, Weiss G (eds) From small molecules to system biology and drug design. Wiley-VCH, Weinheim
5. Wang T, Yan P, Squier TC, Mayer MU (2007) Prospecting the proteome: identification of naturally occurring binding motifs for biarsenical probes. Chembiochem 8:1937–1940
6. Spagnuolo CC, Vermeij RJ, Jares-Erijman EA (2006) Improved photostable FRET-competent biarsenical-tetracysteine probes based on fluorinated fluoresceins. J Am Chem Soc 128: 12040–12041
7. Cao H, Chen B, Squier TC, Mayer MU (2006) CrAsH: a biarsenical multi-use affinity probe with low non-specific fluorescence. Chem Commun 2601–2603
8. Bhunia AK, Miller SC (2007) Labeling tetracysteine-tagged proteins with a SplAsH of color: a modular approach to bis-arsenical fluorophores. Chembiochem 8:1642–1645
9. Madani F, Lind J, Damberg P, Adams SR, Tsien RY, Gräslund AO (2009) Hairpin structure of a biarsenical-tetracysteine motif determined by NMR spectroscopy. J Am Chem Soc 131:4613–4615
10. Chen B, Cao H, Yan P, Mayer MU, Squier TC (2007) Identification of an orthogonal peptide binding motif for biarsenical multiuse affinity probes. Bioconjug Chem 18:1259–1265
11. Cao H, Xiong Y, Wang T, Chen B, Squier TC, Mayer MU (2007) A red cy3-based biarsenical fluorescent probe targeted to a complementary binding peptide. J Am Chem Soc 129:8672–8673
12. Zürn A, Klenk C, Zabel U, Reiner S, Lohse MJ, Hoffmann C (2010) Site-specific, orthogonal labeling of proteins in intact cells with two small biarsenical fluorophores. Bioconjug Chem 21:853–859
13. Getz EB, Xiao M, Chakrabarty T, Cooke R, Selvin PR (1999) A comparison between the sulfhydryl reductants tris(2-carboxyethyl)phosphine and dithiothreitol for use in protein biochemistry. Anal Biochem 273:73–80
14. Soh N (2008) Selective chemical labeling of proteins with small fluorescent molecules based on metal-chelation methodology. Sensors 8:1004–1024
15. Adams SR, Tsien RY (2008) Preparation of the membrane-permeant biarsenicals FlAsH-EDT_2 and ReAsH-EDT_2 for fluorescent labeling of tetracysteine-tagged proteins. Nat Protocols 3:1527–1534

16. Gaietta G, Deerinck TJ, Adams SR, Bouwer J, Tour O, Laird DW, Sosinsky GE, Tsien RY, Ellisman MH (2002) Multicolor and electron microscopic imaging of connexin trafficking. Science 296:503–507
17. Park H, Hanson GT, Duff SR, Selvin PR (2004) Nanometre localization of single ReAsH molecules. J Microsc 216:199–205
18. Taguchi Y, Shi ZD, Ruddy B, Dorward DW, Greene L, Baron GS (2009) Specific biarsenical labeling of cell surface proteins allows fluorescent- and biotin-tagging of amyloid precursor protein and prion proteins. Mol Biol Cell 20:233–244
19. Nakanishi J, Nakajima T, Sato M, Ozawa T, Tohda K, Umezawa Y (2001) Imaging of conformational changes of proteins with a new environment-sensitive fluorescent probe designed for site-specific labeling of recombinant proteins in live cells. Anal Chem 73: 2920–2928
20. Nakanishi J, Maeda M, Umezawa Y (2004) A new protein conformation indicator based on biarsenical fluorescein with an extended benzoic acid moiety. Anal Sci 20:273–278
21. Urano Y, Kamiya M, Kanda K, Ueno T, Hirose K, Nagano T (2005) Evolution of fluorescein as a platform for finely tunable fluorescence probes. J Am Chem Soc 127: 4888–4894
22. Tour O, Adams SR, Kerr RA, Meijer RM, Sejnowski TJ, Tsien RW, Tsien RY (2007) Calcium Green FlAsH as a genetically targeted small-molecule calcium indicator. Nat Chem Biol 3: 423–431
23. Liu B, Archer CT, Burdine L, Gillette TG, Kodadek T (2007) Label transfer chemistry for the characterization of protein-protein interactions. J Am Chem Soc 129:12348–12349
24. Thorn KS, Naber N, Matuska M, Vale RD, Cooke R (2000) A novel method of affinity-purifying proteins using a bis-arsenical fluorescein. Protein Sci 9:213–217
25. Feldman G, Bogoev R, Shevirov J, Sartiel A, Margalit I (2004) Detection of tetracysteine-tagged proteins using a biarsenical fluorescein derivative through dry microplate array gel electrophoresis. Electrophoresis 25:2447–2451
26. Sciara MI, Spagnuolo CC, Jares-Erijman EA, Garcia Véscovi E (2008) Cytolocalization of the PhoP response regulator in *Salmonella enterica*: modulation by extracellular Mg^{2+} and by the SCV environment. Mol Microbiol 70:479–493
27. Senf F, Tommassen J, Koster M (2008) Polar secretion of proteins via the Xcp type II secretion system in Pseudomonas aeruginosa. Microbiology 154:3025–3032
28. Panchal RG, Ruthel G, Kenny TA, Kallstrom GH, Lane D, Badie SS, Li L, Bavari S, Aman MJ (2003) In vivo oligomerization and raft localization of Ebola virus protein VP40 during vesicular budding. Proc Natl Acad Sci USA 100:15936–15941
29. Ignatova Z, Gierasch LM (2004) Monitoring protein stability and aggregation in vivo by real-time fluorescent labeling. Proc Natl Acad Sci USA 101:523–528
30. Muñoz-Pinedo C, Guío-Carrión A, Goldstein JC, Fitzgerald P, Newmeyer DD, Green DR (2006) Different mitochondrial intermembrane space proteins are released during apoptosis in a manner that is coordinately initiated but can vary in duration. Proc Natl Acad Sci USA 103:11573–11578
31. Ju W, Morishita W, Tsui J, Gaietta G, Deerinck TJ, Adams SR, Garner CC, Tsien RY, Ellisman MH, Malenka RC (2004) Activity-dependent regulation of dendritic synthesis and trafficking of AMPA receptors. Nat Neurosci 7:244–253
32. Marek KW, Davis GW (2002) Transgenically encoded protein photoinactivation (FlAsH-FALI): acute inactivation of synaptotagmin I. Neuron 36:805–813
33. Heerssen H, Fetter R, Davis G (2008) Clathrin dependence of synaptic vesicle formation at the Drosophila neuromuscular junction. Curr Biol 18:401–409
34. Kasprowicz J, Kuenen S, Miskiewicz K, Habets RL, Smitz L, Verstreken P (2008) Inactivation of clathrin heavy chain inhibits synaptic recycling but allows bulk membrane uptake J. Cell Biol 182:1007–1016

35. Venken KJ, Kasprowicz J, Kuenen S, Yan J, Hassan BA, Verstreken P (2008) Recombineering-mediated tagging of Drosophila genomic constructs for in vivo localization and acute protein inactivation. Nucleic Acids Res 36:e114
36. Gaietta GM, Giepmans BN, Deerinck TJ, Smith WB, Ngan L, Llopis J, Adams SR, Tsien RY, Ellisman MH (2006) Golgi twins in late mitosis revealed by genetically encoded tags for live cell imaging and correlated electron microscopy. Proc Natl Acad Sci USA 103:17777–17782
37. Jares-Erijman EA, Jovin TM (2003) FRET imaging. Nat Biotechnol 21:1387–1395
38. Domingo B, Sabariegos R, Picazo F, Llopis J (2007) Imaging FRET standards by steady-state fluorescence and lifetime methods. Microsc Res Tech 1021:1010–1021
39. Hoffmann C, Gaietta G, Bünemann M, Adams SR, Oberdorff-Maass S, Behr B, Vilardaga JP, Tsien RY, Ellisman MH, Lohse MJ (2005) A FlAsH-based FRET approach to determine G protein coupled receptor activation in living cells. Nat Methods 2:171–176
40. Evans NJ, Walker JW (2008) Endothelin receptor dimers evaluated by FRET, ligand binding, and calcium mobilization. Biophys J 95:483–492
41. Liu R, Hu D, Tan X, Lu HP (2006) Revealing two-state protein-protein interactions of calmodulin by single-molecule spectroscopy. J Am Chem Soc 128:10034–10042
42. Granier S, Kim S, Shafer AM, Ratnala VR, Fung JJ, Zare RN, Kobilka B (2007) Structure and conformational changes in the C-terminal domain of the beta2-adrenoceptor: insights from fluorescence resonance energy transfer studies. J Biol Chem 282:13895–13905
43. Robia SL, Flohr NC, Thomas DD (2005) Phospholamban pentamer quaternary conformation determined by in-gel fluorescence anisotropy. Biochemistry 44:4302–4311
44. Roberti MJ, Bertoncini CW, Klement R, Jares-Erijman EA, Jovin TM (2007) Fluorescence imaging of amyloid formation in living cells by a functional, tetracysteine-tagged alpha-synuclein. Nat Methods 4:345–351
45. Kerppola T (2006) Visualization of molecular interactions by fluorescence complementation. Nat Rev 7:449–456
46. Hu C, Kerppola T (2003) Simultaneous visualization of multiple protein interactions in living cells using multicolor fluorescence complementation analysis. Nat Biotechnol 21:539–545
47. Luedtke NW, Dexter RJ, Fried DB, Schepartz A (2007) Surveying polypeptide and protein domain conformation and association with FlAsH and ReAsH. Nat Chem Biol 3:779–784
48. Krishnan B, Gierasch L (2008) Cross-strand split tetra-Cys motifs as structure sensors in a β-sheet protein. Chem Biol 20:1104–1115
49. Goodman J, Fried D, Schepartz A (2009) Bipartite tetracysteine display requires site flexibility for ReAsH coordination. Chembiochem 10:1644–1647

Labeling of Oligohistidine-Tagged Proteins

Jacob Piehler

Abstract Complexation of Histidine residues by chelated transition metal ions can be exploited for noncovalent, reversible labeling of His-tagged proteins. While the affinity of individual transition metal ions complexed by nitrilotriacetic acid (NTA) is not sufficient for stable fluorescence labeling, this problem has been overcome by multivalent NTA, which bind His-tagged proteins with subnanomolar affinity and complex lifetimes of >1 h. Selective labeling with a defined stoichiometry is possible in cell lysates and on the surface of living cells. Thus, rapid labeling in situ with these relatively small probes at low concentration is possible, which can be reversed under mild conditions.

Keywords Cyanine dye · Förster resonance energy transfer (FRET) · His-tag · HisZiFit · Metal chelator · Multivalency · Nitrilotriacetic acid · Quantum dot · Transition metal ion

Contents

J. Piehler
University of Osnabrück, Barbarastr. 11, 49068 Osnabrück, Germany
e-mail: piehler@uos.de

A.P. Demchenko (ed.), *Advanced Fluorescence Reporters in Chemistry and Biology III: Applications in Sensing and Imaging*, Springer Ser Fluoresc (2011) 10: 297–310,
DOI 10.1007/978-3-642-18035-4_9,

1 Introduction

Peptides consisting of 6–12 consecutives histidine residues – commonly called "His-tags" – have become the most prominent affinity tags for protein purification, protein immobilization, protein detection and many other biochemical applications [1]. Recognition of the His-tag is based on the strong coordination of the imidazole side chain of Histidine by transition metal ions such as Zn^{2+}, Cu^{2+}, Co^{2+} or Ni^{2+}. For His-tag-specific capturing of proteins these transition metal ions are complexed by chelating compounds such as iminotriacetic acid (IDA) or nitrilotriacetic acid (NTA) so that two coordination sites remain free for complexation with Histidine residues (Fig. 1a). Thus, simultaneous coordination of two histidine residues of the His-tag by a single transition metal ion is achieved. Specificity is accomplished due to the relative low abundance of Histidines and other transition metal ion-coordinating residues such as cysteine in natural proteins. Compared to traditional biochemical binders of short affinity tags such as antibodies or other proteins, this interaction has two important advantages:

1. The small, chemically defined chelating compound enables for controlled synthetic conjugations as well as attachment of the chromophore close to the protein of interest.
2. The interaction between the His-tag and the transition metal ion can be switched off by competing agents such as imidazole or by complexing of the transition metal ion with EDTA (Fig. 1b).

For these advantages, transition metal ion mediated attachment of fluorescence probes to His-tagged proteins is highly attractive, e.g., for FRET experiments, which

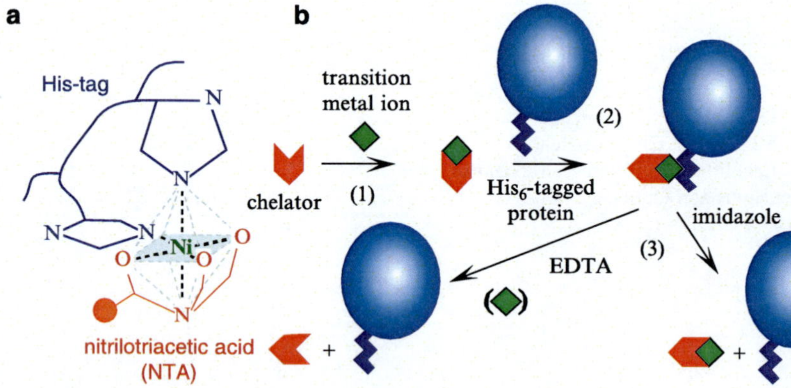

Fig. 1 Noncovalent interaction of the His-tag with chelated transition metal ions. (**a**) Schematic structure of NTA complexed with Ni^{2+}, which is coordinated by two adjacent histidine residues from a His-tagged protein. The site for chemical attachment is indicated by a *red dot*. (**b**) Switchable, reversible binding of chelated transition metal ions to His-tagged proteins. By loading transition metal ions (1), affinity toward the His-tag is triggered (2), which can be reversed by competition with imidazole or removal of complexed transition metal ions by ethylene diamine tetracetate (EDTA) (3)

require close proximity and defined position of fluorescence donor and acceptor dyes. However, the binding affinities of the His-tag toward individual transition metal ions are relatively low. For Ni-NTA, binding affinities of 10–20 μM have been determined [2–4], corresponding to a complex lifetime (binding stability) of a few seconds [2]. Such affinities and binding stabilities are not sufficient for most applications in vitro and in living cells. Stable, high affinity binding of His-tagged proteins to affinity resins or solid supports are based on simultaneous interaction of a single His-tag with several immobilized transition metal ion, which is possible due to a high density of chelating moieties. For labeling with fluorescence dyes, however, such surface-mediated multivalent interactions cannot be exploited. For this reason, efficient approaches for specific fluorescence labeling of His-tagged proteins are based on supramolecular entities, which incorporate more than one chelator moiety. While some of these approaches exploit fluorescence dyes with multiple functionalities, more generic labeling strategies have been implemented based on multivalent chelator heads.

2 NTA-Based Labeling

Despite the low affinity of individual metal ions, fluorescence dyes conjugated to single NTA moieties have been used for fluorescence labeling in vitro and on the surface of living cells [3–6]. Vogel and coworkers synthesized two rhodamine derivatives conjugated with NTA, which were loaded with Ni(II) ions [3]. These probes were shown to bind to purified hexahistidine-tagged GFP in vitro as detected by FRET, yielding a binding affinity of 3 and 6 μM, respectively. This dye was successfully used for reversible labeling of the 5HT3 serotonin receptor (5HT3R) on the surface of living cells. Interestingly, different binding affinities were observed for different positions of the His-tag within the 5HT3R, as well as an increase in binding affinity with increasing numbers of histidine residues in the His-tag fused to the 5HT3R, though only two His residues are involved in recognition of the NTA-complexed Ni(II) ions. Possibly, different structural constraints given by the environment of the His-tag affect their binding affinity. Specific cell surface labeling was confirmed by FRET between acceptor-labeled 5HT3R and a donor-labeled 5HT3R antagonist [3]. Moreover, detection of individual cell surface receptors was possible in living cells [7]. A similar approach toward cell surface labeling based on fluorescein conjugated to NTA has also been reported [4]. However, in both reports the severe drawbacks of the low binding affinities become obvious.

3 Multivalent Recognition

3.1 *Dye-Specific Approaches*

In order to increase the stability and the affinity of His-tag-specific fluorescence probes, some chromophores have been modified with two chelating moieties. Cyanine dyes are particularly attractive for this approach, because bifunctionality

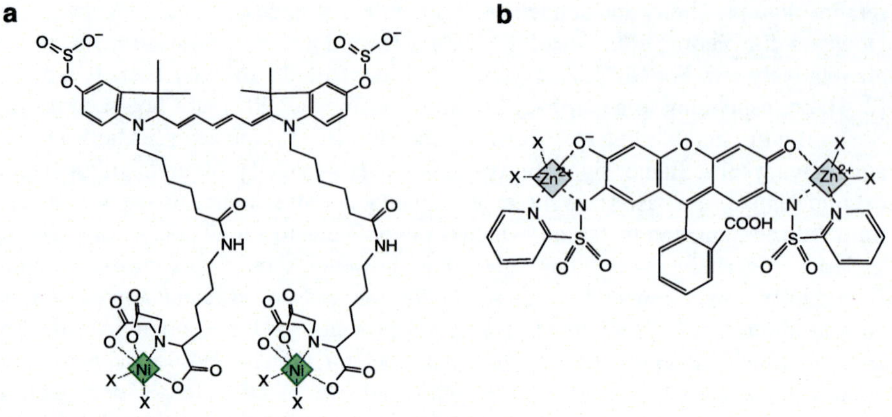

Fig. 2 Fluorescence dyes with two coordinated metals ions. (**a**) Cy5 conjugated with two NTA residues. (**b**) Dimethylfluoresceine derivative HisZiFiT

is readily achieved (Fig. 2a). This approach was pioneered by Ebright and coworkers, who synthesized Cy3 and Cy5 derivatives conjugated to two NTA moieties [8]. Binding to hexahistidine-tagged proteins was probed by fluorescence anisotropy in vitro, yielding submicromolar binding affinities. Substantial loss of the quantum yield (by 75%) was observed for these probes upon complexing the NTA moieties with Ni(II)-ions, which was ascribed to photoinduced electron transfer (PET) due to unpaired electrons in the *d*-orbitals of Ni(II), This problem, though less pronounced, has also been observed for probes conjugated with single NTA moieties.

While the binding affinities of these probes were substantially higher compared probes with a single NTA moiety, they have not been tested for protein labeling on cell surfaces.

Tsien and coworker synthesized a dimethylfluoresceine derivative, which integrated the chromophore itself into the chelating moieties (Fig. 2b) [9]. Similar approaches have been used to design metal-ion sensitive fluorescence probes. In this case, the two 3-dentate chelating moietes were complexed with Zn(II) ions, yielding a His-tag-specific probe termed "HisZiFit" (Fig. 2b) with a binding affinity of 40 nM toward the hexahistidine-tag [9]. In contrast to approaches based on Ni-NTA, Zn(II) has the tremendous advantage to be a diamagnetic, colorless transition metal ion. Thus, the fluorescence of this probe is not quenched by PET or FRET as observed for Ni(II)-based probes (see also below).

The HisZiFit probe was successfully applied for rapid and stable labeling of membrane proteins in the plasma membrane. Orthogonality of this approach with tetracysteine-specific labeling with biarsenical probes (cf. [10]) was demonstrated, enabling for dual color tag-specific labeling with small probes. Moreover, the importance of small tags for appropriate membrane protein trafficking was demonstrated. Thus, HisZiFit was proved to be a very powerful compound for membrane protein labeling on the surface of live cells. However, this approach is dye-specific and thus not applicable to the majority powerful organic dyes with drastically improved photophysical properties compared to fluoresceine.

3.2 *Multivalent Chelator Heads*

For more generic application of multivalent recognition of His-tagged proteins, multivalent chelators have been synthesized [2, 11]. These are based on NTA moieties, which have higher metal binding affinity than 3-dentate chelators such as IDA or HisZiFit. Both linear and cyclic scaffolds have been used for obtaining 2, 3 and 4 metal ion binding sites within close proximity (Fig. 3) [2]. With increasing

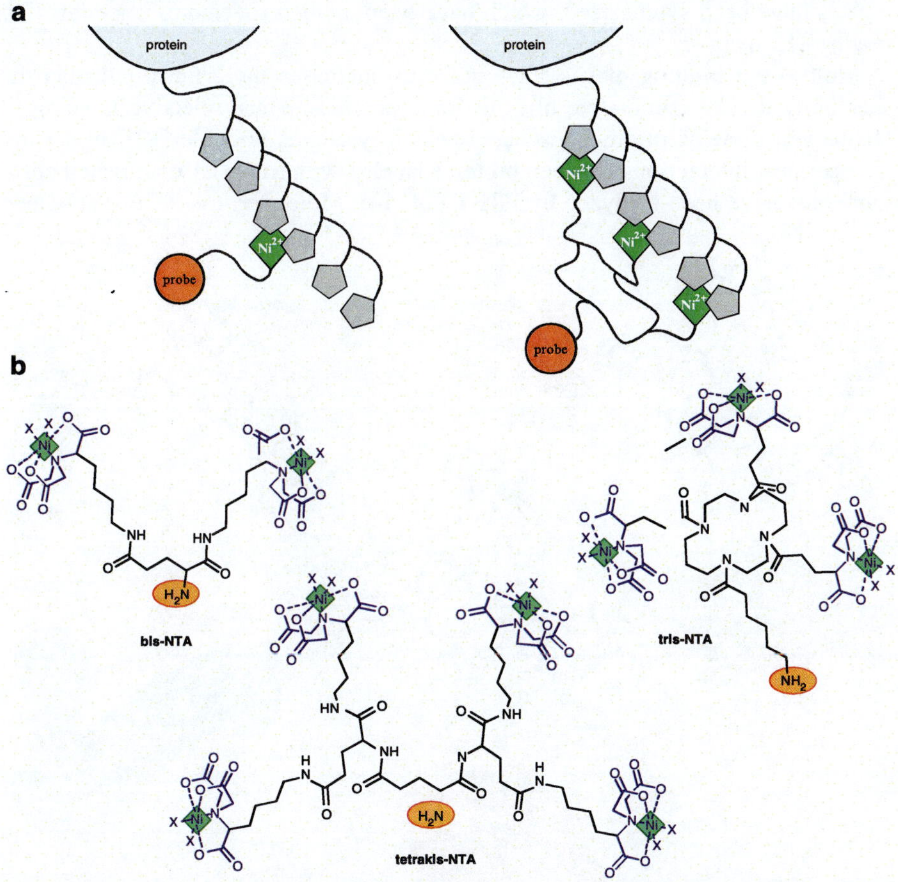

Fig. 3 Multivalent chelators. (**a**) Concept: recognition of a H6-tag by a single Ni-NTA and tris-NTA in comparison. (**b**) Structures of some multivalent chelator heads based on linear (bis-NTA, tetrakis-NTA) as well as cyclic (tris-NTA) scaffolds. The orthogonal functional group for attachments of spectroscopic probes is highlighted in *orange*

number of NTA moieties, substantially increased binding affinities were obtained by this approach. Highly stable, stoichiometric labeling of His-tagged proteins with fluorescence probes attached to tris-NTA or tetrakis-NTA was demonstrated by size exclusion chromatography [12]. Depending on the length of the His-tag (six or ten histidine residues), binding affinities in the subnanomolar regime were estimated for tris-NTA, the most potent multivalent chelator. Tris-NTA offers six coordination sites, perfectly matching the classic hexahistidine tag (H6), which is still very frequently applied. However, up to fivefold stronger binding of tris-NTA to the decahistidine-tag (H10) was observed.

Complexes of tris-NTA with His-tagged proteins have a lifetime of >1 h. Thus, tris-NTA-labeled proteins can be purified prior to further applications. Tris-NTA is readily conjugated to NHS-functionalized derivatives of fluorescence dyes, which are available for all relevant probes used for fluorescence imaging. Various conjugates have been synthesized, which have been applied for fluorescence labeling in vitro [13] (Fig. 4).

Multivalent binding of tris-NTA not only increases the affinity but also the specificity toward cumulative oligohistidine peptides. Thus, selective labeling of His-tagged proteins in crude Escherichia coli lysates is possible [13]. In several applications, interactions between proteins labeled with tris-NTA conjugated chromophores have been analyzed by FRET [13, 14]. Moreover, tris-NTA conjugated

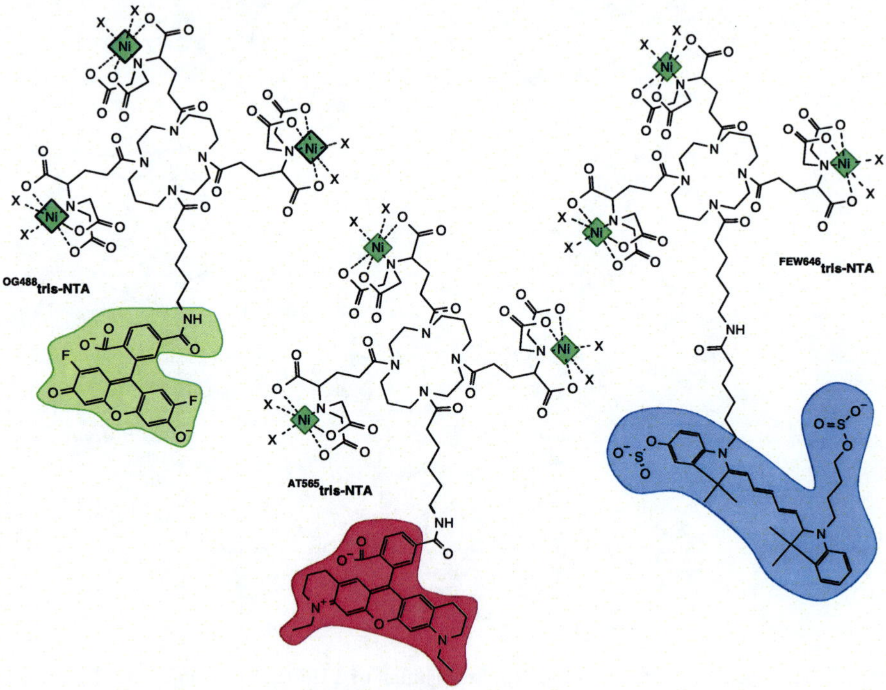

Fig. 4 Structure of some tris-NTA chromophore conjugates, which have been successfully used for labeling His-tagged proteins in vitro and on the cell surface

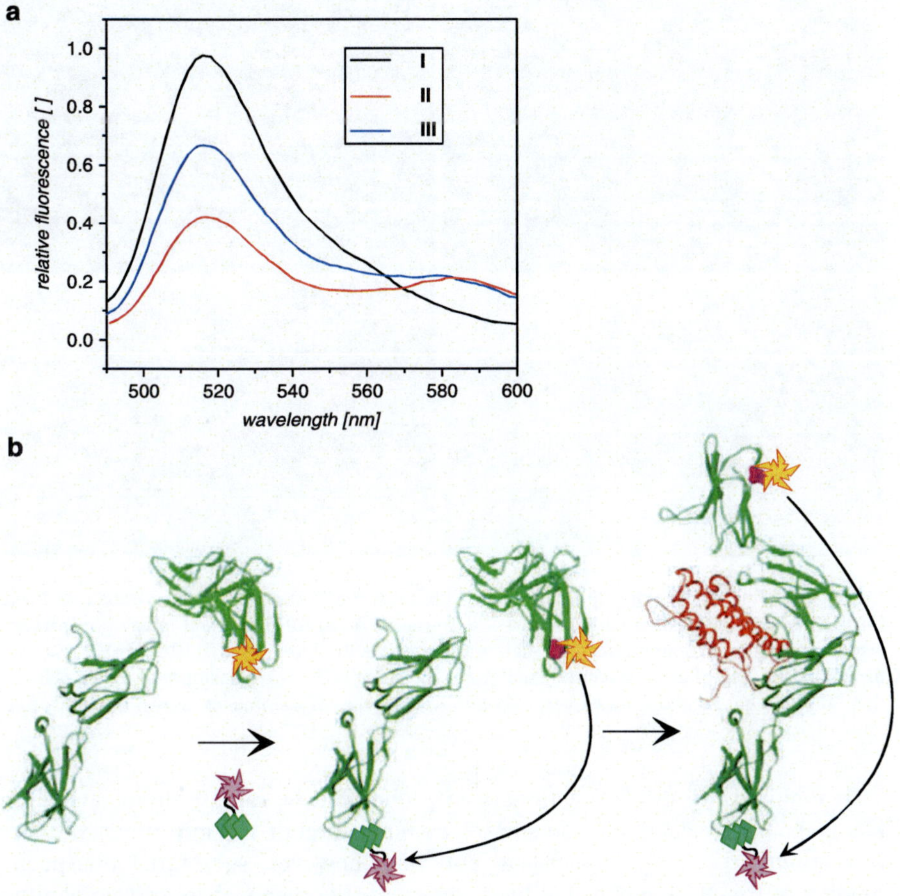

Fig. 5 In situ labeling with a tris-NTA conjugated chromophores as FRET acceptor for probing protein conformations and conformational changes. (**a**) IFNAR1 labeled with Oregon Green 488 was incubated with ^{AT565}tris-NTA, followed by binding of a nonfluorescent ligand. (**b**) The corresponding fluorescence

chromophores are particularly interesting for dual-color protein labeling for probing protein conformation (Fig. 5). High affinity and rapid, His-tag-specific binding at stoichiometric concentrations of the probe enables for incorporating FRET acceptor [15] or quencher molecules [16] in situ. Thus, FRET is readily detected and FRET efficiencies can be quantified very reliably. Owing to the paramagnetic complexed Ni(II) ions, however, very low sensitized fluorescence from the acceptor probe is detected (cf. Fig. 5) [13, 14].

Fluorescence probes conjugated to tris-NTA have also been successfully employed for labeling membrane proteins on the surface of living insect cells [13]. Also for mammalian cells, rapid and selective labeling was achieved (unpublished result, Fig. 6).

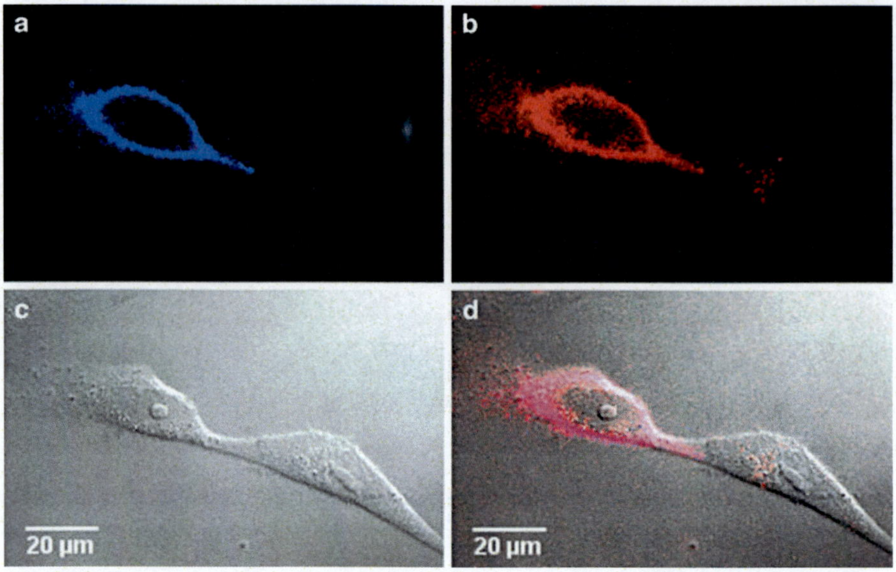

Fig. 6 Labeling of His-tagged IFNAR2 in live cells. HeLa cells transiently transfected with H10-IFNAR2 were labeled with 100 nM ^{FEW646}tris-NTA (**a**) for 5 min and imaged by confocal laser-scanning microscopy after thorough washing. Specific labeling of H10-IFNAR2 was confirmed by binding of a fluorescent ligand (**b**) to the same cells. (**c**) Transmission image showing that only one out of two cells was transfected. (**d**) Overlay of the images shown in panels **a**–**c**

If required, tris-NTA functionalized probes can be rapidly dissociated from His-tagged proteins on the cell surface by incubation of 100 mM imidazole for a few minutes [13]. For a neuronal cell line, however, substantial nonspecific binding of tris-NTA probes has been observed (Maxime Dahan, private communication). Specificity may be optimized by washing with lower imidazole concentrations (10–20 mM) as also done in affinity chromatography of His-tagged proteins.

A critical issue for multivalent NTA-based labeling is the strong loss of fluorescence quantum yield upon coordination of Ni(II) ions (by 70–80% for tris-NTA chromophores). This effect has been ascribed to the paramagnetism of Ni(II), which facilitates radiation-less deactivation from the excited state based on photoinduced electron transfer (cf. [17]). Since very close proximity is required for electron transfer, increasing the distance between the multivalent NTA and the chromophore by a spacer should drastically increase the quantum yield of such probes. With flexible oligoethylene-glycol linker, only a small increase in quantum yield was obtained [13], while with more rigid polyproline linkers substantially improved quantum yields were achieved [14]. However, the brightness of tris-NTA chromophore conjugates without additional spacers are still sufficient for probing protein–protein interactions on the single molecule level [18]. Reversible His-tag-specific labeling with tris-NTA functionalized fluorophores has recently been exploited for single molecule localization-based super-resolution imaging of cell surface receptors in living cells [19].

3.3 Labeling with Fluorescent Nanoparticles

Fluorescent nanoparticles such as semiconductor are powerful labels. The His-tag has also been successfully applied for targeting of fluorescent nanoparticles – e.g., quantum dots (QD) – to cell surface receptors. Since the inorganic materials used for QD synthesis involve transition metal ions with affinity toward the imidazole moiety (e.g., the ZnS shell of a CdSe/ZnS core/shell nanoparticle), His-tag-specific protein binding to QDs is readily achieved. This feature has been exploited for decorating the surface of QD with recombinant proteins [20–23]. Interestingly, this interaction cannot be reversed by imidazole. However, the specificity of this approach is not sufficient for selectively targeting His-tagged proteins on the cell surface, because several other amino acid side chains also interact with the ZnS surface, thus mediating nonspecific binding. Moreover, only very stable proteins do not unfold and lose their function upon direct adsorption on the inorganic surface of a quantum dot. For this reason, covering the QD surface with a protein-repellent coating followed by surface functionalization with NTA chelating transition metal ions is a more promising approach. Since multiple NTA moieties are readily incorporated on a nanoparticle surface with multiple functional groups, stable binding of His-tagged proteins by multivalent interaction is readily achieved. Indeed, efficient capturing of His-tagged proteins to such surface architectures has been reported [24, 25].

These approaches were demonstrated to be useful for capturing His-tagged protein in vitro and on the surface of living cells. For biophysical applications, e.g., tracking of individual membrane receptors, stable binding in a 1:1 ratio is required, which cannot be achieved by dense surface functionalization with NTA. Again, multivalent NTA can be used in order to achieve high stability and selectivity, and to control the stoichiometry. Commercially available QDs functionalized with streptavidin have been targeted to His-tagged proteins by using tris-NTA conjugated to biotin (^{BT}tris-NTA, Fig. 7) [26]. These QDs were successfully applied for labeling cell surface receptors in HeLa cells [26]. The average degree of particle functionalization is readily controlled by the relative concentration of ^{BT}-tris-NTA.

Direct surface coating of QDs and their functionalization with tris-NTA were achieved by using amphiphilic gallic acid derivatives depicted in Fig. 8. Hydrophobic QDs were solubilized in micelles containing tris-NTA functionalized and nonfunctionalized gallates [27]. Thus, the average number of tris-NTA functionalities per QD was controlled by the mixing ratio. Again, selective targeting of these tris-NTA functionalized QDs to His-tagged receptors at the cell surface was demonstrated, enabling for dual color single particle tracking in living cells [27]. Both these techniques allow to control the average number of tris-NTA groups on

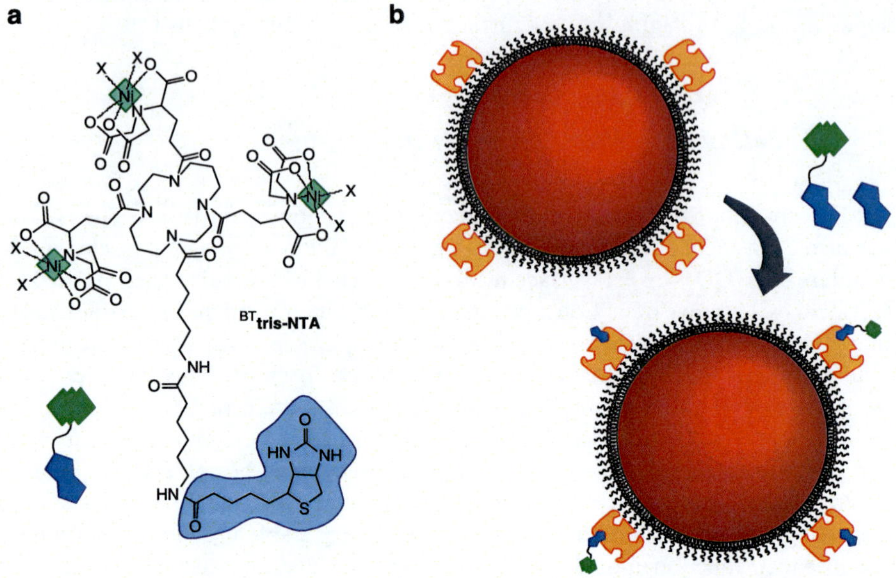

Fig. 7 Streptavidin/biotin interaction for functionalization of QD surfaces tris-NTA. (**a**) Structure of ^{BT}tris-NTA. (**b**) Commercially available streptavidin-QDs can be functionalized with tris-NTA by using ^{BT}tris-NTA. For lowering the functionalization degree, free biotin can be mixed with ^{BT}tris-NTA

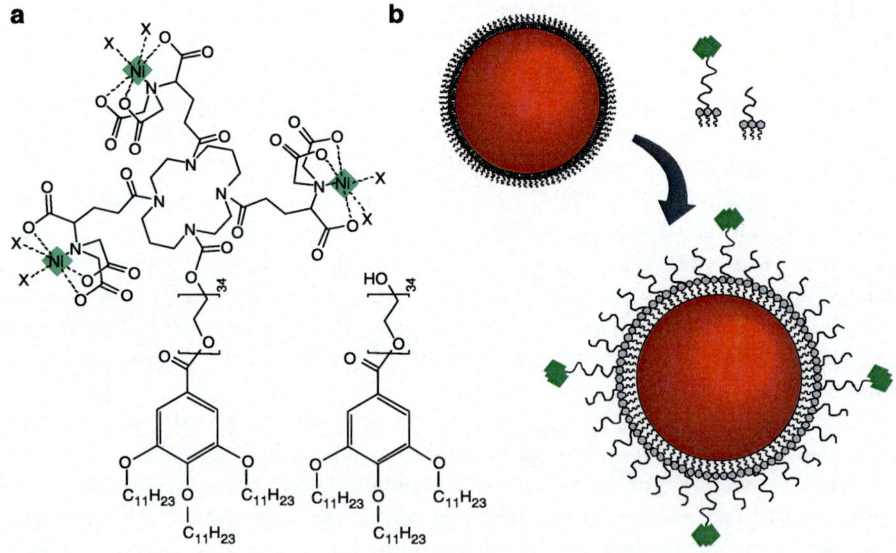

Fig. 8 QD surface functionalization using gallate –tris-NTA. (**a**) Structure of the amphiphilic gallate derivatives. (**b**) Tris-NTA-functionalized QDs obtained by encapsulation of QDs into gallate micelles

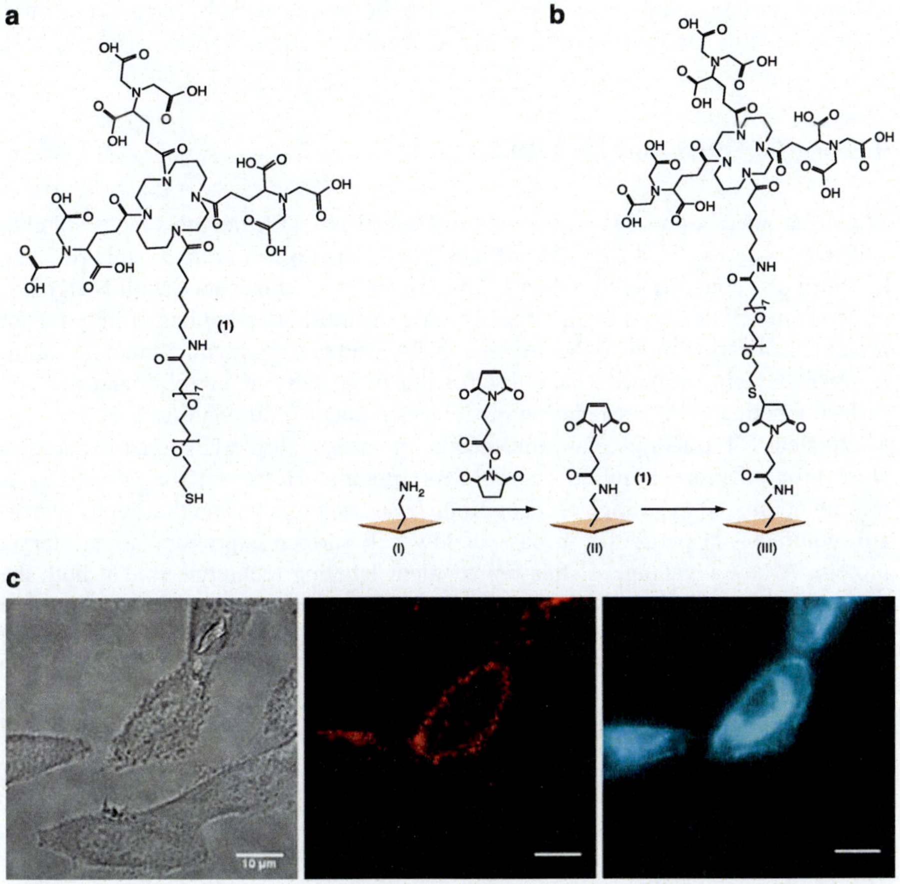

Fig. 9 Monofunctional tris-NTA QDs. **(a, b)** Surface modification of commercial quantum dots with tris-NTA thiol. **(c)** Labeling of HeLa cells transiently transfected with membrane-anchored H10-CFP (*cyan*) with QDs (*red*). Only transfected cells show staining by the QDs

QDs, yielding a statistic distribution around this average value. For avoiding experimental bias due to protein cross-linking on the cell surface, a single tris-NTA on each QD is required.

Monofunctionalization was achieved in a simple chemical functionalization approach depicted in Fig. 9 using commercially available, PEG-coated QDs [28]. This reaction by itself provides a statistic distribution as other coupling techniques. However, when this reaction was carried out at a very low ionic strength, monofunctionalization was observed, which was ascribed to electrostatic control: once a tris-NTA group was coupled to the QD surface, the strong negative charge of the first tris-NTA strongly reduces the probability for coupling of the second tris-NTA moiety. This approach probably depends on the charge of the nanoparticle before the coupling reaction (relatively low) and a kinetically controlled coupling reaction.

These monofunctional tris-NTA QDs have been successfully used for protein labeling in vitro and for reversible protein labeling on the cell surface (Fig. 9) [28].

4 Conclusions and Outlook

While the interaction of His-tags with individual Ni-NTA moieties lacks stability and selectivity required for efficient labeling of His-tagged proteins, this problem has been overcome by multivalent NTA. Tris-NTA groups loaded with Ni(II) ions perfectly match the hexahistdine-tag, which is the most often applied affinity tag for protein purification in vitro. Multivalent NTA bind His-tags with nanomolar affinity, which yields sufficient stability for purifying labeled proteins and for cell surface labeling. Yet, dissociation of the probe under mild conditions is possible. Multivalent NTA also increase the specificity since multiple Histidine residues in close proximity are required for efficient binding. However, the specificity is inferior compared to biological recognition based on proteins and thorough controls are required – in particular in case of low cell surface expression of the target protein. A key advantage of this noncovalent labeling technique is that both the recognized tag and the multivalent NTA are small molecules, and thus the chromophore is localized very close to the labeled protein. This is important not only for minimizing interfering with protein function, but also for analyzing interaction and conformational changes by FRET. Simple and rapid labeling in vitro and on the surface at low concentrations is possible, often not requiring the removal of excess dye. The application of this elegant labeling scheme, however, is limited by the sensitivity of Ni-NTA complexes to reducing conditions, chelating compounds (including amino acids at high concentrations) and bivalent metal ions. Moreover, Ni–Histidine complexes are not stable at a pH below 7 due to protonation of the coordinating imidazole nitrogen. Another critical issue is the loss in fluorescence quantum yield by the proximity of paramagnetic Ni(II) ions. For these reasons, combining rapid and selective targeting by the His-tag/Ni-NTA interaction with an irreversible covalent reaction (cf. [29]) could be a future solution for improving the performance of His-tag-specific labeling schemes.

References

1. Ueda EK, Gout PW, Morganti L (2003) Current and prospective applications of metal ion-protein binding. J Chromatogr A 988:1–23
2. Lata S, Reichel A, Brock R, Tampé R, Piehler J (2005) High-affinity adaptors for switchable recognition of histidine-tagged proteins. J Am Chem Soc 127:10205–10215
3. Guignet EG, Hovius R, Vogel H (2004) Reversible site-selective labeling of membrane proteins in live cells. Nat Biotechnol 22:440–444
4. Goldsmith CR, Jaworski J, Sheng M, Lippard SJ (2006) Selective labeling of extracellular proteins containing polyhistidine sequences by a fluorescein-nitrilotriacetic acid conjugate. J Am Chem Soc 128:418–419

5. Peneva K, Mihov G, Herrmann A, Zarrabi N, Borsch M, Duncan TM, Mullen K (2008) Exploiting the nitrilotriacetic acid moiety for biolabeling with ultrastable perylene dyes. J Am Chem Soc 130:5398–5399
6. Soh N, Seto D, Nakano K, Imato T (2006) Methodology of reversible protein labeling for ratiometric fluorescent measurement. Mol Biosyst 2:128–131
7. Guignet EG, Segura JM, Hovius R, Vogel H (2007) Repetitive reversible labeling of proteins at polyhistidine sequences for single-molecule imaging in live cells. Chemphyschem 8:1221–1227
8. Kapanidis AN, Ebright YW, Ebright RH (2001) Site-specific incorporation of fluorescent probes into protein: hexahistidine-tag-mediated fluorescent labeling with (Ni(2+):nitrilotriacetic Acid (n)-fluorochrome conjugates. J Am Chem Soc 123:12123–12125
9. Hauser CT, Tsien RY (2007) A hexahistidine-Zn2+-dye label reveals STIM1 surface exposure. Proc Natl Acad Sci USA 104:3693–3697
10. Spagnuolo C, Joselevich M, Leskow CF, Jares-Erijman EA (2010) Tetracysteine and bipartite tags for biarsenical organic fluorophores. In: Demchenko A (ed) Fluorescence reporters in chemistry and biology III: methods and applications. Springer Series on Fluorescence 10. Springer, Heidelberg, pp 263–295
11. Huang Z, Park JI, Watson DS, Hwang P, Szoka FC Jr (2006) Facile synthesis of multivalent nitrilotriacetic acid (NTA) and NTA conjugates for analytical and drug delivery applications. Bioconjug Chem 17:1592–1600
12. Lata S, Piehler J (2005) Stable and functional immobilization of histidine-tagged proteins via multivalent chelator head-groups on a molecular poly(ethylene glycol) brush. Anal Chem 77:1096–1105
13. Lata S, Gavutis M, Tampé R, Piehler J (2006) Specific and stable fluorescence labeling of histidine-tagged proteins for dissecting multi-protein complex formation. J Am Chem Soc 128:2365–2372
14. van der Does C, Presenti C, Schulze K, Dinkelaker S, Tampe R (2006) Kinetics of the ATP hydrolysis cycle of the nucleotide-binding domain of Mdl1 studied by a novel site-specific labeling technique. J Biol Chem 281:5694–5701
15. Strunk JJ, Gregor I, Becker Y, Li Z, Gavutis M, Jaks E, Lamken P, Walz T, Enderlein J, Piehler J (2008) Ligand binding induces a conformational change in ifnar1 that is propagated to its membrane-proximal domain. J Mol Biol 377:725–739
16. Wruss J, Pollheimer PD, Meindl I, Reichel A, Schulze K, Schofberger W, Piehler J, Tampe R, Blaas D, Gruber HJ (2009) Conformation of receptor adopted upon interaction with virus revealed by site-specific fluorescence quenchers and FRET analysis. J Am Chem Soc 131:5478–5482
17. Cheng-Chih Hsieh, Mei-Lin Ho, Pi-Tai Chou (2010) Organic dyes with excited-state transformations (electron, charge, and proton transfers). In: Demchenko A (ed) Advanced fluorescence reporters in chemistry and biology I: fundamentals and molecular design. Springer Series on Fluorescence 8. Springer, Heidelberg, pp 225–266
18. DeRocco VC, Anderson T, Piehler J, Erie DA, Weninger K (2010) Four-color single molecule fluorescence with noncovalent dye labeling to monitor dynamic multimolecular complexes. Biotechniques 49:807–816
19. Giannone G, Hosy E, Levet F, Constals A, Schulze K, Sobolevsky AI, Rosconi MP, Gouaux E, Tampe R, Choquet D et al (2010) Dynamic superresolution imaging of endogenous proteins on living cells at ultra-high density. Biophys J 99:1303–1310
20. Howarth M, Liu W, Puthenveetil S, Zheng Y, Marshall LF, Schmidt MM, Wittrup KD, Bawendi MG, Ting AY (2008) Monovalent, reduced-size quantum dots for imaging receptors on living cells. Nat Methods 5:397–399
21. Liu W, Howarth M, Greytak AB, Zheng Y, Nocera DG, Ting AY, Bawendi MG (2008) Compact biocompatible quantum dots functionalized for cellular imaging. J Am Chem Soc 130:1274–1284
22. Sapsford KE, Pons T, Medintz IL, Higashiya S, Brunel FM, Dawson PE, Mattoussi H (2007) Kinetics of metal-affinity driven self-assembly between proteins or peptides and CdSe-ZnS quantum dots. J Phys Chem C 111:11528–11538

23. Medintz IL, Clapp AR, Brunel FM, Tiefenbrunn T, Uyeda HT, Chang EL, Deschamps JR, Dawson PE, Mattoussi H (2006) Proteolytic activity monitored by fluorescence resonance energy transfer through quantum-dot-peptide conjugates. Nat Mater 5:581–589
24. Kim J, Park HY, Ryu J, Kwon do Y, Grailhe R, Song R (2008) Ni-nitrilotriacetic acid-modified quantum dots as a site-specific labeling agent of histidine-tagged proteins in live cells. Chem Commun 1910–1912
25. Gupta M, Caniard A, Touceda-Varela A, Campopiano DJ, Mareque-Rivas JC (2008) Nitrilotriacetic acid-derivatized quantum dots for simple purification and site-selective fluorescent labeling of active proteins in a single step. Bioconjug Chem 19:1964–1967
26. Reichel A, Schaible D, Al Furoukh N, Cohen M, Schreiber G, Piehler J (2007) Noncovalent, site-specific biotinylation of histidine-tagged proteins. Anal Chem 79:8590–8600
27. Roullier V, Clarke S, You C, Pinaud F, Gouzer GG, Schaible D, Marchi-Artzner V, Piehler J, Dahan M (2009) High-affinity labeling and tracking of individual histidine-tagged proteins in live cells using Ni2+ tris-nitrilotriacetic acid quantum dot conjugates. Nano Lett 9:1228–1234
28. You C, Wilmes S, Beutel O, Lochte S, Podoplelowa Y, Roder F, Richter C, Seine T, Schaible D, Uze G et al (2010) Self-controlled monofunctionalization of quantum dots for multiplexed protein tracking in live cells. Angew Chem Int Ed Engl 49:4108–4112
29. Plass T, Schultz C (2010) Covalent labeling of biomolecules in living cells. In: Demchenko A (ed) Fluorescence reporters in chemistry and biology III: methods and applications. Springer Series on Fluorescence 10. Springer, Heidelberg, pp 225–261

Part V
Tissue and Whole Body Imaging

In Vivo Imaging of Vascular Targets Using Near-Infrared Fluorescent Probes

Jan Klohs and Markus Rudin

Abstract The development of new optical imaging devices over the past few years has facilitated the translation of fluorescence imaging from the microscopic imaging of cellular processes to the macroscopic imaging of intact animals. Novel near-infrared fluorescent (NIRF) probes have been designed and synthesized for in vivo application. The probes come with different targeting moieties and possess different mechanism of contrast generation, determining their specific use. Herein, we present examples for the application of NIRF probes for imaging thrombus formation, expression of vascular and cell adhesion molecules, angiogenesis and proteolytic activity in animal models of cancer and cardiovascular diseases.

Keywords Fluorescent dyes · Near-infrared fluorescence imaging · Optical molecular imaging

Contents

J. Klohs (✉) and M. Rudin
Institute for Biomedical Engineering, University of Zurich and ETH Zurich, AIC – ETH HCI D426, Wolfgang-Pauli-Str. 10, 8093 Zürich, Switzerland
e-mail: klohs@biomed.ee.ethz.ch

A.P. Demchenko (ed.), *Advanced Fluorescence Reporters in Chemistry and Biology III: Applications in Sensing and Imaging*, Springer Ser Fluoresc (2011) 10: 313–328,
DOI 10.1007/978-3-642-18035-4_10,

1 Introduction

Cancer, diabetes, and cardiovascular diseases, such as atherosclerosis, myocardial infarction, and stroke account collectively for approximately two-thirds of all deaths in the Western world and cause enormous suffering, disability, and socio-economic costs [1]. Despite the remarkable progress that has been made in the last decades in understanding the pathologic processes underlying these diseases and in developing treatments to alleviate symptoms and improve quality of life in these patients, there is a high medical need for new, disease-modifying therapies. The development of such therapies critically depends on the availability of novel diagnostic approaches with improved sensitivity and specificity.

In the past few years, imaging technologies such as positron emission computed tomography, single-photon emission computed tomography, magnetic resonance imaging (MRI), ultrasound and near-infrared fluorescence (NIRF) imaging using specific probes have emerged which enable the imaging of physiological, metabolic and molecular function in experimental models and patients [2]. Due to the non-invasiveness of these methods, repetitive imaging over the course of the disease is feasible, enabling disease staging, assessing the role of a specific pathophysiological process, and monitoring the response to therapeutic intervention. Among the existing noninvasive imaging methods, NIRF imaging combines the advantage of high detection sensitivity, the use of nonionizing radiation with a demand for relatively simple and inexpensive instrumentation. Today, NIRF imaging has grown into an important tool in biomedical research and drug discovery [3–5].

While cancer and cardiovascular diseases are distinctively different diseases with complex disease etiologies, they share common pathophysiological processes. These include endothelial activation, infiltration of leukocytes, remodeling of the extracellular matrix with an increase of vascular permeability and angiogenesis. Most of these processes occur within or in the vicinity of blood vessels which renders the vasculature a suitable target for therapeutic drugs and imaging probes. In addition, the endothelium is the most accessible compartment for imaging probes. The probes need to be injected and delivery to targets distal to the vascular wall is often hampered by the presence of blood–tissue barriers. Given the potentially generic applicability of the probes and the target accessibility it is not surprising that major efforts have been invested in developing imaging probes for vascular targets. In this chapter, we give an overview on currently used NIRF imaging approaches and their application to study vascular pathophysiology in animal models of cancer and cardiovascular diseases.

1.1 The Use of NIR Light as Basis for In Vivo Fluorescence Imaging

In the NIR spectral domain, ranging from 700 to 900 nm, major absorbers in the tissue like hemoglobin, lipids, and water have their lowest absorption coefficient,

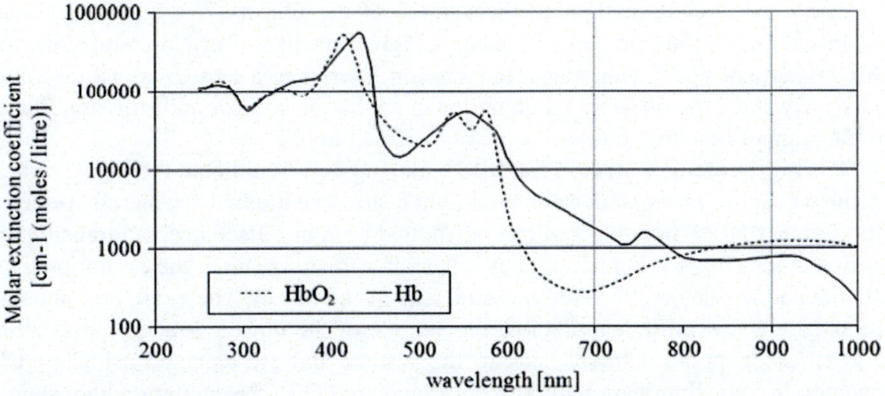

Fig. 1 Absorption coefficient of oxyhemoglobin and deoxyhemoglobin as a function of wavelength. The absorption coefficients of these endogenous absorbers are at the minimum between 700 and 900 nm

enabling photons to propagate deep into the tissue (Fig. 1). In addition, metabolites and structural components in animal tissue, such as elastin, collagen, tryptophan, NADH, porphyrins, lipofuscins, and flavins have high autofluorescence emission in the visible and midinfrared spectrum, but not in the NIR range [6]. This enhances the sensitivity for detecting exogenously administered fluorochromes against reduced background fluorescence. It has been shown that NIRF imaging can detect picomole amounts of NIR dyes noninvasively in mice [3, 7].

When applying NIRF imaging to living animals one faces various methodological constraints: The interaction of the photons with tissue components, such as membranes and cell organelles, leads to an attenuation of the photons by absorption and scattering. The absorption limits the ultimate depth photons can penetrate into the tissue. Depth penetration can range from several millimeters to 5 cm, depending on the optical properties of the tissue [3]. The scattering of photons increases exponentially with depth in tissue. The result is a reduction of the achievable image resolution. The overall photon attenuation increases nonlinearly with increasing depth of the fluorescent object. This nonlinear relationship hampers the ability to absolutely quantify the fluorochrome concentration. Therefore, the in vivo application of NIRF imaging relies on the technical ability to spatially resolve the fluorescence emission from a specific NIRF probe and requires complex image acquisition schemes and reconstruction algorithm for quantification of the signal.

1.2 NIRF Instrumentation and Techniques

A number of NIRF imaging devices with different set-ups and mode of data acquisition have become available. The standard NIRF imaging instrumentation consists of an excitation light source, most commonly a laser diode, and fiber optics to direct the excitation light to the surface of the object (Fig. 2). The emitted fluorescence photons are captured using a highly sensitive charged-coupled device (CCD) camera fitted with adequate emission filters and a lens to adjust the focal plane. By choosing different combination of excitation sources and emission filters, multichannel imaging in the same object is feasible [8].

The most common method for NIRF imaging is to illuminate the object with a plane wave, i.e., an expanded coherent light beam. The method is generally referred to as planar NIRF imaging and can be applied both in reflectance and transillumination mode (Fig. 3). In fluorescence reflectance imaging (FRI), the excitation light source and the detector are on the same side of the object. The excitation photons propagate a few millimeters below the surface of the object. Scattered excitation and emission photons travel back to the surface and are detected using a CCD camera. In transillumination fluorescence imaging (TFI), the excitation light source

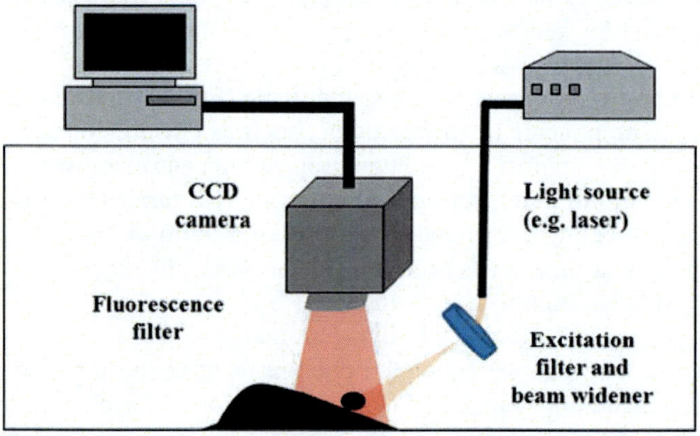

Fig. 2 NIRF imaging instrumentation. Typical planar NIRF imaging set-ups consist of an excitation light source (e.g., a laser diode) and fiber optics to direct the excitation light to the surface of the object. Fluorescence is detected using a CCD camera fitted with adequate emission filters

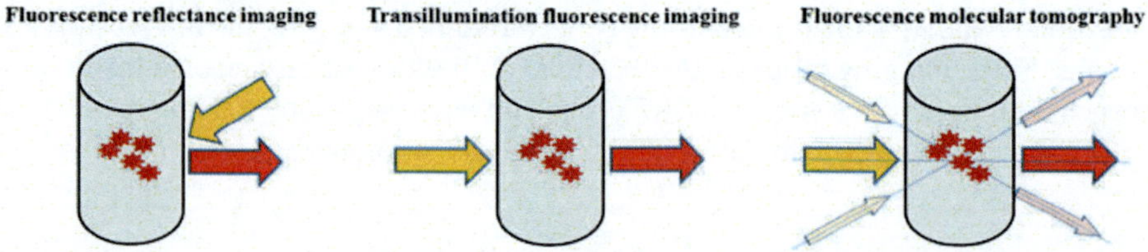

Fig. 3 Modes of data acquisition in NIRF imaging. In FRI the excitation light is applied and the fluorescence emission is collected from the same side of the object. In TFI the excitation has to travel through the object before it gets collected by the detector on the opposite site of the object. In FMT the object is sequentially illuminated at different locations. The fluorescence emission is collected by the detector and a three-dimensional image is reconstructed. Modified after [3], with permission

and the detector are on opposite sides. The advantage of TFI is that the entire volume of the object is sampled upon the passage of the excitation light. Furthermore, autofluorescence in the skin is reduced in TFI because the photon fluence (i.e., flux of photons per unit area) impinging on the surface of the animal is significantly lower compared to the FRI mode [9].

Planar NIRF imaging methods have the advantage that they require relatively inexpensive instrumentation, are easy to operate and facilitate high-throughput imaging of animals. However, planar NIRF imaging methods have some limitations: The detected fluorescence emission depends linearly on the fluorochrome concentration and nonlinearly on the depth of the fluorescent object and the optical properties of the surrounding tissue. As a consequence, planar NIRF images do not allow for absolute quantification of the detected fluorescence. Hence, in most studies the fluorescence emission is expressed in terms of a contrast, i.e., as a ratio of the fluorescence intensity measured over the disease-affected region to the intensity of measured over a nonaffected region, termed target-to-background ratio (TBR).

Quantitative analysis of fluorescence emission can be achieved using fluorescence molecular tomography (FMT). In FMT, the excitation light is coupled sequentially to the object at different locations, while the position of the detector array is fixed. In hybrid FMT/X-ray computerized tomography (CT) system both source and detector array are rotated using gantry [10]. Recording multiple projections yields a sufficient number of source–detector pair combinations for constructing a synthetic measurement and to apply to an inversion code for three-dimensional reconstruction of the fluorochrome distribution [11]. The reconstruction leads also to an improved image resolution of the order of 1 mm in all three dimensions as compared to planar NIRF techniques, where image resolution is about 2–3 mm with no information of depth [11].

1.3 Near-Infrared Fluorescent Probes

Today, a large number of fluorochromes emitting in the NIR including cyanine dyes, lanthanide chelates, squaraines, tetrapyrrole-based probes, quantum dots, and others are available for the generation of NIRF probes [12]. NIRF probes can be classified by their mechanism of contrast generation as nonspecific, targeted, and activatable probes.

Nonspecific fluorescent probes are usually NIR dyes which achieve contrast by distributing differentially in tissue based on the pharmacokinetic properties. This kind of probes can be used for imaging physiological processes such as vascular perfusion and for detecting changes in vascular permeability. Since there is always a fraction of circulating fluorescent probe, the degree of background fluorescence is in general high, leading to low TBRs.

Targeted NIRF probes consist of a NIR dye and a targeting moiety such as a peptide, antibody, or antibody fragment. Contrast is achieved when the probe binds to the target, e.g., a receptor, while the unbound probe fraction is cleared from the circulation. The higher the proportion is between the target-bound probe and unbound probe fraction, the higher the resulting TBRs.

Activatable probes change their biophysical properties upon interaction with the target, typically an enzyme. Most activatable probes are designed in such a way that the fluorescent dye molecules are in close proximity to each other, leading to dark quenching of the fluorescence. The fluorescence emission is recovered by enzyme-mediated cleavage, which liberates the dye molecules. Other activation principles such as addition reaction or molecular rearrangements are feasible. Since activatable probes have almost no fluorescence emission in the inactivated state, but generate a strong fluorescence signal upon activation, high TBRs can be achieved.

Regardless of the mechanism of contrast generation of the probe, factors such as the degree of target expression, the biodistribution and clearance kinetics of the probe, the ability of the probe to penetrate blood–tissue barriers and its metabolic stability are essential in determining the TBR that can be ultimately achieved in vivo.

2 Imaging Vascular Targets with Near-Infrared Fluorescent Probes

An increasing number of imaging probes, including probes for NIRF imaging, that probe molecular targets and processes at the vasculature have been developed (Table 1). They have been used to detect of thrombus formation and the activity of proteolytic enzymes, to assess the expression of vascular and cell adhesion molecules and the formation of new blood vessels during angiogenesis.

2.1 Thrombus Formation

Coagulation is an important process to restore vascular integrity and prevent excessive loss of blood when a blood vessel becomes damaged. The blood vessel wall is then covered by a clot containing platelets and fibrin to stop bleeding and to initiate repair of the damaged vessel. Coagulation can be impaired during inflammation and atherosclerosis which may lead to thrombus formation, ischemia, and tissue injury. Noninvasive assessment of this process might therefore be of considerable interest.

The coagulation cascade is complex and involves a number of mediators (i.e., coagulation factors), which can be visualized with NIRF probes. The first step in coagulation is the formation of a stable fibrin clot. The initiator of this process is the tissue factor protein, which is expressed in subendothelial tissue, platelets, and leukocytes. NIRF probes to visualize tissue factor accumulation, platelet

Table 1 Examples of imaging vascular targets in cancer and inflammatory diseases

Imaging application	Target	References
Thrombus formation		
Targeted	Tissue factor, fibrin-IIβ chain and CD41	[13]
Targeted	FXIIIa	[14–16]
Targeted	Thrombin	[17, 18]
Activatable	Glycoprotein IIb/IIIa	[19]
Adhesion molecules		
Targeted	VCAM-1	[20–22]
Targeted	E-selectin	[23]
Targeted	CD40	[24]
Targeted	$\alpha_V\beta_3$ Integrin	[25–31]
Targeted	α3 Integrin subunit	[32, 33]
Angiogenesis		
Targeted	ED-B fibronectin	[34–36]
Targeted	VEGF receptor	[37, 38]
Targeted	PECAM-1	[39]
Proteolytic activity		
Activatable	Cathepsin D	[40]
Targeted	Aminopeptidase N	[41]
Activatable	Cathepsin B	[42–44]
Activatable	Cathepsin K	[45]
Activatable	MMP	[22, 46–50]

deposition, and fibrin generation were developed by labeling antibodies against a tissue factor, fibrin-IIβ chain and CD41 receptor with Alexa dyes emitting at different wavelengths [13]. These processes were monitored in mice after laser-induced endothelial injury in real-time using intravital microscopy. In a different approach, an A15 probe peptide directed against fibrin stabilizing factor a (FXIIIa) was labeled with Alexa680 [14]. The use of the A15 probe for studying thrombus formation with intravital microscopy and FRI has been demonstrated in a mouse model of cerebral venous formation and intravascular thrombus formation [15, 16].

Thrombin is a serine protease that cleaves fibrinogen to form fibrin monomers that polymerize to form networks. Hence, thrombin plays a central role in thrombus formation. An activatable NIRF probe was developed by conjugation of a NIR fluorochrome to an oligopeptide substrate as a cleaving motif for thrombin [17]. The probe showed an 18-fold increase in fluorescence intensity upon in vitro activation with thrombin and detected acute and subacute thrombus formation in a mouse intravascular thrombosis model [18].

The glycoprotein IIb/IIIa receptor is expressed at the surface of activated thrombocytes, binds to fibrinogen and leads to platelet aggregation. The expression of the receptor has been monitored using a glycoprotein IIb/IIIa peptide ligand that was conjugated to the NIR dye Cy5.5 [19]. The probe showed a 16-fold increase in fluorescence intensity when incubated with clots generated from platelet-rich plasma. This study demonstrated the general potential of the probe for visualizing activated platelets during thrombosis.

Tumor cells also posses the capacity to directly activate the coagulation cascade by producing procoagulant factors and the methods described could be readily applied for tumor studies. Yet, to our knowledge, NIRF imaging of thrombus formation has not yet been described for mouse models of cancer.

2.2 Expression of Vascular Adhesion Molecules

A hallmark of most cardiovascular diseases is the local infiltration of leukocytes, which on the one hand contribute to tissue damage, e.g., in nascent atherosclerotic lesions and on the other hand are involved in repair and regeneration of injured tissue. Tumors usually promote an immunosuppressive environment and are devoid of leukocytes. The infiltration of leukocytes has been implicated to partake in the spontaneous regression in some form of cancers.

A common mechanism underlying leukocyte infiltration is the expression of adhesion molecules at the vasculature, which direct leukocytes to the site of inflammation. Vascular cell adhesion molecule 1 (VCAM-1) and E-selectin expression causes leukocytes to roll at the vascular wall and intercellular adhesion molecule 1 (ICAM-1) expression results in a firm adherence. Ultimately, the expression of platelet endothelial cell adhesion molecule 1 (PECAM-1) or CD31 leads to a transmigration of the leukocytes through the endothelial cells. Imaging the expression of the different adhesion molecules is therefore of interest when studying the processes of leukocyte infiltration in different diseases.

VCAM-1 adheres to very late antigen 4 (VLA4), a receptor expressed on the plasma membrane of activated leukocytes. A peptide was identified as VCAM-1 binder from a phage display screen [20]. A VCAM-1 targeted probe was created by conjugation of the peptide segment with a cross-linked iron oxide (CLIO) nanoparticle and the NIR dye Cy5.5 for detection with MRI and intravital microscopy. The probe was found to be internalized in VCAM-1 expressing cells in atherosclerotic lesions [20]. Similarly, the probe was used in a mouse model of atherosclerosis and in a model of aortic valve disease [21, 22].

E-selectin recognizes and binds to sialylated carbohydrates on the surface of certain leukocytes. A targeted probe was synthesized by conjugation of E-selectin binding peptide sequence to a CLIO nanoparticle and Cy5.5 [23]. E-selectin expression was visualized with FRI in a mouse xenograft model of Lewis lung carcinoma.

CD40, belonging to the tumor necrosis factor receptor superfamily, is found on the surface of endothelial cells, platelets and immune cells like B cells, monocytes, and dendritic cells. CD40 is mediating a variety of inflammatory and immune responses. An antibody against the CD40 receptor labeled with Cy5.5 was used to detect stroke-induced inflammation with FRI in a mouse model of cerebral ischemia [24]. Some of the fluorescence detected within the ischemic lesion was attributed to the presence of blood-derived cells which infiltrated the site of injury after binding and internalization of the Cy5.5-labeled antibody.

2.3 Expression of Cell Adhesion Molecules

Cell adhesion molecules are transmembrane proteins involved in the adhesion of cells to other cells or to the extracellular matrix. The family comprises calcium-dependent adhesion molecules (cadherins), selectins, and lymphocyte homing receptors. The cell adhesion molecules are located at the cell surface and hence constitute attractive targets for NIRF probes.

For example, the expression of the cell adhesion molecule $\alpha_V\beta_3$ integrin is involved in mediating migration of endothelial cells into the basement membrane and in regulating endothelial cell growth, survival and differentiation during angiogenesis. In cancer, $\alpha_V\beta_3$ integrin expression promotes the development, invasion and metastasis of the tumor cells. A variety of fluorescent peptides probes based on the arginine–glycine–aspartic acid (RGD) binding motif have been developed as ligands for targeting $\alpha_V\beta_3$ integrin expression in small animal models of cancer. The exact structure of the RGD peptide is determining the metabolic stability, the selectivity for a subtype of $\alpha_V\beta_3$ integrin, the affinity, and hence the TBR that can be achieved in vivo.

Several studies assessing $\alpha_V\beta_3$ integrin expression in tumors with different RGD-based NIRF probes have been reported. For example, monomeric RGD peptides were conjugated to Cy5.5 for NIRF imaging $\alpha_V\beta_3$ integrin expression in a glioblastoma xenograft and Kaposi's sarcoma [25, 26]. A RGD peptide labeled with IRDye800 and a disulfide bridged RGD peptide labeled with Cy5.5 were used for FRI and FMT of melanoma xenograft, respectively [27, 28]. A cyclic pentapeptide containing the RGD motif was conjugated with Cy5 for FRI of human embryonic kidney cells engrafted subcutaneously in mice [29]. Subcutaneously inoculated tumors were detected with FRI after either injecting a mono-, di-, and tetrameric RGD peptide conjugated with Cy7 [30]. The tetrameric peptide showed the highest accumulation in the tumor and gave the highest TBR. A probe for detecting $\alpha_V\beta_3$ integrin expression in BT-20 tumor with FRI, FMT, and MRI was created by conjugating the RGD peptide with CLIO nanoparticles and Cy5.5 dye [31].

NIRF probes with selectivity for the $\alpha3$ integrin subunit have been developed. A $\alpha3$ binding peptides were conjugated to Cy5.5 or Alexa680 as well as to biotin and were complexed with streptavidin-Cy5.5 for FRI of ovarian cancer xenografts [32]. Similarly, a cyclic peptide was biotinylated and complexed with streptavidin-Cy5.5 for FRI of subcutaneous and orthotopic U-87MG xenograft mouse model [33].

2.4 Angiogenesis

The formation of new functional blood vessels from pre-existing vessels is a physiological process. In cancer, angiogenesis is a part of tumor development. As the tumor outgrowths, oxygen and nutrients are supplied through molecular diffusion. When the tumor reaches a certain size it becomes hypoxic and initiates signaling programs that trigger the formation of new blood vessels. This so-called "angiogenic switch" is a key element in promoting tumor growth and is associated with high malignancy and poor prognosis. Angiogenesis is also involved in inflammatory processes occurring for example in certain cardiovascular diseases.

During angiogenesis, marker molecules such as integrins (see Sect. 2.3), growth factor receptors and G-protein coupled receptors are expressed de novo in the stromal compartment associated with newly formed blood vessels or on the endothelial cell membrane where they can be targeted for imaging.

Fibronectin is a glycoprotein that binds to integrins and extracellular matrix components such as collagen, fibrin, and proteoglycans. The protein is highly expressed on activated endothelial cells during angiogenesis. A NIRF probe against the extra domain-B (ED-B) of fibronectin was created by conjugation of a human antibody fragment against ED-B to the NIR dyes Cy5 and Cy7 for NIRF imaging [34]. It has been shown that these probes can detect neovasculature in different mouse models of angiogenesis, including cancer. In a related study, an ED-B fibronectin targeting minibody labeled with Cy5 and Cy7, with improved pharmacokinetic properties as compared to the full antibody, was used to visualize plaque angiogenesis in a mouse model of atherosclerosis. A high TBR of 18 was achieved in the atherosclerotic plaque allowing the identification of the molecular target with high sensitivity [35]. Similarly, a single chain antibody against the ED-B fibronectin was labeled with tetrasulfonated carbocyaninemaleimide for FRI of atheroscelotic plaques [36].

Vascular endothelial growth factor (VEGF) is a proangiogenic factor. The protein is involved in the mitogenesis and cell migration of endothelial cells. VEGF leads also to an increase in vascular permeability. A VEGF receptor targeted probe was produced by coupling the NIR dye Cy5.5 to single-chain VEGF [37]. FRI detected angiogenesis in tumor bearing mice injected with the conjugate (Fig. 4). The specificity of the approach was demonstrated by using an inactive form of VEGF, which showed no binding to the VEGF receptor, as control. With this approach it was demonstrated that tumor uptake was most likely due to

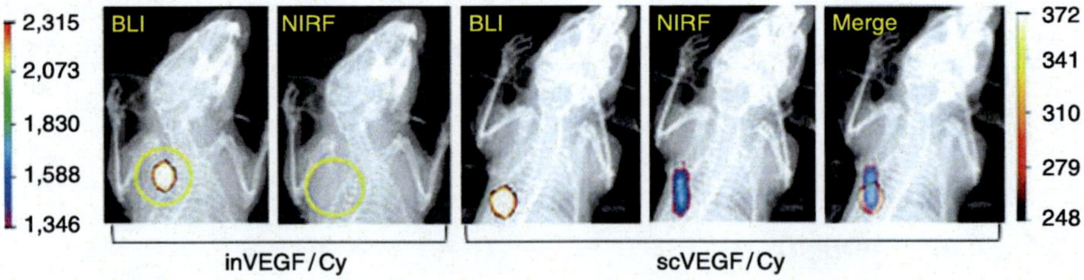

Fig. 4 Multimodel tumor imaging (reproduced from [37]; with permission). The tumor was stably transfected with luciferase for detection with bioluminescence imaging. Mice were mice injected with Cy5.5 labeled VEGF (scVEGF) for FRI imaging of angiogenesis. Inactive VEGF labeled with Cy5.5 (inVEGF) was used as a control for the specificity of the approach. The FRI system was also equipped with X-ray for visualization of the mouse anatomy

specific binding to the VEGF receptor and not due to extravasation from leaky tumor vessels. Alternatively, VEGF receptor expression was demonstrated using a monoclonal antibody against the VEGF receptor 2 labeled with the IRdye 800CW. FRI detected NIRF signals in tumor xenografts in mice [38].

Inhibiting angiogenesis as treatment strategy in cancer is a widely studied. An attractive application of NIRF imaging is the evaluation of therapy efficacy of an antiangiogenic drug by directly labeling the therapeutic compound with a NIR dye. For example, the angiogenesis inhibitor Endostatin was labeled with Cy5.5 for NIRF imaging of mice bearing Lewis lung carcinomas [39]. The probe was found to colocalize with its target receptor PECAM-1, indicating that in fact the drug is interacting with this vascular adhesion molecule inhibiting and its endogenous ligands.

2.5 *Proteolytic Activity*

Proteases such as cathepsins and matrix metalloproteinases (MMP) cleave peptides and proteins and take part in a multitude of physiological reactions. They are capable of degrading extracellular matrix proteins and tissue remodeling. Proteases are involved in the disruption of blood–tissue barriers, angiogenesis, tumor invasion and metastasis, regeneration and wound repair, processes which are implicated in the pathophysiology of cancer and in many cardiovascular diseases.

The peptide cleavage function of proteases has led to the development of so-called activatable NIRF probes, which produce a substantial fluorescent signal when the peptide substrate is cleaved. The detected fluorescent signal is not only indicative for the presence of a particular protease but also for its particular degree of activity. And because a single protease can process multiple substrate molecules (imaging probes) over time the NIRF signal can be amplified.

The principle of activatable NIRF probes targeting proteases was first demonstrated in mouse models of carcinomas [51]. In a follow-up study, a NIRF probe that is specifically activated by cathepsin D was developed and tested in mice bearing tumors with and without having cathepsin D activity [40]. FRI detected significantly higher NIRF signal in cathepsin D positive tumor, demonstrating the specificity of the probe. Similarly, a NIRF probe activated by cathepsin B was evaluated in a mouse model of human breast cancer [42]. MMP specific probes have been used for NIRF imaging of mice bearing human fibrosarcoma [46, 47]. FRI did not detect fluorescence emission in mice bearing mammary adenocarcinoma which have little specific MMP activity [47].

Similarly, a peptide containing the asparagine-glycine-arginine-motive labeled with Cy5.5 was used to target CD13/aminopeptidase N expression, a cell surface metalloproteinase, which is associated with tumor malignancy [41]. The probe was detected in tumor xenografts both with FRI and FMT.

While initial characterization of protease probes was achieved in tumor models, they soon found application in other disease conditions involving inflammation.

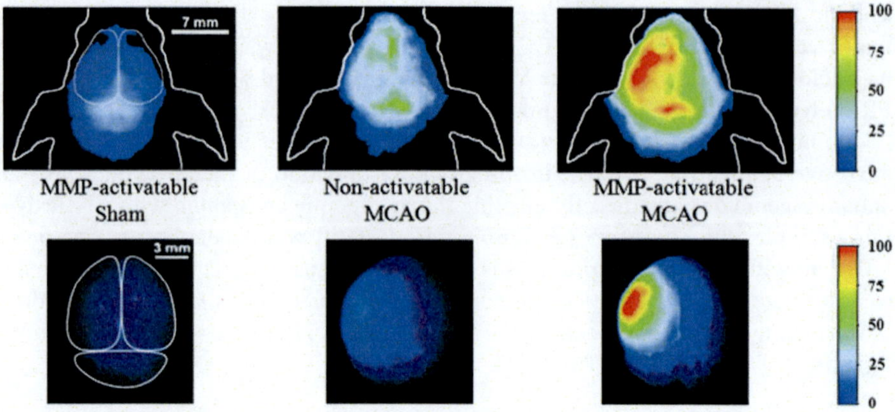

Fig. 5 NIRF imaging of MMP activity after cerebral ischemia (reproduced from [50]; with permission). FRI was performed 24 h after induction of cerebral ischemia in vivo and after removal of the brain ex vivo. While no differences between the hemispheres were in sham-operated mice injected with the MMP-activatable probe, slightly higher fluorescence intensities over the ischemic hemisphere were detected in mice after cerebral ischemia injected with the nonactivatable probe as control and strong fluorescence was seen in mice injected with the MMP-activatable probe

For example, a cathepsin B specific NIRF probe has been used in a mouse model of myocardial infarction and atherosclerosis [43, 44]. FMT localized the NIRF signal in the atherosclerotic aorta and microscopy demonstrated colocalization of the NIRF signal with the lyosomal cysteine protease cathepsin B [44]. Alternatively, a cathepsin K specific NIRF probe has been used for ex vivo imaging of carotid plaques with a laser scanning microscope [45]. Also, MMP specific NIRF probes were used in mouse models of cardiovascular diseases. FRI detected fluorescence emission due to MMP activity in a mouse model of myocardial infarction and aortic valve disease [22, 48]. FRI and FMT were used to detect NIRF signal in the aortic root, arch and thoracic aorta of ApoE$^{-/-}$ mice on a high-cholesterol diet [49].

Finally, MMP activity was detected with FRI in the brain of mice after focal cerebral ischemia [50]. NIRF was confined to the ischemic hemisphere and was significantly higher than in mice that received a nonactivatable NIRF probe as control (Fig. 5). Since MMP activity in the brain leads to an impairment of blood–brain barrier (BBB) function this can be visualized with unspecific NIRF probes. Albumin was labeled with Cy5.5 and an indotricarbocyanine dye to serve as NIRF probes [52, 53]. Extravasation of conjugates, which under physiological conditions do not cross the BBB, was observed to occur after focal cerebral ischemia.

3 Summary and Conclusion

Recent progress in the development of optical instrumentation and reconstruction algorithms together with the design and synthesis of fluorescent probes that absorb and emit in the NIR spectral domain has enabled researchers to study physiological, metabolic, and molecular function in intact, living animals. The examples presented here demonstrated the successful application of NIRF probes to assess relevant pathophysiological processes at the vasculature. These comprise the visualization of thrombus formation, endothelial activation, infiltration of leukocytes, proteolytic activity, and the formation of new blood vessels. Since these processes are crucially involved in cancer and cardiovascular diseases, NIRF imaging with specific probes can help to give new insight into the role of these processes during the course of the diseases and to develop specific therapies.

Compared to other existing noninvasive imaging modalities NIRF imaging offers several advantages. The techniques share with SPECT and PET the high detection sensitivity in vivo, but the fluorochromes are stable tracer and nonionizing radiation is used for probe detection. Of all available imaging modalities, NIRF imaging requires relatively simple and inexpensive instrumentation and the ease of use enables high-throughput imaging of animals. An added advantage of using NIRF probes is that microscopic techniques can be used to track the distribution of the injected probes ex vivo with high spatial resolution, thus enabling to evaluate probe specificity. However, the application of NIRF probes in vivo also possesses several limitations. Unlike in MRI, PET, and SPECT imaging, the ultimate penetration depth of light is limited. While the application of NIRF imaging is most often not restricted in small animals, the use of the technique in humans is confined to applications on tissue surfaces, for studies of low absorbing tissue such as the female breast, and to use in intraoperative systems [54, 55]. The nonlinear dependence of fluorescence signal and depth makes quantification technically difficult, in particular for planar NIRF imaging. In addition, the scattering of the light limits the achievable resolution of the method. Although further research effort is needed before fluorescent probes reach a significant stage in clinical practice, it is expected that NIRF imaging will continue to play a pivotal role in studying animal models of human disease and for the preclinical development of drugs.

References

1. Anderson RN, Smith BL (2003) Deaths: leading causes for 2001. Natl Vital Stat Rep 52 (9):1–85
2. Rudin M, Weissleder R (2003) Molecular imaging in drug discovery and development. Nat Rev Drug Discov 2(2):123–131
3. Ntziachristos V, Ripoll J, Wang LV, Weissleder R (2005) Looking and listening to light: the evolution of whole-body photonic imaging. Nat Biotechnol 23(3):313–320
4. Klohs J, Wunder A, Licha K (2008) Near-infrared fluorescent probes for imaging vascular pathophysiology. Basic Res Cardiol 103(2):144–151
5. Shah K, Jacobs A, Breakefield XO, Weissleder R (2004) Molecular imaging of gene therapy for cancer. Gene Ther 11(15):1175–1187

6. Billinton N, Knight AW (2001) Seeing the wood through the trees: a review of techniques for distinguishing green fluorescent protein from endogenous autofluorescence. Anal Biochem 291(2):175–197
7. Klohs J, Steinbrink J, Nierhaus T, Bourayou R, Lindauer U, Bahmani P, Dirnagl U, Wunder A (2006) Noninvasive near-infrared imaging of fluorochromes within the brain of live mice: an in vivo phantom study. Mol Imaging 5(3):180–187
8. Kobayashi H, Koyama Y, Barrett T, Hama Y, Regino CA, Shin IS, Jang BS, Le N, Paik CH, Choyke PL, Urano Y (2007) Multimodal nanoprobes for radionuclide and five-color near-infrared optical lymphatic imaging. ASC Nano 1(4):258–264
9. Ntziachristos V, Turner G, Dunham J, Windsor S, Soubret A, Ripoll J, Shih HA (2005) Planar fluorescence imaging using normalized data. J Biomed Opt 10(6):64007
10. Schulz RB, Ale A, Sarantopoulos A, Freyer M, Soehngen E, Zientkowska M, Ntziachristos V (2010) Hybrid system for simultaneous fluorescence and X-ray computed tomography. IEEE Trans Med Imaging 29(2):465–473
11. Ntziachristos V, Tung CH, Bremer C, Weissleder R (2002) Fluorescence molecular tomography resolves protease activity in vivo. Nat Med 8(7):757–760
12. Patsenker LD, Tatarets AL, Terpetschnig EA (2010) Long-wavelength probes and labels based on cyanines and squaraines. In: Demchenko A (ed) Advanced fluorescence reporters in chemistry and biology I: fundamentals and molecular design. Springer Series on Fluorescence 8. Springer, Heidelberg, pp 65–104
13. Falati S, Gross P, Merrill-Skoloff G, Furie BC, Furie B (2002) Real-time in vivo imaging of platelets, tissue factor and fibrin during arterial thrombus formation in the mouse. Nat Med 8 (10):1175–1181
14. Tung CH, Ho NH, Zeng Q, Tang Y, Jaffer FA, Reed GL, Weissleder R (2003) Novel factor XIII probes for blood coagulation imaging. Chembiochem 4(9):897–899
15. Kim DE, Schellingerhout D, Jaffer FA, Weissleder R, Tung CH (2005) Near-infrared fluorescent imaging of cerebral thrombi and blood-brain barrier disruption in a mouse model of cerebral venous sinus thrombosis. J Cereb Blood Flow Metab 25(2):226–233
16. Jaffer FA, Tung CH, Wykrzykowska JJ, Ho NH, Houng AK, Reed GL, Weissleder R (2004) Molecular imaging of factor XIIIa activity in thrombosis using a novel, near-infrared fluorescent contrast agent that covalently links to thrombi. Circulation 110(2):170–176
17. Tung CH, Gerszten RE, Jaffer FA, Weissleder R (2002) A novel near-infrared fluorescence sensor for detection of thrombin activation in the blood. Chembiochem 3(2–3): 207–211
18. Jaffer FA, Tung CH, Gerszten RE, Weissleder R (2002) In vivo imaging of thrombin activity in experimental thrombi with thrombin-sensitive near-infrared molecular probe. Arterioscler Thromb Vasc Biol 22(11):1929–1935
19. Tung CH, Quinti L, Jaffer FA, Weissleder R (2005) A branched fluorescent peptide probe for imaging of activated platelets. Mol Pharm 2(1):92–95
20. Kelly KA, Allport JR, Tsourkas A, Shinde-Patil VR, Josephson L, Weissleder R (2005) Detection of vascular adhesion molecule-1 expression using a novel multimodal nanoparticle. Circ Res 96(3):327–336
21. Nahrendorf M, Jaffer FA, Kelly KA, Sosnovik DE, Aikawa E, Libby P, Weissleder R (2006) Noninvasive vascular cell adhesion molecule-1 imaging identifies inflammatory activation of cells in atherosclerosis. Circulation 114(14):1504–1511
22. Aikawa E, Nahrendorf M, Sosnovik D, Lok VM, Jaffer FA, Aikawa M, Weissleder R (2007) Multimodality molecular imaging identifies proteolytic and osteogenic activities in early aortic valve disease. Circulation 115(3):377–386
23. Funovics M, Montet X, Reynolds F, Weissleder R, Josephson L (2005) Nanoparticles for the optical imaging of tumor E-selectin. Neoplasia 7(10):904–911
24. Klohs J, Gräfe M, Graf K, Steinbrink J, Dietrich T, Stibenz D, Bahmani P, Kronenberg G, Harms C, Endres M, Lindauer U, Greger K, Stelzer EH, Dirnagl U, Wunder A (2008) In vivo

imaging of the inflammatory receptor CD40 after cerebral ischemia using a fluorescent antibody. Stroke 39(10):2845–2852
25. Chen X, Conti PS, Moats RA (2004) In vivo near-infrared fluorescence imaging of integrin alphavbeta3 in brain tumor xenografts. Cancer Res 64(21):8009–8014
26. Wang W, Ke S, Wu Q, Charnsangavej C, Gurfinkel M, Gelovani JG, Abbruzzese JL, Sevick-Muraca EM, Li C (2004) Near-infrared optical imaging of integrin alphavbeta3 in human tumor xenografts. Mol Imaging 3(4):343–51
27. Kwon S, Ke S, Houston JP, Wang W, Wu Q, Li C, Sevick-Muraca EM (2005) Imaging dose-dependent pharmacokinetics of an RGD-fluorescent dye conjugate targeted to alpha v beta 3 receptor expressed in Kaposi's sarcoma. Mol Imaging 4(2):75–87
28. Von Wallbrunn A, Höltke C, Zühlsdorf M, Heindel W, Schäfers M, Bremer C (2007) In vivo imaging of integrin alpha v beta 3 expression using fluorescence-mediated tomography. Eur J Nucl Med Mol Imaging 34(5):745–754
29. Jin ZH, Josserand V, Foillard S, Boturyn D, Dumy P, Favrot MC, Coll JL (2007) In vivo optical imaging of alphaV-beta3 in mice using multivalent or monovalent cRGD targeting vectors. Mol Cancer 6:41
30. Wu Y, Cai W, Chen X (2006) Near-infrared fluorescence imaging of tumor integrin alpha v beta 3 expression with Cy7-labeled RGD multimers. Mol Imaging Biol 8(4):226–236
31. Montet X, Montet-Abou K, Reynolds F, Weissleder R, Josephson L (2006) Nanoparticle imaging of integrins on tumor cells. Neoplasia 8(3):214–222
32. Aina OH, Marik J, Gandour-Edwards R, Lam KS (2005) Near-infrared optical imaging of ovarian cancer xenografts with novel alpha 3-integrin binding peptide "OA02". Mol Imaging 4(4):439–447
33. Xiao W, Yao N, Peng L, Liu R, Lam KS (2009) Near-infrared optical imaging in glioblastoma xenograft with ligand-targeting alpha 3 integrin. Eur J Nucl Med Mol Imaging 36(1):94–103
34. Birchler M, Neri G, Tarli L, Halin C, Viti F, Neri D (1999) Infrared photodetection for the in vivo localization of phage-derived antibodies directed against angiogenic marker. J Immunol Methods 231(1–2):239–248
35. Matter CM, Schuler PK, Alessi P, Meier P, Ricci R, Zhang D, Halin C, Castellani P, Zardi L, Hofer CK, Montani M, Neri D, Lüscher TF (2004) Molecular imaging of atherosclerotic plaques using a human antibody against the extra-domain B of fibronectin. Circ Res 95 (12):1225–1233
36. Dietrich T, Perlitz C, Licha K, Stawowy P, Atrott K, Tachezy M, Meyborg H, Stocker C, Gräfe M, Fleck E, Schirner M, Graf K (2007) ED-B fibronectin (ED-B) can be targeted using a novel single chain antibody conjugate and is associated with macrophage accumulation in atherosclerotic lesions. Basic Res Cardiol 102(4):298–307
37. Backer MV, Levashova Z, Patel V, Jehning BT, Claffey K, Blankenberg FC, Backer JM (2007) Molecular imaging of VEGF receptors in angiogenic vasculature with single-chain VEGF-based probes. Nat Med 13(4):504–509
38. Virostko J, Xie J, Hallahan DE, Arteaga CL, Gore JC, Manning HC (2009) A molecular imaging paradigm to rapidly profile response to angiogenesis-directed therapy in small animals. Mol Imaging Biol 11(3):204–212
39. Citrin D, Lee AK, Scott T, Sproull M, Menard C, Tofilon PJ, Camphausen K (2004) In vivo tumor imaging in mice with near-infrared fluorescence endostatin. Mol Cancer Ther 3(4): 481–488
40. Tung CH, Mahmood U, Bredow S, Weissleder R (2000) In vivo imaging of proteolytic enzyme activity using a novel molecular reporter. Cancer Res 60(17):4953–4958
41. Von Wallbrunn A, Waldeck J, Höltke C, Zühlsdorf M, Mesters R, Heindel W, Schäfers M, Bremer C (2008) In vivo optical imaging of CD13/APN-expression in tumor xenografts. J Biomed Opt 13(1):011007
42. Bremer C, Tung CH, Bogdanov A Jr, Weissleder R (2002) Imaging of differential protease expression in breast cancers for detection of aggressive tumor phenotypes. Radiology 222 (3):814–818

43. Nahrendorf M, Sosnovik DE, Waterman P, Swirski FK, Pande AN, Aikawa E, Figueiredo JL, Pittet MP, Weissleder R (2007) Dual channel optical tomographic imaging of leukocyte recruitment and protease activity in the healing myocardial infarct. Circ Res 100(8): 1218–1225
44. Chen J, Tung CH, Mahmood U, Ntziachristos V, Gyurko R, Fishman MC, Huang PL, Weissleder R (2002) In vivo imaging of proteolytic activity in atherosclerosis. Circulation 105(23):2766–2771
45. Jaffer FA, Kim DE, Quinti L, Tung CH, Aikawa E, Pande AN, Kohler RH, Shi GP, Libby P, Weissleder R (2007) Optical visualization of cathepsin K activity in atherosclerosis with a novel protease-activatable fluorescence sensor. Circulation 115(17):2292–2298
46. Bremer C, Tung CH, Weissleder R (2001) In vivo molecular target assessment of matrix metalloproteinase inhibition. Nat Med 7(6):743–748
47. Bremer C, Bredow S, Mahmood U, Weissleder R, Tung CH (2001) Optical imaging of matrix metalloproteinase-2 activity in tumors: feasibility study in a mouse model. Radiology 221 (2):523–529
48. Chen J, Tung CH, Allport JR, Chen S, Weissleder R, Huang PL (2005) Near-infrared fluorescent imaging of matrix metalloproteinase activity after myocardial infarction. Circulation 111(14):1800–1805
49. Deguchi JO, Aikawa M, Tung CH, Aikawa E, Kim DE, Ntziachristos V, Weissleder R, Libby P (2006) Inflammation in atherosclerosis: visualizing matrix metalloproteinase action in macrophages in vivo. Circulation 114(1):55–62
50. Klohs J, Baeva N, Steinbrink J, Bourayou R, Boettcher C, Royl G, Megow D, Dirnagl U, Priller J, Wunder A (2009) In vivo near-infrared fluorescence imaging of matrix metalloproteinase activity after cerebral ischemia. J Cereb Blood Flow Metab 29(7):1284–1292
51. Weissleder R, Tung CH, Mahmood U, Bogdanov A Jr (1999) In vivo imaging of tumors with protease-activated near-infrared fluorescent probes. Nat Biotechnol 17(4):375–378
52. Abulrob A, Brunette E, Slinn J, Baumann E, Stanimirovic D (2008) Dynamic analysis of the blood-brain barrier disruption in experimental stroke using time domain in vivo fluorescence imaging. Mol Imaging 7(6):248–62
53. Klohs J, Steinbrink J, Bourayou R, Mueller S, Cordell R, Licha K, Schirner M, Dirnagl U, Lindauer U, Wunder A (2009) Near-infrared fluorescence imaging with fluorescently labeled albumin: a novel method for non-invasive optical imaging of blood-brain barrier impairment after focal cerebral ischemia in mice. J Neurosci Methods 180(1):126–132
54. Ntziachristos V, Yodh AG, Schnall M, Chance B (2000) Concurrent MRI and diffuse optical tomography of breast after indocyanine green enhancement. Proc Natl Acad Sci 97(6): 2767–72
55. Aoki T, Yasuda D, Shimizu Y, Odaira M, Niiya T, Kusano T, Mitamura K, Hayashi K, Murai N, Koizumi T, Kato H, Enami Y, Miwa M, Kusano M (2008) Image-guided liver mapping using fluorescence navigation system with indocyanine green for anatomical hepatic resection. World J Surg 32(8):1762–1767

Whole-Body Imaging of Hematopoietic and Cancer Cells Using Near-Infrared Probes

Vyacheslav Kalchenko, Michal Neeman, and Alon Harmelin

Abstract Noninvasive in vivo monitoring of cell behavior presents a major challenge to imaging researchers. Cells labeled with fluorescent probes or expressing bioluminescent reporter genes offer opportunities for in vivo tracking. However, it is important to verify that labeling remains stable once it is in the cell, does not impair cell phenotype or influence engrafting behavior. Herein, we provide a brief outline of current approaches to whole-body fluorescence imaging of cancer and hematopoietic cells by means of near-infrared (NIR) fluorescence-emitting agents. Cell-labeling approaches will be reviewed, with particular focus on the use of NIR lipophilic membrane dyes, nontargeted NIR cyanine-based fluorescent markers, NIR esterase-activatable fluorescent probes, and NIR Quantum Dots. Each approach features both advantages and limitations, typically related to ease of use, sensitivity, specificity, toxicity or stability. The beneficial aspects of NIR imaging are maximized when used in conjunction of complementary imaging modalities. Thus, the final section of this chapter is devoted to the use of NIR imaging in the context of multimodal imaging.

Keywords Cancer cells · Hematopoietic cells · Near-infrared fluorescence imaging · Whole-body optical imaging

Contents

V. Kalchenko and A. Harmelin (✉)
Department of Veterinary Resources, The Weizmann Institute of Science, Rehovot 76100, Israel
e-mail: Alon.harmelin@weizmann.ac.il

M. Neeman
Department of Biological Regulation, The Weizmann Institute of Science, Rehovot 76100, Israel

A.P. Demchenko (ed.), *Advanced Fluorescence Reporters in Chemistry and Biology III: Applications in Sensing and Imaging*, Springer Ser Fluoresc (2011) 10: 329–346,
DOI 10.1007/978-3-642-18035-4_11,

1 Introduction

Imaging technologies have revolutionized medicine by significantly improving noninvasive diagnosis, monitoring of disease progression, and the effectiveness of therapy. In basic and preclinical research, noninvasive imaging systems provide dynamic and multiparametric access to complex anatomical, physiological, and molecular information. Such technologies offer important tools for investigating the biological basis of diseases and facilitating the development of novel therapeutic approaches. The ability to simultaneously follow multiple cell lineages, analyze their patterns of migration, determine their sites and rates of proliferation and subsequently monitor their differentiation and function at the terminal site, have significantly broadened scientific research capacities.

More specifically, changes in tissue organization and organ growth can be derived from anatomical imaging protocols. Functional imaging provides real time information on hemodynamics, vascular permeability, muscle contraction, cardiac function, and cognitive brain activity. Molecular imaging allows for in vivo detection of gene or cell surface marker expression monitoring and analysis of enzymatic activity. Cellular imaging methods allow for tracking of cell migration, differentiation, proliferation, and survival within the living organism. Tight monitoring of cell fate is critical during fetal growth, pathological processes, and therapeutic processes, such as tissue repair.

The ever-growing demand for temporal evaluation of cell distribution, pharmacokinetics and pharmacodynamics of administered agents, progression of infectious agents, and expression of cell surface antigens, with simultaneous recording of anatomical, hemodynamic, or other tissue function parameter changes has outlined the need for multifaceted imaging techniques. Integration of diverse imaging tools and coregistration of their output can provide imaging systems with unprecedented capacities. To date, a number of imaging approaches concurrently incorporate various instruments for whole-body in vivo cell imaging. In most cases, cell-labeling contrast materials, each designed to be detected by a unique imaging technique, are simultaneously applied to allow for detection and integration of multiple signals from a single live sample. Cell-labeling methods enabling cell visualization in vivo include introduction of exogenous contrast media (fluorescent dyes or nanoparticles) or use of reporter genes (fluorescent protein such as GFP/RFP or bioluminescent enzyme/luciferin).

In this chapter we will provide an overview of the state-of-the-art of whole-body near-infrared (NIR) fluorescence-based imaging techniques applied toward in vivo monitoring of cell fate.

2 Levels of Spatial Resolution for In Vivo Fluorescent Cellular Imaging

Optical imaging provides a unique opportunity to observe cellular behavior within the intact organism at different levels of resolution. Cellular behavior could be observed both on the level of single cell or at the level of groups of cells. In order to achieve different scales such as microscopic magnification or panoramic view different approaches should be chosen. For example, intravital microscopy (IM) and especially fluorescent intravital video microscopy (fIVM) is a technique that enables visualizing single cell or small groups of cells inside live animals at high spatial and temporal resolution (Fig. 1). This technique requires specific adaptations of the microscope so as to optimize it for in vivo imaging applications [1]. The major limitation of fIVM is the small depth of penetration (up to 1 mm when using two-photon microscopy) [2]. IM systems are optimized for high resolution imaging of superficial cells, but have limited capabilities for panoramic imaging of relatively large objects such as live experimental animal. To overcome those limitations whole-body fluorescence imaging applies sensitive

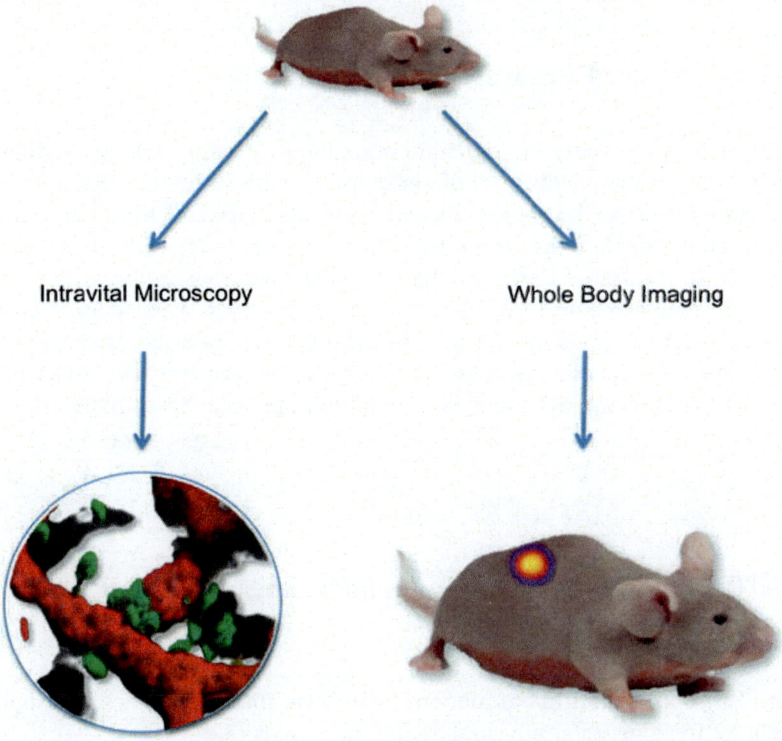

Fig. 1 Two levels of fluorescent cellular imaging (Intravital Microscopy and Whole-Body Optical Imaging)

cameras, without microscopic magnification, allowing capturing images from larger objects ranging from millimeters up to ten or more centimeters [3]. Combined macroscopic imaging systems adapted with an in vivo microscope are available and provide a continuum of spatial resolution with a reciprocal shift in depth of penetration [4].

Translation from superficial intravital microscopy to whole organism (whole body) raises various challenges including the need to overcome fluorescent light absorption, scattering and higher auto-fluorescence. Fluorescent probes that provide good images for superficial intravital microscopy imaging are frequently not optimal for whole-body imaging. For example, the most popular fluorescent probes, dyes, and fluorescent reporter genes emit in the visual spectrum of 400–650 nm. At this visible light range, the emitted fluorescence is highly absorbed by the endogenous body chromophores [5]. Increasing sensitivity, specificity, three-dimensional reconstruction, and spatial resolution are the major challenges for whole-body optical imaging. Imaging cellular behavior using whole-body optical imaging, and particularly visualizing signal from labeled cells located in deep tissue, requires combination of state-of-the-art instruments with advanced cell-labeling technologies.

3 Near-Infrared Fluorescence Imaging

Hemoglobin, the primary biological chromophore, strongly absorbs visible light, thereby limiting the penetration of short wavelength light to a few millimeters. Other components of biological tissues, such as water and lipids, are optically transparent at wavelengths within the visible to NIR range (600–1,000 nm), but strongly absorb infrared light. In addition, light scattering and auto-fluorescence of biological fluorophores [6, 7] are lower in the NIR when compared to the visible light range (<600 nm), where they can strongly interfere with detection of short wavelength-emitting probes [8]. Thus, this transparent window, where the absorption coefficient of tissue is at a minimum, can be utilized for optical imaging.

4 NIR Fluorescence Imaging Approaches for Small Animal Imaging

Table 1 presents a short list of modern fluorescent imaging approaches applicable for cellular imaging, their advantages and limitations [3, 9–16].

Table 1 NIR fluorescent imaging approaches for small animal whole-body imaging

Type of instrumentation	Advantages	Disadvantages
Two-dimensional single channel epifluorescence	Simple operation; fast	Quantitative only for superficial signals; offers no anatomical information
Two-dimensional spectral imaging	Multiplexing	Only semiquantitative and only for superficial signals; offers no anatomical information
Time-domain fluorescence imaging	Multiplexing	Quantitative only for superficial signals; offers no anatomical information
Fluorescence tomography	Three-dimensional rendering; quantitative also for deeper signals	Complex operation; computation of light dispersion through the tissue is based on many approximations
Fluorescence endoscopy/intravital microscopy	High spatial and temporal resolution	Limited to superficial organs

5 General Approaches for Flourescent Labeling of Hematopoietic and Cancer Cell

Cells can be fluorescently labeled either in vivo or ex vivo, using either endogenously expressed reporter genes with constitutive or induced expression, or using exogenously applied targeted fluorescent probes (targeted to specific chemicals including antigens) or untargeted probes, depending on the aim of the study. Exogenously applied cell-labeling probes allow rapid labeling of all cell types, but rapid cell proliferation will lead to dye dilution and corresponding loss of signal. Long term cell lineage tracking frequently requires the use of endogenously and constitutively expressed reporter genes. In vivo cell labeling for the purpose of cell tracking is a common practice in intravital microscopy [17]. For example, intravenous administration of the cationic fluorescent dye Rhodamine 6G (Rh6G) results in selective accumulation in nuclei and mitochondria of living cells such as leukocytes. In vivo administration of NIR fluorescent (NIRF) dyes for the purpose of hematopoietic or cancer cell labeling (following whole-body optical imaging) is not yet a common practice. Nevertheless the use of NIRF probes targeted (and/or conjugated with antibodies) might have a good potential. For example anticancer antibodies conjugated with NIRF dye like Cy7 could be administrated in vivo and their accumulation in site of tumor growth could be visualized using whole-body fluorescence imaging [18]. On the other hand, ex vivo cell labeling by NIRF dyes and probes is a common and widespread practice as will be described below.

6 Ex Vivo Cell Labeling Using Nonspecific NIR Fluorescent Dyes

The standard procedure for ex vivo cell labeling is based on introducing a contrast agent (fluorescent dye) to cell culture or cell suspension. Schematically it is shown in the Fig. 2.

A number of infrared dyes can be used for labeling of cancer and hematopoietic cells. However, as any labeling procedure induce cellular stress, validation of cell viability is critical in determination of the labeling protocol most appropriate for the cell model of interest.

To ensure effective labeling, the fluorescent molecule must cross the cell membrane, either spontaneously or with the aid of membrane permeabilizing treatments (such as the use transfection agents), and remain within the cell for an extended period. For example, the FDA-approved NIRF Indocyanine green (ICG) contrast agent is commonly used in ophthalmological applications, such as retinal and choroidal vasculature imaging, or for measuring and monitor cardiac output, liver function and various parameters related to peripheral circulation. Its absorption and emission in the NIR region allow for deep tissue penetration (absorption maximum: 780 nm and emission maximum: 820 nm in aqueous solution) [19]. ICG concentration and solvent largely influence its optical properties, where aggregation leads to self-quenching and a consequential reduction in fluorescence while micellar organizations improve signals and are often used as delivery vehicles [9]. In aqueous solutions, ICG forms dimers and oligomers at dye concentrations >3.9 mg/L [20] or large J-aggregates at concentrations >103 mg/L ICG with an absorption maximum

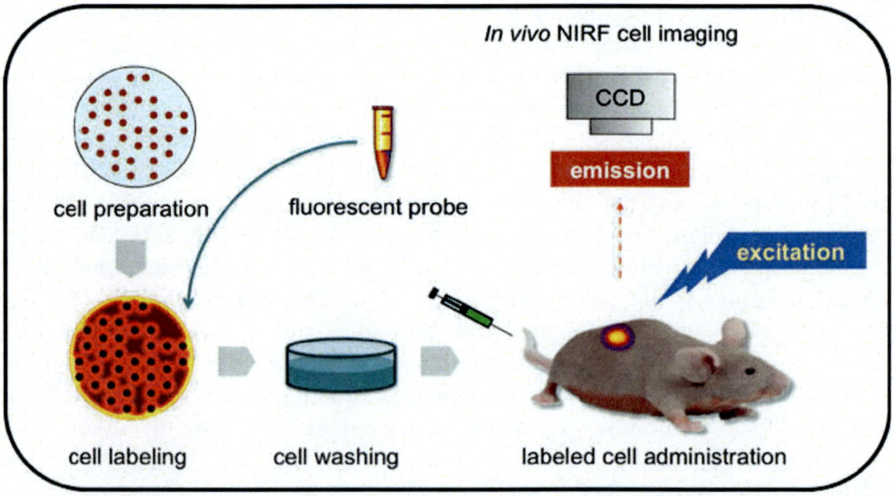

Fig. 2 A schematic representation of whole-body near-infrared imaging of ex vivo fluorescently labeled hematopoietic or cancer cells

at $\lambda = 900$ nm after 7 days [21]. Alternatively, organic solvents such as DMSO and or the Protamine sulfate polycation transfection agent improve ICG solubility, in vitro cellular uptake and fluorescence emission [22].

In addition to ICG, dyes such as Cy7, Cy5.5, Alexa Fluor 680, Alexa Fluor 750, IrD700, IrD800, which fluoresce in NIR region, can be used for ex vivo cell labeling and whole-body fluorescence imaging.

7 Cell Labeling Using Lipophilic NIR Fluorescent Probes

The cell membrane with its lipid bilayer provides an important target for labeling cells using lipophilic fluorescent probes. Lipophilic carbocyanine dyes with long hydrocarbon chains and lipophilic styryl fluorescent dyes, each with its unique chemical and optical specifications, are often used for in vivo imaging. Cell labeling with lipophilic carbocyanines is generally performed by incubating the cells ex vivo with the fluorescent molecule prior to transplantation [23–25]. The cells remain labeled for extended periods, providing a powerful tool for routine cell labeling and whole-body NIRF optical imaging.

Due to its suitable optical properties, DiR (1,1-dioctadecyl-3,3,3,3-tetramethylindotricarbocyanine iodide, Invitrogen), excitation maximum: ~750 nm, emission maximum: 782 nm [26] is one of the most widely applied NIRF lipophilic dyes in cell-labeling techniques and in vivo imaging in Figs. 3 and 4.

Some relevant examples for the use of lipophilic dyes in human cancer and murine hematopoietic cell models are described below and illustrated in Figs. 5–9.

Imaging of trans-endothelial migration of macrophages in a cutaneous granuloma model by means of two-dimensional fluorescence imaging and by three-dimensional fluorescence-mediated tomography are versatile techniques to monitor and quantify cellular inflammatory response in vivo [27]. An example of recruitment of fibroblasts into s.c. tumors detected by NIR fluorescence imaging is shown in Fig. 10 [28].

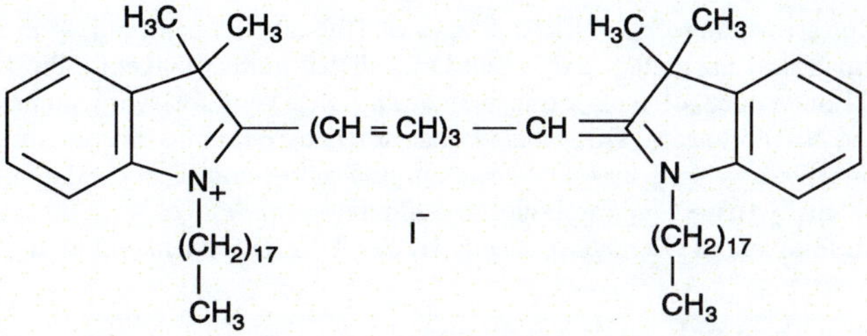

Fig. 3 (DiR) 1 "DiR"; DiIC$_{18}$(7)1,1′-dioctadecyl-3,3,3′,3′-tetramethylindotricarbocyanine iodide. Molecular Formula: $C_{63}H_{101}IN_2$, Molecular Weight: 1,013.41

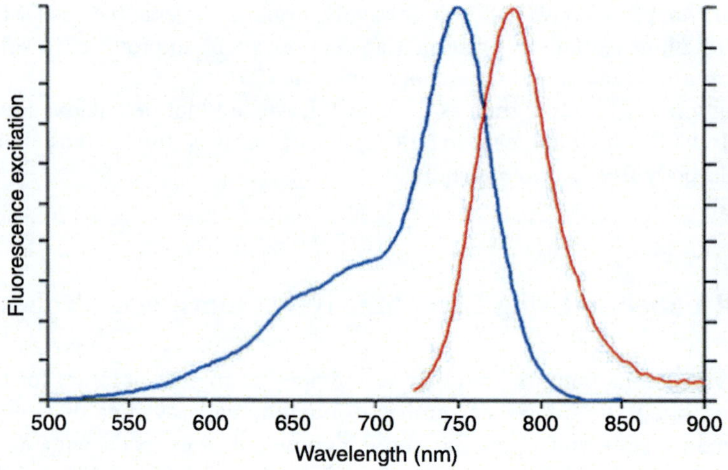

Fig. 4 Fluorescence excitation and emission spectra of $DiIC_{18}(7)$ ("DiR") bound to phospholipid bilayer membranes

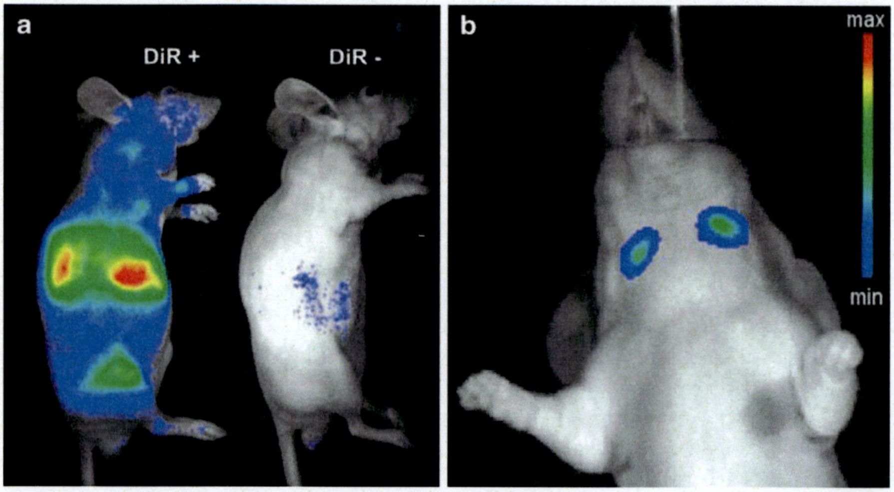

Fig. 5 Near-infrared-fluorescent (NIRF) images of DiR-labeled cell homing in mice. (**a**) Color coded NIRF images of unlabeled (DiR-) and DiR-labeled human leukemic Pre-B ALL G2 cells homing in CD1 nude mice, 48 h after transplantation (DiR+) and following noninvasive imaging. (**b**) Color coded NIRF image of DiR-labeled mice splenocyte homing overlaid on a photographic image. The threshold was set above the background auto-fluorescence. The *color bar* defines fluorescence intensity (from *blue*/minimum to *red*/maximum). Images were obtained by IVIS 100 from Caliper Life Sciences (Xenogen), Alameda, CA, USA (Kalchenko et al. unpublished)

However, a thorough analysis demonstrated that in addition to signal arising from the labeled cells and their progeny, some of the fluorescence could be attributed to dye transfer to neighboring cells either by cell–cell contacts or by

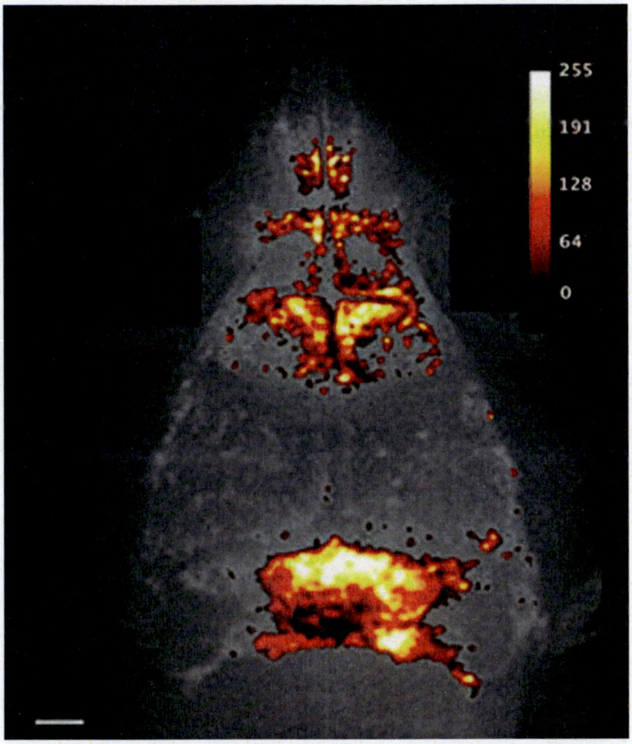

Fig. 6 Color-coded NIRF image of the cranium of sacrificed CD1 nude mouse 24 h after transplantation of 1×10^7 DiR-labeled human leukemic Pre-B ALL G2 cells. The cell homing pattern shows that the G2 cells reached the bone marrow. *Bar* is 1 mm (adapted from [23])

released microparticles, resulting in "microenvironmental contamination" of the regions penetrated by labeled cells [29]. Such contamination was observed for various cell types and various lipophylic dyes. Thus, tracking cells labeled with lipophilic dyes must be performed with caution and independent tools should critically validate the data.

8 Esterase-Sensitive Fluorescent Probes for Cell Tracking

Esterase-sensitive fluorescent probes show effective intracellular accumulation, and are therefore widely used for cell labeling. After cellular uptake and ester bond hydrolysis by intracellular esterases, these probes are activated to their fluorescent derivatives and are trapped within the cell. A widely used esterase-sensitive probe is carboxyfluorescein succinimidyl ester (CFSE) [30]. CFSC or

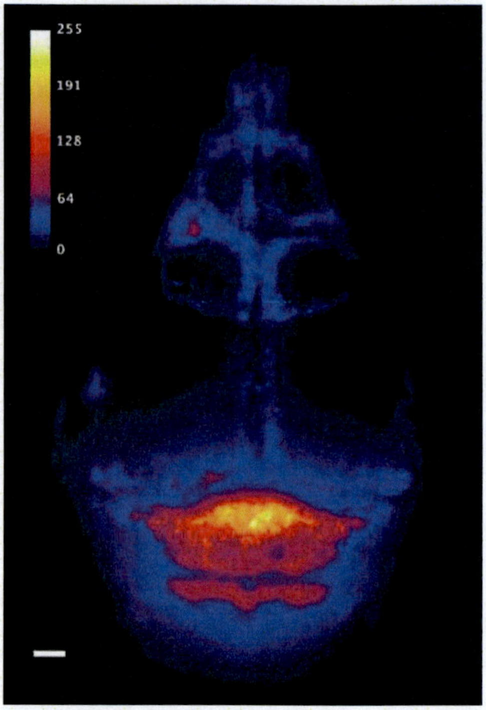

Fig. 7 Color-coded NIRF image of the cranium of a sacrificed NOD/SCID mouse 48 h after transplantation of 1×10^7 DiR-labeled ARH-77, Epstein–Barr plasma cells. *Bar* is 1 mm (Kalchenko et al. unpublished)

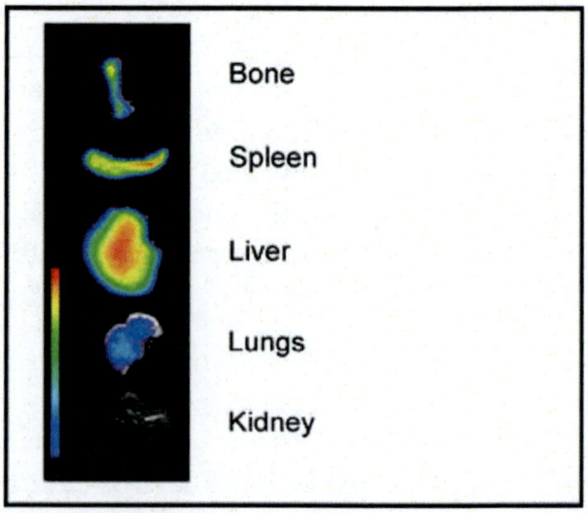

Fig. 8 Color-coded NIRF image of the internal organs of CD1 nude mice overlaid on a photographic image 24 h after transplantation of 1×10^7 DiR-labeled G2L (human leukemic) cells. The *color bar* defines fluorescence intensity (from *blue*/minimum to *red*/maximum). Images were obtained by IVIS 100 from Caliper Life Sciences (Xenogen), Alameda, CA, USA (adapted from [23])

analogs are long acting probe and biocompatible and are acceptable for intravital microscopy and in vitro studies, however, the use in whole-body imaging is limited because CFSC maximum emission is in the range which is highly absorbed by

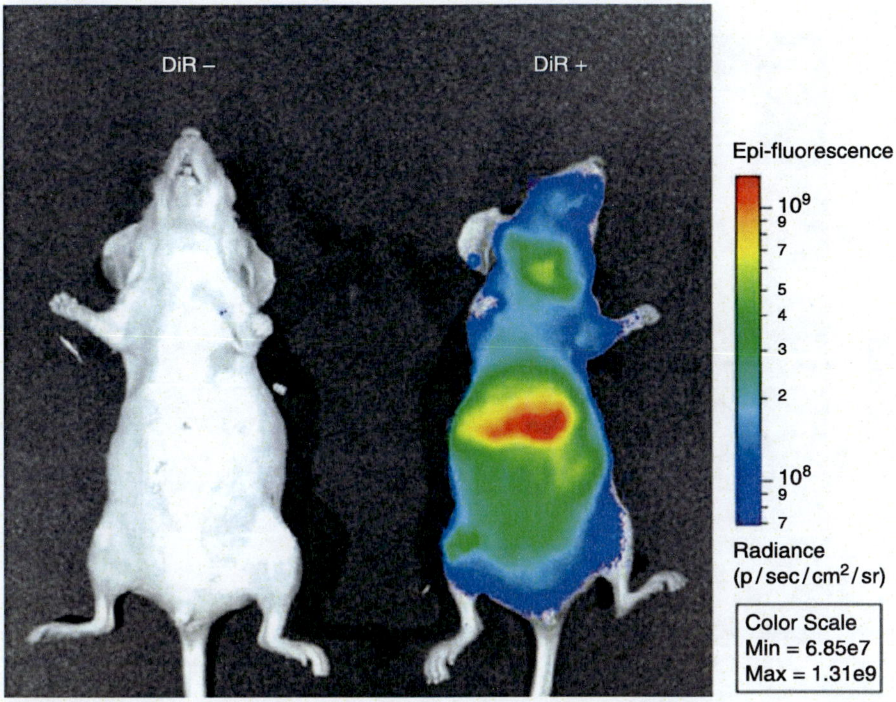

Fig. 9 Color-coded NIRF images overlaid on photographic images of Cd1 Nude mice 5 min after i.v. implantation of 1×10^7 DiR-labeled human prostatic cancer cells. The *color bar* defines fluorescence intensity (from *blue*/minimum to *red*/maximum). Images were obtained by IVIS 100 from Caliper Life Sciences (Xenogen), Alameda, CA, USA (Kalchenko et al. unpublished)

tissue chromophores. Additional esterase-triggered NIRF probes, such as Alexa Fluor 750 carboxylic acid, succinimidyl ester (AF750) were recently reported [31, 32]. An example for noninvasive NIRF imaging of Alexa Fluor 750-labeled donor cells homing into bone marrow is shown in the Fig. 11. Esterase-sensitive NIRF probes appear to provide a convenient alternative for labeling hematopoietic and cancer cells for whole-body NIRF optical imaging.

9 NIR Imaging by Means of Reporter Genes

Development of fluorescent proteins with bright emission in NIR region will dramatically change whole-body imaging approach of hematopoietic and cancer cells. The currently known infrared-fluorescent proteins (IFPs) can be used for cell

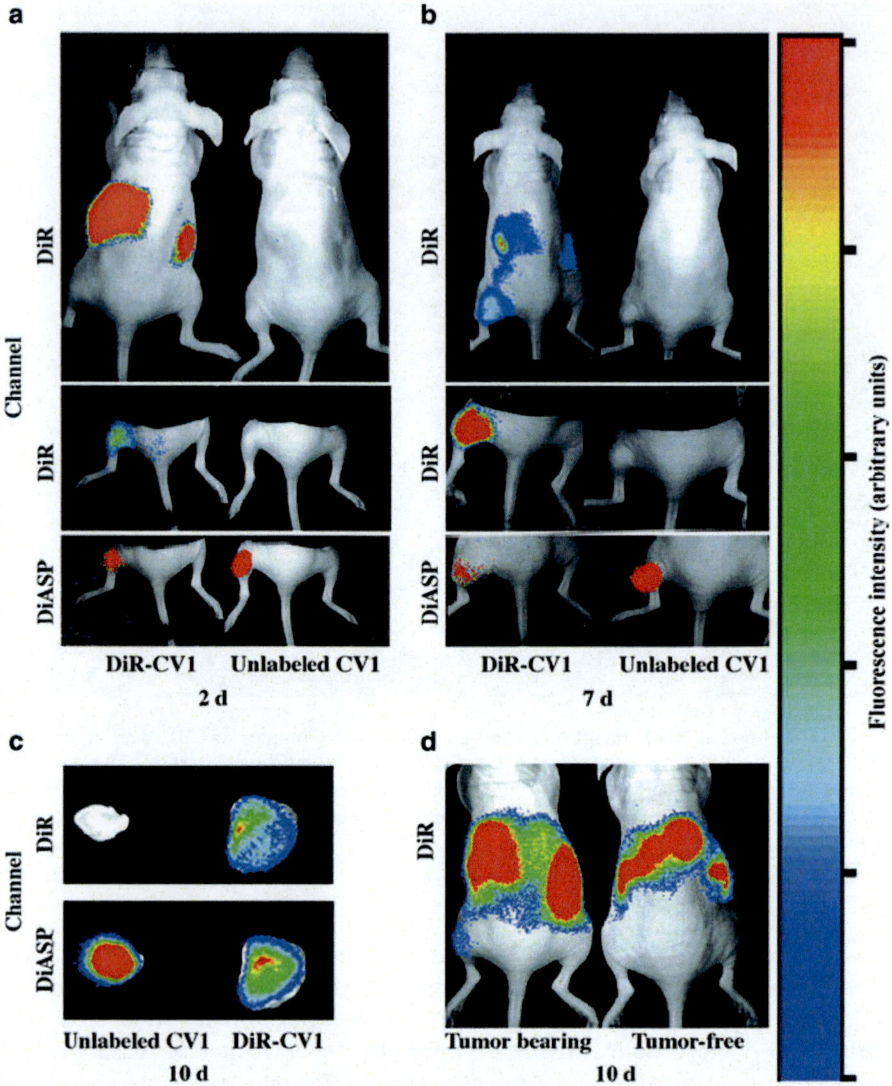

Fig. 10 Recruitment of fibroblasts into s.c. tumors detected by near-IR fluorescence imaging. Tumors were initiated in the left hind limb of CD-1 nude mice by s.c. inoculation of MLS cells labeled with DiASP followed by a second i.p. injection of CV-1 fibroblasts (2×10^6) labeled with DiR or unlabeled cells (*left* and *right* mice, respectively). Color-coded near-IR fluorescent images were acquired by IVIS 100 (Xenogen) system and were overlaid on photographic images of s.c. tumors. (**a**) and (**b**), mice were viewed using near-IR fluorescence (*top*, no masking; *middle*, black tape covering on the abdomen) as well as DiASP fluorescence imaging (**a** and **b**, *bottom*). Images were acquired 2 days (**a**) and 7 days (**b**) after fibroblast injection. (**c**), excised tumors. *Top*, near-IR fluorescence imaging of MLS tumors derived from mice injected with DiR-labeled (*right*) and unlabeled (*left*) CV-1 cells. *Bottom*, DiASP fluorescence of the labeled MLS tumor cells. (**d**) CV-1 fibroblasts labeled with DiR were injected into the peritoneum of a tumor-bearing mouse (*left* mouse, tumor located at the left limb) or to a control tumor-free mouse (*right* mouse). Mice were viewed 10 days after inoculation (**c** and **d**) (adapted from [28])

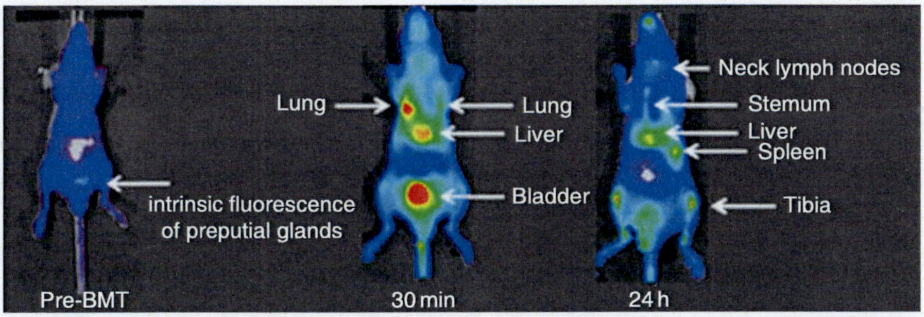

Fig. 11 Noninvasive NIRF imaging of Alexa Fluor 750-labeled donor cells homing into bone marrow. In vivo imaging was performed using an IVIS Spectrum system (Xenogen, Alameda, CA, USA). Alexa Fluor 750 carboxylic acid, succinimidyl ester (AF750) (peak excitation: 749 nm, peak emission: 775 nm; Invitrogen, Eugene, OR, USA). Pre-BMT – Before bone marrow transplantation (Alexa Fluor 750-labeled donor cells administration). Adopted under Creative Commons Attribution License (CCAL) from [32]

imaging using whole-body imaging but application is limited by their low quantum yield [33].

Reporter genes provide a stable signal level for the cells and their progeny thus facilitating lineage tracking, as signals are not diluted with cell proliferation. Furthermore, molecular tools now allow to study specific differentiation pathways. Beta-galactosidase (betaGal) is one of the most popular reporter genes, and its activity can be detected by fluorescence imaging after NIR activation of the X-Gal reaction product. Far red fluorescence signals can be provided by using an alternative betaGal substrate [34].

10 Quantum Dots in NIR Imaging Applications

Fluorescent quantum dots (QDs) are inorganic nanometer-sized, luminescent semiconductor crystals featuring unique chemical and physical properties due to their size and highly compact structure. QDs are characterized by a very well-defined absorption and a narrow emission spectrum, but on the other hand, QDs are available for a broad range of emission wavelengths, from the visible to infrared, depending on their size and chemical composition [35]. Use of QDs in imaging applications, bypasses many limitations associated with organic fluorescent dyes such as photobleaching and chemical instability [36]. QDs are particularly useful for in vivo studies requiring long-term and multicolor imaging of cellular and molecular interactions [37]. A number of reports demonstrate feasibility of QD usage in in vivo cellular imaging of hematopoietic and cancer cells. Figure 12 shows an example of uptake of QDs by dendritic cells (DC) [38].

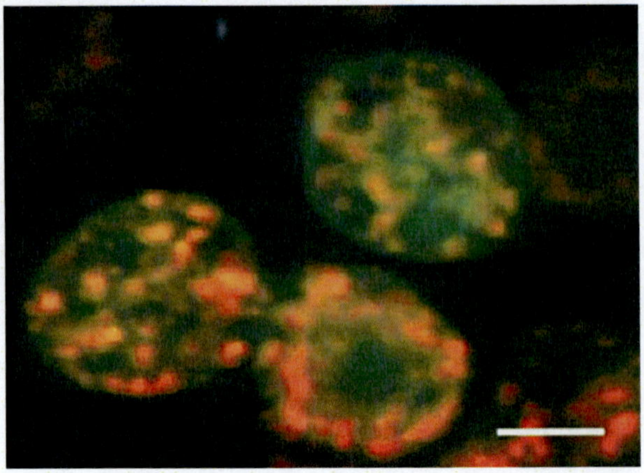

Fig. 12 Two-photon image showing endocytosed QDs (*red*) inside CD11c+DCs (*green*) cultured from bone marrow of transgenic EYFP-CD11c reporter mice. Image adopted under Creative Commons Attribution License (CCAL) from [38]

NIR-emitting QDs have also been applied for noninvasive in vivo tracking of DC migration into lymph nodes [39] and for targeting cancer cells for NIRF imaging [40, 41]. The relatively large size of QDs (15–30 nm), along with their low biocompatibility and biostability [42], and their high cost, limit their broad application.

11 Multimodal Imaging

Whole-body NIR imaging alone lacks anatomical information. Therefore, unless specific anatomical structures are targeted with fluorescence probes, it can be difficult to assign a signal to a specific location. Integration of NIR fluorescence imaging with complementary optical and nonoptical imaging modalities can enhance the molecular information provided by the fluorescent probes and can aid in interpretation of the data, especially for macroscopic imaging systems. Multimodality imaging approaches combine two or more imaging modalities. NIRF imaging can be combined with visual fluorescence [43], bioluminescence (BLI), X-ray or X-ray computed thomography (CT) [44], magnetic resonance imaging (MRI) (Fig. 13) [9], [28, 45], and positron emission tomography (PET), FMT version of NIRF can be combined with MRI and PET [46]. Beside gaining anatomical localization of the NIRF signal in the body it simultaneously allow visualization of multiple functions. For example, it might allow the uncovering of the functions of different components of the microenvironment while cancer or inflammation forms. Furthermore, the ability to noninvasively follow these cells may potentially be beneficial for targeted image-guided cellular therapy.

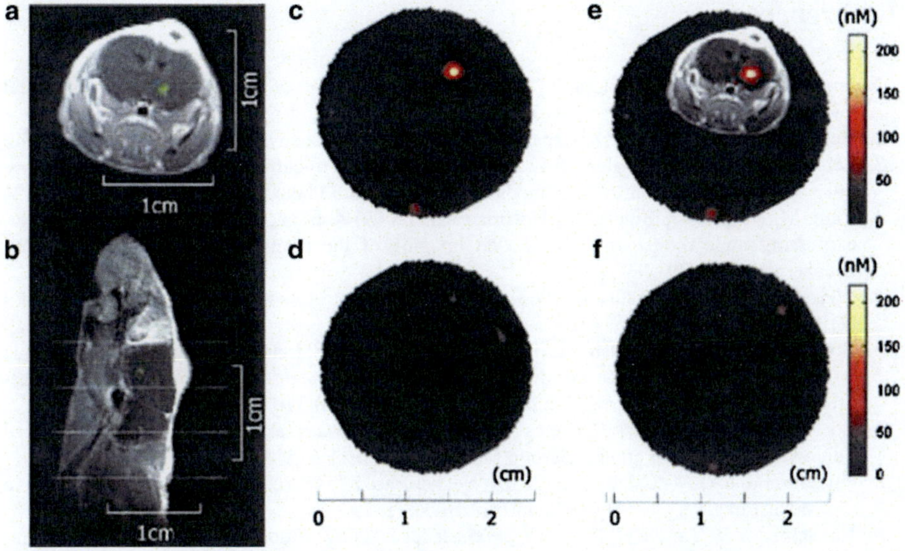

Fig. 13 In vivo FMT of cathepsin B expression levels in 9L gliosarcomas stereotactically implanted into unilateral brain hemispheres of nude mice. (**a** and **b**) Axial and sagittal MR slices of an animal implanted with a tumor, which is shown in *green* after gadolinium enhancement. (**c**, **d**, and **f**) Consecutive FMT slices obtained from *top* to *bottom* from the volume of interest shown on (**b**) by *thin white horizontal lines*. (**e**) Superposition of the MR axial slice passing through the tumor (**a**) onto the corresponding FMT slice (**c**) after appropriately translating the MR image to the actual dimensions of the FMT image (adopted from [9])

12 Summary and Conclusions

Whole-body imaging of cancer and hematopoietic cells facilitates monitoring of complex biological processes, disease progression, and response to therapy. Visualization of cancer cell in living animals reveals tumor growth and metastasis, allowing researchers to monitor tumor burden and assess the efficacy of candidate antitumor and antimetastatic agents. Similarly, observation of hematopoietic cell rolling, homing and interactions within vessels and tissues such as bone marrow, spleen, lymph-nodes, and liver can contribute toward an in-depth understanding of the cells in their natural and disease environments.

The use of NIRF dyes for labeling hematopoietic and cancer cells enables in vivo imaging of cell fate at microscopic and whole-body levels. Further innovations in probe design, multichannel techniques, and streamlined toxicity screening, will result in the development of safer and more versatile imaging agents. The added value obtained by combining molecular information provided by optical imaging agents with anatomical data collected from other imaging modalities will expand the use and potential applications of optical imaging in the coming years.

References

1. Jain RK, Munn LL, Fukumura D (2002) Dissecting tumour pathophysiology using intravital microscopy. Nat Rev Cancer 2(4):266–276
2. Helmchen F, Denk W (2005) Deep tissue two-photon microscopy. Nat Methods 2(12):932–940
3. Leblond F, Davis SC, Valdés PA, Pogue BW (2010) Pre-clinical whole-body fluorescence imaging: review of instruments, methods and applications. J Photochem Photobiol B 98(1):77–94
4. Yang M, Reynoso J, Bouvet M, Hoffman RM (2009) A transgenic red fluorescent protein-expressing nude mouse for color-coded imaging of the tumor microenvironment. J Cell Biochem 106(2):279–284
5. Weissleder R, Ntziachristos V (2003) Shedding light onto live molecular targets. Nat Med 9 (1):123–128
6. Boumaza S, Arribas SM, Osborne-Pellegrin M, McGrath JC, Laurent S, Lacolley P, Challande P (2001) Fenestrations of the carotid internal elastic lamina and structural adaptation in stroke-prone spontaneously hypertensive rats. Hypertension 37(4):1101–1107
7. Lee WK, Bell J, Kilpatrick E, Hayes M, Lindop GB, Dominiczak MH (1993) Collagen-linked fluorescence in human atherosclerotic plaques. Atherosclerosis 98(2):219–227
8. Hilderbrand SA, Weissleder R (2010) Near-infrared fluorescence: application to in vivo molecular imaging. Curr Opin Chem Biol 14(1):71–79
9. Ntziachristos V, Tung CH, Bremer C, Weissleder R (2002) Fluorescence molecular tomography resolves protease activity in vivo. Nat Med 8(7):757–760
10. Tung CH, Mahmood U, Bredow S, Weissleder R (2000) In vivo imaging of proteolytic enzyme activity using a novel molecular reporter. Cancer Res 60(17):4953–4958
11. Ntziachristos V, Schellenberger EA, Ripoll J, Yessayan D, Graves E, Bogdanov A Jr, Josephson L, Weissleder R (2004) Visualization of antitumor treatment by means of fluorescence molecular tomography with an annexin V-Cy5.5 conjugate. Proc Natl Acad Sci USA 101(33):12294–12299
12. Jaffer FA, Tung CH, Gerszten RE, Weissleder R (2002) In vivo imaging of thrombin activity in experimental thrombi with thrombin-sensitive near-infrared molecular probe. Arterioscler Thromb Vasc Biol 22(11):1929–1935
13. Moon WK, Lin Y, O'Loughlin T, Tang Y, Kim DE, Weissleder R, Tung CH (2003) Enhanced tumor detection using a folate receptor-targeted near-infrared fluorochrome conjugate. Bioconjug Chem 14(3):539–545
14. Pham W, Choi Y, Weissleder R, Tung CH (2004) Developing a peptide-based near-infrared molecular probe for protease sensing. Bioconjug Chem 15(6):1403–1407
15. Nahrendorf M, Sosnovik DE, Waterman P, Swirski FK, Pande AN, Aikawa E, Figueiredo JL, Pittet MJ, Weissleder R (2007) Dual channel optical tomographic imaging of leukocyte recruitment and protease activity in the healing myocardial infarct. Circ Res 100(8):1218–1225
16. Bremer C, Ntziachristos V, Weitkamp B, Theilmeier G, Heindel W, Weissleder R (2005) Optical imaging of spontaneous breast tumors using protease sensing 'smart' optical probes. Invest Radiol 40(6):321–327
17. Saetzler RK, Jallo J, Lehr HA, Philips CM, Vasthare U, Arfors KE, Tuma RF (1997) Intravital fluorescence microscopy: impact of light-induced phototoxicity on adhesion of fluorescently labeled leukocytes. J Histochem Cytochem 45(4):505–513
18. Zou P, Xu S, Povoski SP, Wang A, Johnson MA, Martin EW Jr, Subramaniam V, Xu R, Sun D (2009) Near-infrared fluorescence labeled anti-TAG-72 monoclonal antibodies for tumor imaging in colorectal cancer xenograft mice. Mol Pharm 6(2):428–440
19. Slakter JS, Yannuzzi LA, Guyer DR, Sorenson JA, Orlock DA (1995) Indocyanine-green angiography. Curr Opin Ophthalmol 6(3):25–32
20. Cardillo JA, Jorge R, Costa RA, Nunes SM, Lavinsky D, Kuppermann BD, Tedesco AC, Farah ME (2008) Experimental selective choriocapillaris photothrombosis using a modified indocyanine green formulation. Br J Ophthalmol 92(2):276–280

21. Landsman ML, Kwant G, Mook GA, Zijlstra WG (1976) Light-absorbing properties, stability, and spectral stabilization of indocyanine green. J Appl Physiol 40(4):575–583
22. Boddington S, Henning TD, Sutton EJ, Daldrup-Link HE (2008) Labeling stem cells with fluorescent dyes for non-invasive detection with optical imaging. J Vis Exp 2(14):686. doi:10.3791/686
23. Kalchenko V, Shivtiel S, Malina V, Lapid K, Haramati S, Lapidot T, Brill A, Harmelin A (2006) Use of lipophilic near-infrared dye in whole-body optical imaging of hematopoietic cell homing. J Biomed Opt 11(5):050507
24. Nadav L, Kalchenko V, Barak MM, Naparstek E, Geiger B, Katz BZ (2008) Tumorigenic potential and disease manifestations of malignant B-cell variants differing in their fibronectin adhesiveness. Exp Hematol 36(11):1524–1534
25. Varol C, Landsman L, Fogg DK, Greenshtein L, Gildor B, Margalit R, Kalchenko V, Geissmann F, Jung S (2007) Monocytes give rise to mucosal, but not splenic, conventional dendritic cells. J Exp Med 204(1):171–180
26. Menon MB, Ronkina N, Schwermann J, Kotlyarov A, Gaestel M (2009) Fluorescence-based quantitative scratch wound healing assay demonstrating the role of MAPKAPK-2/3 in fibroblast migration. Cell Motil Cytoskeleton 66(12):1041–1047
27. Eisenblätter M, Ehrchen J, Varga G, Sunderkötter C, Heindel W, Roth J, Bremer C, Wall A (2009) In vivo optical imaging of cellular inflammatory response in granuloma formation using fluorescence-labeled macrophages. J Nucl Med 50(10):1676–1682
28. Granot D, Addadi Y, Kalchenko V, Harmelin A, Kunz-Schughart LA, Neeman M (2007) In vivo imaging of the systemic recruitment of fibroblasts to the angiogenic rim of ovarian carcinoma tumors. Cancer Res 67(19):9180–9189
29. Lassailly F, Griessinger E, Bonnet D (2010) "Microenvironmental contaminations" induced by fluorescent lipophilic dyes used for noninvasive in vitro and in vivo cell tracking. Blood 115(26):5347–5354
30. Hsieh AS, Winet H, Bao JY, Stevanovic M (1999) Model for intravital microscopic evaluation of the effects of arterial occlusion-caused ischemia in bone. Ann Biomed Eng 27(4):508–516
31. Kim Y, Choi Y, Weissleder R, Tung CH (2007) Membrane permeable esterase-activated fluorescent imaging probe. Bioorg Med Chem Lett 17(18):5054–5057
32. Ushiki T, Kizaka-Kondoh S, Ashihara E, Tanaka S, Masuko M, Hirai H, Kimura S, Aizawa Y, Maekawa T, Hiraoka M (2010) Noninvasive tracking of donor cell homing by near-infrared fluorescence imaging shortly after bone marrow transplantation. PLoS ONE 5(6):e11114
33. Shu X, Royant A, Lin MZ, Aguilera TA, Lev-Ram V, Steinbach PA, Tsien RY (2009) Mammalian expression of infrared fluorescent proteins engineered from a bacterial phytochrome. Science 324(5928):804–807
34. Tung CH, Zeng Q, Shah K, Kim DE, Schellingerhout D, Weissleder R (2004) In vivo imaging of beta-galactosidase activity using far red fluorescent switch. Cancer Res 64(5):1579–1583
35. Ghasemi Y, Peymani P, Afifi S (2009) Quantum dot: magic nanoparticle for imaging, detection and targeting. Acta Biomed 80(2):156–165
36. Walling MA, Novak JA, Shepard JR (2009) Quantum dots for live cell and in vivo imaging. Int J Mol Sci 10(2):441–491
37. Jaiswal JK, Goldman ER, Mattoussi H, Simon SM (2004) Use of quantum dots for live cell imaging. Nat Methods 1(1):73–78
38. Sen D, Deerinck TJ, Ellisman MH, Parker I, Cahalan MD (2008) Quantum dots for tracking dendritic cells and priming an immune response in vitro and in vivo. PLoS ONE 3(9):e3290
39. Noh YW, Lim YT, Chung BH (2008) Noninvasive imaging of dendritic cell migration into lymph nodes using near-infrared fluorescent semiconductor nanocrystals. FASEB J 22(11):3908–3918
40. He X, Wang K, Cheng Z (2010) In vivo near-infrared fluorescence imaging of cancer with nanoparticle-based probes. Wiley Interdiscip Rev Nanomed Nanobiotechnol 2(4):349–366
41. Stroh M, Zimmer JP, Duda DG, Levchenko TS, Cohen KS, Brown EB, Scadden DT, Torchilin VP, Bawendi MG, Fukumura D, Jain RK (2005) Quantum dots spectrally distinguish multiple species within the tumor milieu in vivo. Nat Med 11(6):678–682

42. Lovrić J, Cho SJ, Winnik FM, Maysinger D (2005) Unmodified cadmium telluride quantum dots induce reactive oxygen species formation leading to multiple organelle damage and cell death. Chem Biol 12(11):1227–1234
43. Goldshaid L, Rubinstein E, Brandis A, Segal D, Leshem N, Brenner O, Kalchenko V, Eren D, Yecheskel T, Salitra Y, Salomon Y, Scherz A (2010) Novel design principles enable specific targeting of imaging and therapeutic agents to necrotic domains in breast tumors. Breast Cancer Res 12(3):R29
44. Cao L, Breithaupt M, Peter J (2010) Geometrical co-calibration of a tomographic optical system with CT for intrinsically co-registered imaging. Phys Med Biol 55(6):1591–1606
45. Pittet MJ, Swirski FK, Reynolds F, Josephson L, Weissleder R (2006) Labeling of immune cells for in vivo imaging using magnetofluorescent nanoparticles. Nat Protoc 1(1):73–79
46. Nahrendorf M, Keliher E, Marinelli B, Waterman P, Feruglio PF, Fexon L, Pivovarov M, Swirski FK, Pittet MJ, Vinegoni C, Weissleder R (2010) Hybrid PET-optical imaging using targeted probes. Proc Natl Acad Sci USA 107(17):7910–7915

Index

A.P. Demchenko (ed.), *Advanced Fluorescence Reporters in Chemistry and Biology III: Applications in Sensing and Imaging*, Springer Ser Fluoresc (2011) 10: 347–352,
DOI 10.1007/978-3-642-18035-4, © Springer-Verlag Berlin Heidelberg 2011